ALLE ZEIT WACH 1842

Glutamine Metabolism in Mammalian Tissues

Edited by
D. Häussinger and H. Sies

With 89 Figures

Springer-Verlag
Berlin Heidelberg New York Tokyo 1984

Dr. DIETER HÄUSSINGER
Medizinische Universitätsklinik
Hugstetterstraße 55
7800 Freiburg, FRG

Professor Dr. HELMUT SIES
Institut für Physiologische Chemie I
Universität Düsseldorf
Moorenstraße 5
4000 Düsseldorf, FRG

ISBN-13:978-3-642-69756-2 e-ISBN-13:978-3-642-69754-8
DOI: 10.1007/978-3-642-69754-8

Library of Congress Cataloging in Publication Data. Main entry under title. Glutamine metabolism in mammalian tissues. Bibliography: p. Includes index. 1. Glutamine – Metabolism. 2. Mammals – Physiology. I. Häussinger, D. (Dieter), 1951– II. Sies, H. (Helmut), 1942– QP562.G55G58 1984 599'.0133 84-5517.

Softcover reprint of the hardcover 1st edition 1984

2131/3130-543210

Preface

Glutamine is a key aminoacid for the synthesis of numerous biologically important compounds in mammalian cells and is a carrier form of ammonia. The advance in knowledge on the metabolic significance of this amino acid is presented in in-depth treatments by experts in this active research field. This includes the enzymology of glutamine synthetase and glutaminase activities in different mammalian organs, notably liver, kidney and brain; properties of glutamine transport across biological membranes; role of glutamine metabolism in the liver, with emphasis on the recent discovery of intercellular heterogeneity with respect to enzyme distribution and its functional consequences for ammonia/urea metabolism and pH regulation; renal and intestinal glutamine metabolism; cerebral glutamine/glutamate interrelationships; skeletal muscle; role of glutamine in cell culture; and finally the clinical aspects, including the new outlook of glutamine antagonists in cancer therapy as well as the role of glutamine in hepatic coma and encephalopathy.

Some, but not all, of the contributors to this work attended the 48th Conference of the Gesellschaft für Biologische Chemie on glutamine metabolism held in Göttingen, W. Germany, in September 1983. This conference was supported by the Stiftung Volkswagenwerk, Dr. H. Falk Foundation, Freiburg, and the J. Pfrimmer Co., Erlangen. The abstracts of the contributions to the conference were published in Hoppe-Seyler's Z. Physiol. Chem. 364, 1237–1256, 1983, and this book is not intended as the proceedings of that meeting.

Finally, it is a pleasure to thank the many collegues who have contributed to this project by giving their valuable advice and critique; thanks are also due to Peter Graf and Thomas Stehle for their help in compiling the entries to the subject index. We hope that this in-depth account of present knowledge will provide a useful platform for further development in this field.

May 1984 DIETER HÄUSSINGER and HELMUT SIES

Contents

Kidney

Other Tissues

Clinical Aspects

Contributors

You will find the addresses at the beginning of the respective contribution

Enzymology and Transport

Enzymology of Glutamine

A. MEISTER[1]

Introduction

Glutamine is an essential building block of proteins, and occupies a central position in nitrogen metabolism [1]. Thus, it is a storage form and a transport form of both glutamate and ammonia. Glutamine is more effectively transported across the blood-brain barrier and across certain cell membranes than glutamate. Ammonia is toxic to many animal tissues, whereas glutamine serves efficiently as a nontoxic source of ammonia for many reactions. Glutamine provides nitrogen for the synthesis of N^3 and N^9 of the purine ring, the amide nitrogen atoms of asparagine and NAD, the amino groups of glucosamine 6-phosphate, guanine, cytidine, and (as carbamyl phosphate) N^1 of the pyrimidine ring. The synthesis of glutamine is a reaction of significance in maintaining the normally low levels of ammonia found in tissues and body fluids. The synthesis of glutamine is of particular importance in the central nervous system for removal of ammonia and of glutamate, a neurotransmitter. Glutamine is a major energy source in kidney [2], small intestine [3], bone [4], and in certain cells grown in tissue culture [5, 6]. Hydrolysis of glutamine by renal glutaminase plays a key role in acid-base balance. Glutamine is used in man and in certain higher apes for the detoxication of phenylacetic acid, which is formed in the metabolism of phenylalanine; phenylacetate is coupled, via phenylacetyl adenylate and phenylacetyl coenzyme A, with glutamine to form phenylacetyl glutamine; the daily urinary excretion of this compound in man is about 300 mg [7]. The α-amino group of glutamine is utilized in a number of transamination reactions of metabolic and physiological significance; glutamine thus functions in a homeostatic mechanism for preservation of amino acid balance [8].

The functions of glutamine summarized above are mediated by the actions of a number of enzymes, most of which have been isolated in highly purified form and studied with respect to their structures and mechanisms of action. Glutamine is not a dietary essential amino acid for animals, nor is it required for growth by plants and most microorganisms. Thus, glutamine must be synthesized by a wide variety of cells. It is notable that glutamine is the predominant amino acid of the blood plasma in mammals and accounts for about 20% of the total amino acid content; it comprises about two-thirds of the total amino acid concentration of the cerebrospinal fluid in man. Many tissues, such

1 Department of Biochemistry, Cornell University Medical College, 1300 York Avenue, New York, NY 10021, USA

Glutamine Metabolism in Mammalian Tissues
Edited by D. Häussinger and H. Sies

as the brain, liver, intestine, skeletal muscle, and heart have high concentrations of glutamine. The synthesis of glutamine is catalyzed by the widely distributed and important catalyst glutamine synthetase. A survey of the properties of this enzyme and of the enzymes that catalyze the utilization of glutamine is given below.

Glutamine Synthetase

Glutamine synthetase catalyzes the reversible formation of glutamine from glutamate and ammonia coupled to the cleavage of ATP to ADP and inorganic phosphate [9] (Reaction 1). A divalent cation, such as Mg^{2+}, Mn^{2+}, Co^{2+}

$$\text{L-glutamate} + \text{ATP} + NH_3 \rightleftharpoons \text{L-glutamine} + \text{ADP} + P_i \qquad (1)$$

is required. When the enzyme is incubated with L-glutamate, ammonium ions, and MgATP (at concentrations of 10 m*M*, at pH 7.0 and 37°), equilibrium is attained when about 90% of the glutamate is converted to glutamine. When hydroxylamine is substituted for ammonia, greater than 99% of the glutamate is converted to γ-glutamyl hydroxamate. The apparent free energy change associated with the hydrolysis of glutamine is about 3,500 calories per mole, while that associated with hydrolysis of γ-glutamyl hydroxamate is about 700 calories per mole. The enzyme also interacts with certain other nucleophilic agents, such as hydrazine, monomethyl hydrazine, methylamine, ethylamine, and glycine ethyl ester; in these reactions, the corresponding γ-glutamyl products are formed. As discussed below, the enzyme also acts on a number of glutamate analogs.

Glutamine synthetase was first isolated in highly purified and essentially homogenous form from sheep brain [10]: the enzyme was later isolated from other animal tissues, including liver [9] and muscle [11]. The brain enzyme has a mol. wt. of about 390,000. The enzyme is composed of eight apparently identical subunits which exhibit a cube-like appearance in the electron microscope. The octamer, which possesses D_4 symmetry, is formed by isologous association of two heterologously bonded tetramers. The amino acid compositions of the several mammalian glutamine synthetases that have thus far been obtained are very similar [9].

The mechanism of the reaction catalyzed by glutamine synthetase involves formation of enzyme-bound intermediates including γ-glutamyl phosphate. Information about the mechanism has come from observations on the partial reactions catalyzed by the enzyme, and from studies with ^{18}O-labeled glutamate and with a variety of glutamate analogs. In the γ-glutamyl transfer reaction (Reaction 2), the γ-glutamyl moiety is transferred to hydroxylamine to form

$$\text{L-glutamine} + NH_2OH \rightarrow \gamma\text{-glutamylhydroxamate} + NH_3 \qquad (2)$$

γ-glutamyl hydroxamate. This reaction requires catalytic amounts of either ATP or ADP, divalent metal ions, and either phosphate or arsenate. The enzyme also catalyzes the arsenolysis of glutamine (Reaction 3), which also requires

$$\text{L-glutamine} + H_2O \xrightarrow[As_i]{ADP,\ M^{2+}} \text{L-glutamate} + NH_3 \qquad (3)$$

catalytic quantities of nucleotide. The synthesis of glutamine involves coupling between the exergonic cleavage of ATP to ADP and inorganic phosphate and the endergonic synthesis of glutamine from glutamate and ammonia, but free intermediates, such as amidophosphate and γ-glutamyl phosphate were not found; it became apparent that the intermediates must be bound to the enzyme. Studies with ^{18}O-labeled glutamate showed that the synthesis of glutamine is accompanied by transfer of oxygen from glutamate to inorganic phosphate. Although the physiologically significant amino acid substrate of the enzyme is L-glutamate, it was found that both optical isomers of glutamate are efficient substrates. Although the rates of synthesis of D- and L-γ-glutamyl hydroxamate were found to be similar, the synthesis of D-glutamine occurred at a significantly lower rate than that of L-glutamine [12]. This suggested that the enzyme catalyzes an initial activation of glutamate, a reaction of low optical specificity, followed by a more specific reaction of the activated glutamate with ammonia which becomes rate limiting in the case of D-glutamate. Search for the nature of the activated intermediate led to studies in which the enzyme was incubated with ATP and glutamate in the absence of ammonia. It was thought that if an enzyme-bound γ-carboxyl activated glutamate derivative were formed, it might exhibit a greater tendency to cyclize than either glutamate or glutamine. It was then found that the enzyme catalyzes a partial reaction in the absence of ammonia in which glutamate is converted stoichiometrically to 5-oxoproline, ADP, and inorganic phosphate [13] (Reaction 4). Interestingly, both isomers of

$$\text{L-(or D-)glutamate} + \text{ATP} \xrightarrow{M^{2+}} \text{5-oxo-L-(or D-)proline} + \text{ADP} + P_i \quad (4)$$

glutamate are cyclized at about the same rate. Evidence for activation of glutamate was also found in isotope trapping studies ("pulse-chase" experiments) in which the enzyme was incubated with MgATP and [^{14}C]glutamate; a mixture containing hydroxylamine and a large excess of unlabeled glutamate was then added. After deproteinization, the formation of labeled γ-glutamyl hydroxamate was determined. It was found that there is preferential conversion of the labeled glutamate to γ-glutamyl hydroxamate; thus, equilibration between the labeled and unlabeled glutamates did not occur. In other approaches, high speed centrifugation and ultrafiltration were used to show that glutamate becomes attached to the enzyme only when Mg^{2+} and ATP are present. Under these conditions, the binding of glutamate to the enzyme is associated with cleavage of ATP to ADP, and the γ-phosphoryl moiety of ATP that is released remains attached to the enzyme complex.

Direct evidence for the participation of an acyl phosphate intermediate was obtained in studies with glutamate analogs. Since both D- and L-glutamate are substrates, it was clear that the amino group of the subtrate does not need to be attached to a specific position on the glutarate carbon chain, and indeed, it was found that β-glutamate (β-aminoglutarate) is a good substrate [14]. The postulated β-aminoglutarylphosphate intermediate was prepared by chemical synthesis and was shown to be used by the enzyme for synthesis of ATP from ADP (Reaction 5). The enzyme is also active with other acyl phosphates, such

$$\beta\text{-glutamyl phosphate} + \text{ADP} \xrightarrow{M^{2+}} \text{ATP} + \beta\text{-glutamate} \quad (5)$$

acetyl phosphate and carbamyl phosphate. When cis-1-amino-1,3-dicarboxycyclohexane (cycloglutamate), which cannot cyclize to form an analog of 5-oxoproline (Fig. 1), was incubated with the enzyme and ATP in the absence of ammonia, an enzyme complex was formed which contained a phosphorylated derivative of the cyclic substrate (Reaction 6).

ENZYME + cycloglutamate + ATP
$\xrightarrow{M^{2+}}$ ENZYME [cycloglutamyl phosphate] [ADP] . (6)

Studies on the mechanism of inhibition of glutamine synthetase by methionine sulfoximine showed that this compound is phosphorylated by glutamine synthetase in the presence of ATP [16–18] (Reaction 7); Fig. 2). Only one of

ENZYME + L-methionine-S-sulfoximine + ATP
$\xrightarrow{M^{2+}}$ ENZYME [L-methionine-S-sulfoximine phosphate] [ADP] (7)

the four isomers of methionine sulfoximine is phosphorylated in this manner and only this same isomer irreversibly inhibits the enzyme. The active site of the enzyme was mapped by studies in which a variety of glutamate analogs were examined [19]. Although the enzyme acts on both L- and D-glutamate, it acts only on the L-isomer of α-methylglutamate to form α-methyl-L-glutamine. On the other hand, the enzyme catalyzes the amidation of β-glutamate to form only D-β-glutamine. Of the four β-methylglutamates, only threo-β-methyl-D-glutamate is a substrate. Sutdies on the four stereoisomers of γ-methylglutamate revealed that only threo-γ-methyl-L-glutamate is a substrate. Thus, of the ten possible monomethyl glutamates only three are substrates of the enzyme, i.e., α-methyl-L-glutamate, threo-β-methyl-D-glutamate, and threo-γ-methyl-L-glutamate. The findings permitted a mapping of the active site of the enzyme; the reader is referred to Meister [20] for details. The findings led to the synthesis of cis-1-amino-1,3-dicarboxycyclohexane, which has a relatively rigid 5-carbon chain that is identical to that of the extended form of L-glutamate (Fig. 1) [21]. This compound is an excellent substrate of the enzyme and confirms conlusions drawn from the studies with the monomethyl derivatives of glutamate.

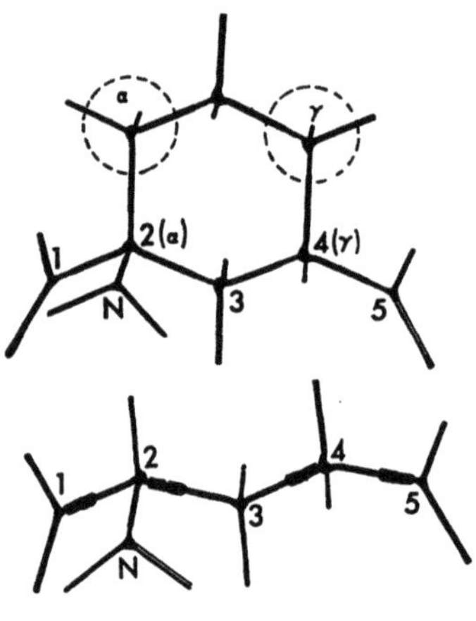

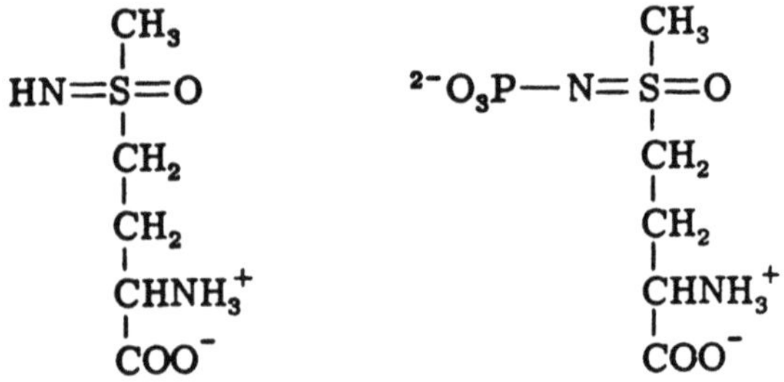

Fig. 2. Methionine sulfoximine and methionine sulfoximine phosphate

◀**Fig. 1.** Dreiding models of L-glutamate and cis-L-1-amino-1,3-dicarboxycyclohexane (cis-L-cycloglutamate) [21]

Extension of this work by use of a computer approach [22] showed that L-glutamate attaches to the active site of the enzyme in an extended conformation in which the carboxyl groups are as far apart (or almost so) as possible, and in which the α-hydrogen atom of L-glutamate is directed away from the enzyme. It was possible to identify five points on the enzyme derived from the positions of particular atoms of L-glutamate, i.e., the α-carboxyl carbon atom, the α-amino nitrogen atom, the γ-carboxyl oxygen atom which is phosphorylated, the γ-carboxyl oxygen atom which is not phosphorylated, and the nitrogen atom of the tetrahedral intermediate derived from L-glutamate. A stereograph of the tetrahedral addition compound formed in the reaction of L-γ-glutamyl phosphate with ammonia is shown in Fig. 3. L-Methionine-S-sulfoximine was found to be an inhibitory analog of the intermediate or transition state formed in the catalytic reaction. This analog is phosphorylated on the sulfoximine nitrogen, which occupies the site on the enzyme that normally binds the γ-carboxyl oxygen atom that is phosphorylated; a stereograph showing enzyme-bound L-methionine-S-sulfoximine is given in Fig. 4. Mapping studies

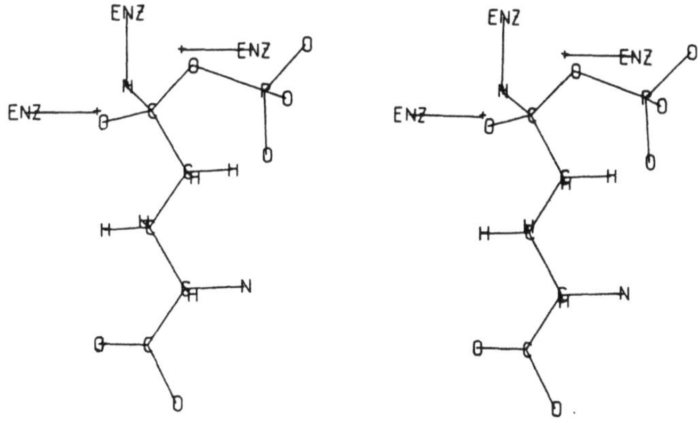

Fig. 3. Stereograph of the tetrahedral addition compound formed in the reaction of L-γ-glutamyl phosphate with ammonia [22]

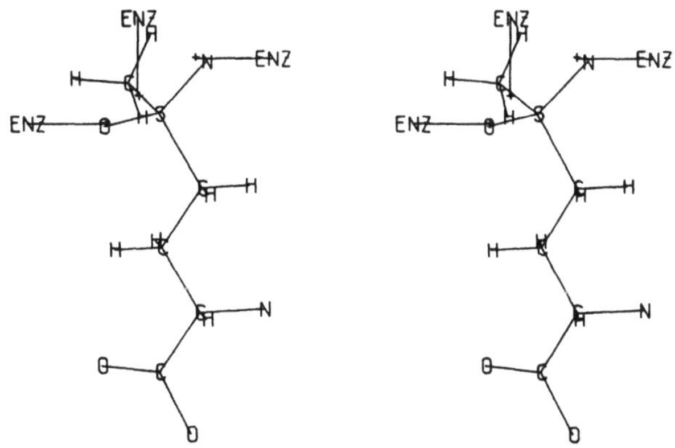

Fig. 4. Stereograph of L-methionine-S-sulfoximine at the active site [22]

thus provided insight into the way in which the enzyme interacts with methionine sulfoximine. The sulfoximine oxygen atom of L-methionine-S-sulfoximine binds to the site that normally binds the glutamate carboxyl oxygen atom that is not phosphorylated.

The methyl group of methionine sulfoximine binds to the ammonia binding site of the enzyme. These studies provided additional support for the view that glutamate is phosphorylated in the normal catalytic reaction. Further evidence for formation of enzyme-bound γ-glutamyl phosphate was obtained by incubating the enzyme with glutamate and ATP labeled with ^{18}O in the $\beta\gamma$-bridge oxygen [23]. Reversible cleavage of the P_β-O-$P_\gamma$bond was detected by appearance of ^{18}O in the β-nonbridge oxygens of ATP. Furthermore and importantly, the relative rate of scrambling was about the same as the maximal velocity of the enzyme-catalyzed reaction. Other evidence indicating intermediate formation of γ-glutamyl phosphate was obtained by trapping this intermediate by treatment with borohydride and subsequent isolation of α-amino-δ-hydroxyvalerate [24].

The mapping studies described above served usefully in the design of selective sulfoximine inhibitors of glutamine synthetase and of γ-glutamyl cysteine synthetase. The reaction catalyzed by a latter enzyme also involves formation of enzyme-bound γ-glutamyl phosphate. Interestingly, the same diastereoisomer of methionine sulfoximine inhibits both enzymes. As discussed in detail elsewhere [25], although both enzymes utilize γ-glutamyl phosphate intermediates, they have substantially different active sites. α-Ethylmethionine sulfoximine was found to bind to glutamine synthetase and to irreversibly inhibit it [26]. However, α-ethylmethionine sulfoximine does not inhibit γ-glutamyl cysteine synthetase, because this enzyme cannot accommodate the α-ethyl moiety. Substitution of the S-methyl moiety of methionine sulfoximine by propyl or butyl moieties (to yield prothionine sulfoximine and buthionine sulfoximine, respectively) gave molecules which could interact with γ-glutamyl cysteine synthetase effectively, but which could not bind to glutamine synthetase [27, 28]. Thus, the bulky propyl and butyl moieties do not attach to the enzyme site that normally binds ammonia; however, the methyl moiety of methionine sulfoximine can be accommodated at this site. These findings are of importance in providing specific sulfoximine inhibitors of the two synthetases; as discussed elsewhere [29], buthionine sulfoximine has been of considerable value for the specific inhibition of glutathione synthesis. It is notable that α-ethylmethionine sulfoximine is a convulsant. That buthionine sulfoximine does not produce convulsions, whereas α-ethylmethionine sulfoximine does, provides strong evidence that the convulsant activity of methionine sulfoximine is due to inhibition of glutamine synthesis rather than to an effect on the synthesis of glutathione.

The convulsant effect of methionine sulfoximine appears to be directly related to inhibition of glutamine synthetase, which is highly concentrated in the glial cells [30]. Decreased synthesis of glutamine would be expected to lead to increases in glutamate concentration; such increases at specific neuronal sites might be expected to lead to increased stimulation. Although conversion of methionine sulfoximine to other products might conceivably lead to convulsions by another mechanism, the finding that α-ethylmethionine sulfoximine, which would not be expected to appreciably metabolized, is a convulsant does not support this idea. Despite these considerations, the conclusion that inhibition of glutamine synthetase is the cause of convulsions has been questioned [31]. Methionine sulfoximine has been reported to have a protective effect against ammonia-induced convulsions [32, 33]. It would be interesting to learn whether α-ethylmethionine sulfoximine has a similar effect. Further studies on the mechanism of this apparent protective effect of methionine sulfoximine and on the compartmentation of ammonia and amino acid metabolism in the brain [31] are needed.

Most of the studies on mammalian glutamine synthetase have been concerned with the brain and liver enzymes. Considerable amounts of glutamine arise from skeletal muscle. Glutamine synthetase has been purified from rat and rabbit skeletal muscle and from rat heart [11]. Highly purified rat skeletal muscle glutamine synthetase is markedly inhibited by ADP; this finding, which is not noted with the brain and liver enzymes, requires further study.

Glutamine Amidotransferases and Glutaminases

There are now 13 enzymes known that catalyze the utilization of the amide nitrogen atom of glutamine [34, 35]. Eight of these enzymes occur in mammalian tissues (Table 1). The glutamine amidotransferases bind glutamine and mediate utilization of its amide nitrogen atom for particular synthesis reactions. Almost all of the glutamine amidotransferases can use ammonia in place of glutamine. Ammonia may appear to be less active than glutamine, but when comparisons

Table 1. Glutamine amidotransferases

Asparagine synthetase	Aspartate + ATP + glutamine + $H_2O \rightarrow$ asparagine + glutamate + AMP + PP_i
Carbamyl phosphate synthetase II	HCO_3^- + 2 ATP + glutamine + $H_2O \rightarrow$ carbamyl phosphate + 2 ADP + P_i + glutamate
CTP synthetase	UTP + ATP + glutamine + $H_2O \rightarrow$ CTP + ADP + P_i + glutamate
Formylglycinamide ribonucleotide synthetase	Formylglycinamide ribonucleotide + glutamine + ATP + $H_2O \rightarrow$ formylglycinamidine ribonucleotide + ADP + P_i + glutamate
Glucosamine-6-phosphate synthase	Fructose-6-P + glutamine $\rightarrow$ glucosamine-6-P + glutamate
GMP synthetase	XMP + ATP + glutamine + $H_2O \rightarrow$ GMP + AMP + PP_i + glutamate
NAD synthetase	Deamido-NAD^+ + ATP + glutamine + $H_2O \rightarrow NAD^+$ + AMP + PP_i + glutamate
5-Phosphoribosylamine synthase	5-Phosphoribosyl pyrophosphate + glutamine + H_2O $\rightarrow$ 5-phosphoribosylamine + PP_i + glutamate

are made on the basis of un-ionized ammonia, this may not be the case. Most of the glutamine amidotransferases exhibit glutaminase activity and are inhibited by glutamine analogs, such as L-2-amino-4-oxo-5-chloropentanoate, L-azaserine, and 6-diazo-5-oxo-L-norleucine. These reagents bind irreversibly to the glutamine binding site of these enzymes, and in many instances it has been shown that an enzyme sulfhydryl group is involved. The glutamine binding ability of these enzymes may be decreased by procedures used for enzyme purification; thus, a modified amidotransferase may be produced which reacts more effectively with ammonia than with glutamine. All of these enzymes catalyze reactions that involve phosphorylated substrates.

Glutamate synthase, a glutamine amidotransferase found in certain microorganisms and plants, catalyzes the reductive amination of α-ketoglutarate by glutamine according to Reaction 8 [36]. This reaction has not as yet been

$$\text{glutamine} + \alpha\text{-ketoglutarate} + \text{NADPH} + \text{H}^+ \rightarrow 2\ \text{glutamate} + \text{NADP}^+ \quad (8)$$

demonstrated in animal tissues. Evidence has been reported that the amide nitrogen atom of glutamine can be transferred directly to pyruvate, oxaloacetate, and α-ketoglutarate by kidney cortex homogenates [37]. The mechanism involved is not yet clear.

Glutaminase activity has been mentioned in the literature since 1904 [38]. Krebs observed that extracts of mammalian brain, retina, liver, and kidney deamidated glutamine [39]. Greenstein and colleagues [40–43] described phosphate-activated glutaminase (glutaminase I), an activity associated with kidney mitochondria and pyruvate-activated glutaminase (glutaminase II); the latter is explained by coupled transaminase-ω-amidase reactions (see below). Greenstein also found glutaminase activity in the sedimentable fraction of rat liver and kidney, which may be identical to that later called "phosphate-independent glutaminase"; this activity, which is stimulated by maleate, is identical to γ-glutamyl transpeptidase [44, 45]. The latter has also been referred to as γ-glutamyl transferase; such use of this term is, however, potentially confusing because a number of enzymes (e.g., glutamine synthetases, glutamine amidotransferases, glutaminases) can catalyze γ-glutamyl transfer. As stated above, many glutamine amidotransferases exhibit glutaminase activity, which under certain conditions can be substantial; see for example, [46].

It is probable that the phosphate-activated glutaminase of kidney is a major catalyst for the formation of urinary ammonia. However, it should be noted that, although glutamine amide nitrogen is a major precursor of urinary ammonia, it is not the sole source. Thus, appreciable amounts of urinary ammonia are derived from the α-amino nitrogen atom of glutamine [47].

Transamination of Glutamine

Glutamine transaminase activity was discovered in the course of studies on the activating effect of α-keto acids on the deamidation of glutamine by liver preparations [8, 48, 49]. Glutamine transaminates with a number of α-keto acids to yield the corresponding amino acids and α-ketoglutaramate; this keto acid amide is hydrolyzed in a separate step by α-keto acid-ω-amidase to α-keto-glutarate and ammonia (Fig. 5).

Two types of glutamine transaminase are now known to occur in rat tissues. Glutamine transaminase L may be considered as a glutamine-methionine transaminase and glutamine transaminase K may be regarded as a glutamine, phenylalanine, methionine transaminase [8, 50, 51]. The two types of enzymes may be conveniently distinguished by their relative activities towards albizziin and phenylpyruvate. Glutamine transaminase L effectively utilizes albizziin in place of glutamine as an amino donor, whereas albizziin is a poor substrate of the K type enzyme. Conversely, phenylpyruvate is a much better substrate of glutamine transaminase K than of the L-type. The L-type is present in the soluble and mitochondrial fractions of liver, and smaller amounts of this enzyme occur in the soluble fraction of the kidney and brain. Glutamine transaminase K activity is found in kidney, liver, brain, and to some extent in other mammalian tissues. These enzymes have been brought to a high state of purity. The L-enzyme is a dimer of mol. wt. 110,000; the K-enzyme is a dimer with a mol. wt. of about 100,000. Although these enzymes can use a number of amino acid substrates, especially phenylalanine and methionine, it appears probable that the physiological substrate is glutamine. The wide distribution of the glutamine transaminases suggests that transamination is a major metabolic function of glutamine. The metabolic breakdown of many amino acids proceeds via the corresponding α-keto acids. The amino acid levels of mammalian tissues probably fluctuate and amino acids may accumulate temporarily in amounts which exceed those necessary for protein synthesis and other metabolic purposes. Accumulated amino acids may be converted to α-keto acids potentially leading to a loss of the corresponding carbon chains. The glutamine transaminases may function to aminate α-keto acids and thereby serve a salvage function for their carbon chains. Loss of methionine sulfur and carbon may occur in association with conversion of S-adenosylmethionine to 5′-methylthioadenosine during the biosynthesis of polyamines. Methionine may be salvaged by formation of 5′-methylthioribose-1-phosphate and α-keto-γ-methiolbutyrate. Conversion of the latter compound to methionine takes place by transamination with glutamine [52]. There is evidence that in the dog [53] the glutamine

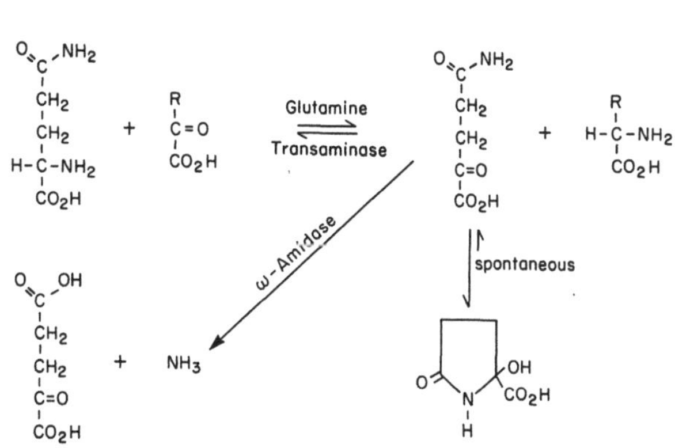

Fig. 5. Reactions involved in transamination of glutamine and in the spontaneous cyclization of α-ketoglutaramate to 2-hydroxy-5-oxoproline

transaminase pathway may contribute significantly to renal ammonia production.

Transamination between glutamine and α-keto acids is driven by removal of α-ketoglutaramate (Fig. 5). Although the transamination reaction is freely reversible, glutamine utilization is favored because α-ketoglutaramate either cyclizes or is hydrolyzed by ω-amidase. It, therefore, appears that the glutamine transaminases function in a homeostatic mechanism for preservation of amino acid balance in which conversion of a α-keto acid to an amino acid is favored. The metabolic balance may be further stabilized by the production of ammonia which is incorporated into the amide group of glutamine by the action of glutamine synthetase (Reactions 9–12). Thus, in such a scheme transamination

$$\text{phenylpyruvate} + \text{glutamine} \rightleftharpoons \text{phenylalanine} + \alpha\text{-ketoglutaramate} \quad (9)$$

$$\alpha\text{-ketoglutaramate} \rightarrow \alpha\text{-ketoglutarate} + NH_3 \quad (10)$$

$$\text{glutamate} + NH_3 + \text{ATP} \rightleftharpoons \text{glutamine} + \text{ADP} + P_i\,. \quad (11)$$

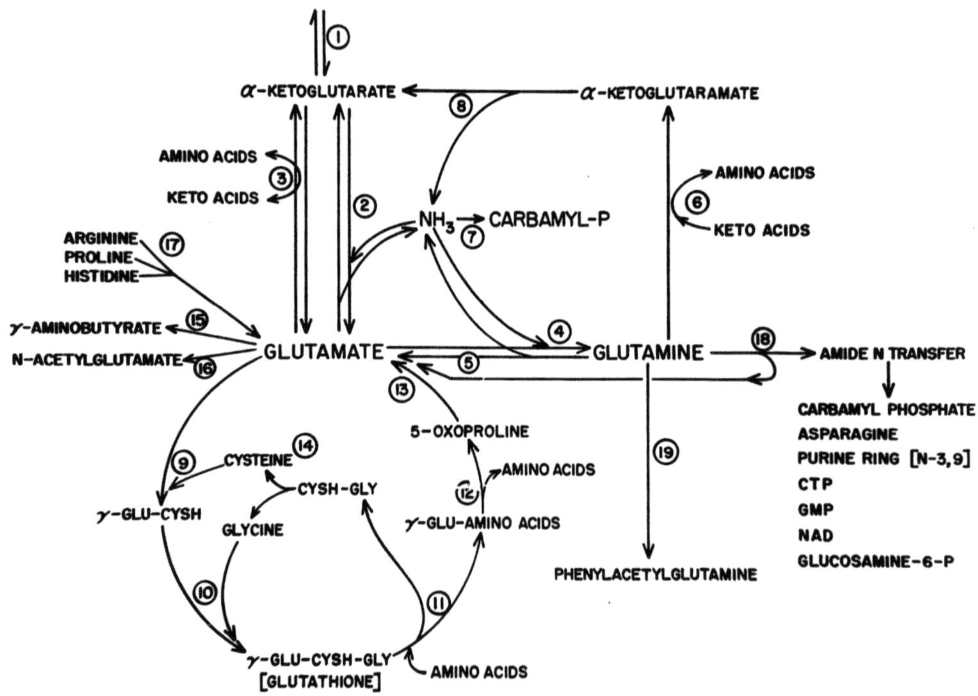

Fig. 6. An outline of glutamate metabolism in mammalina tissues. *1* Reactions of the citric acid cycle; *2* glutamate dehydrogenase; *3* glutamate transaminases; *4* glutamine synthetase; *5* glutaminase; *6* glutamine transaminase; *7* carbamyl phosphate synthetase (liver); *8* α-keto acid ω-amidase; *9* γ-glutamyl cysteine synthetase; *10* glutathione synthetase; *11* γ-glutamyl transpeptidase; *12* γ-glutamyl cyclotransferase; *13* 5-oxoprolinase; *14* cysteinylglycinase; *15* glutamate decarboxylase; *16* glutamate N-acetylase; *17* various enzymes involved in the degradation of these amino acids; *18* glutamine amidotransferases known to occur in mammalian tissues; and *19* phenylacetyl glutamine synthetase (acyl-CoA-L-glutamine N-acyltransferase) [54]

Sum:

$$\begin{aligned}&\text{phenylpyruvate} + \text{glutamate} + \text{ATP}\\&\rightarrow \text{phenylalanine} + \alpha\text{-ketoglutarate} + \text{ADP} + \text{P}_i\end{aligned} \quad (12)$$

reactions would be metabolically coupled with the endergonic synthesis of glutamine, and the equilibrium of the overall reaction (Reaction 12) would be far to the right.

Overview of Glutamine Metabolism

The scheme given in Fig. 6 provides an outline of glutamine metabolism in mammalian tissues, and indicates important interrelationships between the metabolism of glutamine and that of glutamate and α-ketoglutarate. These pathways illustrate significant metabolic connections involving the citric acid cycle, various aspects of the metabolism of individual amino acids, the formation of γ-aminobutyrate, the reactions of the γ-glutamyl cycle, and the utilization of ammonia or the synthesis of glutamate and of carbamyl phosphate.

Although Fig. 6 gives the major reactions and catalysts involved, other important considerations influence glutamine metabolism in different cell types, such as the amounts of the several enzymes present, their intracellular disposition, and factors that affect their synthesis and regulation. Other chapters in this volume deal with these questions.

References

1. Meister A (1980) Catalytic mechanism of glutamine synthetase; overview of glutamine metabolism. In: Palacios R, Mora J (eds) Glutamine: Metabolism, enzymology, and regulation. Academic Press, London New York, pp 1–40
2. Pitts RF (1975) Production of CO_2 by the intact functioning kidney of the dog. In: Baruch S (ed) The medical clinics of North America, vol 59, no 3. Saunders, Philadelphia, p 507
3. Windmueller HG, Spaeth AE (1980) Respiratory fuels and nitrogen metabolism in vivo in small intestine of fed rats. J Biol Chem 255: 107
4. Biltz RM, Letteri JM, Pellegrino ED, Pinkus L (1982) Glutamine: A new metabolic substrate. In: Massry SG, Letteri JM, Ritz E (eds) Regulation of phosphate and mineral metabolism. Adv Exp Med Biol 151: 423
5. Zielke HR, Ozand PT, Tildon JT, Sevdalian DA, Cornblath M (1980) Reciprocal regulation of glucose and glutamine utilization by cultured human diploid fibroblasts. J Cell Physiol 95: 41
6. Reitzer LJ, Wice BM, Kennell D (1979) Evidence that glutamine, not sugar, is the major energy source for cultured hela cells. J Biol Chem 254: 2669
7. Moldave K, Meister A (1957) Synthesis of phenylacetylglutamine by human tissues. J Biol Chem 229: 463
8. Cooper AJL, Meister A (1977) The glutamine transaminase-ω-amidase pathway. Crit Rev Biochem 4: 281–303
9. Meister A (1974) Glutamine synthetase of mammals. Enzymes 10: 699–754
10. Pamiljans V, Krishnaswamy PR, Dumville G, Meister A (1962) Studies on the mechanism of glutamine synthesis; isolation and properties of the enzyme from sheep brain. Biochemistry 1: 153–157
11. Rowe WB (1984) Glutamine synthetase from muscle. Methods Enzymol (in press)

12. Levintow L, Meister A (1953) Enzymatic synthesis of D-glutamine and related hydroxamic acids. J Am Chem Soc 75:3039
13. Krishnaswamy PR, Pamiljans V, Meister A (1962) Studies on the mechanism of glutamine synthesis; evidence for the formation of an enzyme-bound activated glutamic acid. J Biol Chem 237:2932–2940
14. Khedouri E, Wellner VP, Meister A (1964) Enzymatic synthesis of β-aminoglutaramic acid (β-glutamine) by glutamine synthetase: evidence for the utilization of β-aminoglutarylphosphate. Biochemistry 3:824–828
15. Tsuda Y, Stephani RA, Meister A (1971) Direct evidence for the formation of an acyl phosphate by glutamine synthetase. Biochemistry 10:3186–3189
16. Ronzio R, Meister A (1968) Phosphorylation of methionine sulfoximine by glutamine synthetase. Proc Natl Acad Sci USA 59:164–170
17. Ronzio RA, Rowe WB, Wilk S, Meister A (1969) Preparation and studies on the characterization of sheep brain glutamine synthetase. Biochemistry 8:2670–2674
18. Rowe WB, Ronzio RA, Meister A (1969) Inhibition of glutamine synthetase by methionine sulfoximine, studies on methionine sulfoximine phosphate. Biochemistry 8:2674–2680
19. Kagan HM, Meister A (1966) Stereospecific activity of glutamine synthetase toward threo-β-methyl-D-glutamic acid. Biochemistry 5:725–732
20. Meister A (1968) The specificity of glutamine synthetase and its relationship to substrate conformation at the active site. Adv Enzymol 31:183–218
21. Gass JD, Meister A (1970) 1-Amino-1,3-dicarboxycyclohexane (Cycloglutamic acid). A new glutamic acid analog and a substrate of glutamine synthetase. Biochemistry 9:842–846
22. Gass JD, Meister A (1970) Computer analysis of the active site of glutamine synthetase. Biochemistry 9:1380–1390
23. Midelfort CF, Rose IA (1976) A stereochemical method for detection of ATP terminal phosphate transfer in enzymatic reactions. J Biol Chem 251:5881–5887
24. Todhunter JA, Purich DL (1975) Use of the sodium borohydride reduction technique to identify a γ-glutamyl phosphate intermediary in the escherichia coli glutamine synthetase reaction. J Biol Chem 250:3505–3509
25. Meister A (1978) Inhibition of glutamine synthetase and γ-glutamylcysteine synthetase by methionine sulfoximine and related compounds. In: Seiler N, Jung MJ, Koch-Weser J (eds) Enzyme-activated irreversible inhibitors. Elsevier North Holland Biomedical Press, Amsterdam, pp 187–211
26. Griffith OW, Meister A (1978) Differential inhibition of glutamine and γ-glutamylcysteine synthetases by α-alkyl analogs of methionine sulfoximine that induce convulsions. J Biol Chem 253:2333–2338
27. Griffith OW, Anderson ME, Meister A (1979) Inhibition of glutathione biosynthesis by prothionine sulfoximine (S-n-propyl-homocysteine sulfoximine). A selective Inhibitor of γ-glutamylcysteine synthetase. J Biol Chem 254:1205–1210
28. Griffith OW, Meister A (1979) Potent and specific inhibition of glutathione synthesis by buthionine sulfoximine (S-n-butyl homocysteine sulfoximine). J Biol Chem 254:7558–7560
29. Meister A, Anderson ME (1983) Glutathione. Ann Rev Biochem 52:711–760
30. Norenberg MD, Martinez-Hernandez A (1979) Fine structural localization of glutamine synthetase in astrocytes in brain. Brain Res 161:303
31. Cooper AJL, Vergara F, Duffy TE (1984) Cerebral glutamine synthetase. In: Hertz L, Kramme E, McGeer EG, Schousboe A (eds) Metabolic relationships between glutamine, glutamate and GABA in the CNS. Liss, New York (in press)
32. Warren KS, Schenker S (1964) Effect of an inhibitor of glutamine synthesis (Methionine sulfoximine) on ammonia toxicity and metabolism. J Lab Clin Med 64:442
33. Hindfeldt B, Plum F (1975) L-Methionine-DL-sulfoximine and acute ammonia toxicity. J Pharm Pharmacol 27:456
34. Meister A (1975) Structure-function relationships in glutamine amidotransferases; Carbamyl phosphate synthetase. PAABS Rev 4:273–299
35. Buchanan JM (1973) The amidotransferases. Adv Enzymol 39:91–184
36. Tempest DW, Meers JL, Brown CM (1970) Synthesis of glutamate in *Aerobacter aerogenes* by a hitherto unknown route. Biochem J 117:405

37. O'Donovan DJ, Lotspeich WD (1969) The role of the amide group of glutamine in renal biosynthesis of amino acids. Enzymologia 36: 301
38. Lang S (1904) Über Desamidierung im Tierkörper. Beitr Chem Phys Pathol V: 321–345
39. Krebs HA (1935) Metabolism of amino-acids. IV. The synthesis of glutamine from glutamic acid and ammonia, and the enzymic hydroysis of glutamine in animal tissues. Biochem J 29: 1951–1969
40. Carter CE, Greenstein JP (1947) Acceleration of enzymatic desamidation of glutamine by several inorganic anions. J Natl Cancer Inst 7: 433–436
41. Errera M, Greenstein JP (1949) Phosphate-activated glutaminase in kidney and other tissues. J Biol Chem 178: 495–502
42. Greenstein JP, Carter CE (1947) Influence of α-keto acids on desamidation of amino acid amides. J Natl Cancer Inst 7: 55–60
43. Otey MC, Birnbaum SM, Greenstein JP (1958) Solubilized kidney glutaminase. Arch Biochem Biophys 49: 245–246
44. Tate SS, Meister A (1974) Stimulation of the hydrolytic activity and decrease of the transpeptidase activity of γ-glutamyl transpeptidase by maleate; Identity of a rat kidney maleate-stimulated glutaminase and γ-glutamyl transpeptidase. Proc Natl Acad Sci USA 74: 3329–3333
45. Curthoys NP, Kuhlenschmidt T (1975) Phosphate-independent glutaminase from rat kidney. Partial purification and identity with γ-glutamyltranspeptidase. J Biol Chem 250: 2099–2105
46. Wellner VP, Meister A (1975) Enhancement of the glutaminase activity of carbamyl phosphate synthetase by alterations in the interaction between the heavy and light subunits. J Biol Chem 250: 3261–3266
47. Pitts RF, deHaas J, Klein J (1963) Relation of renal amino and amide nitrogen extraction to ammonia production. Am J Physiol 204: 187–191
48. Meister A, Tice SV (1950) Transamination from glutamine to α-keto acids. J Biol Chem 187: 173–187
49. Meister A, Sober HA, Tice SV, Fraser PE (1952) Transamination and associated deamidation of asparagine and glutamine. J Biol Chem 197: 319–330
50. Cooper AJL, Meister A (1972) Isolation and properties of highly purified glutamine transaminase. Biochemistry 11: 661–671
51. Cooper AJL, Meister A (1974) Isolation and properties of a new glutamine transaminase from rat kidney. J Biol Chem 249: 2554–2561
52. Backlund PS Jr, Chang CP, Smith RA (1982) Identification of 2-keto-4-methylthiobutyrate as an intermediate compound in methionine synthesis from 5′-methylthioadenosine. J Biol Chem 257: 4196
53. Fine A, Scott J, Bourke E (1972) Studies on the glutamine aminotransferase-ω-amidase pathway in human kidney in vitro. J Lab Clin Med 80: 591
54. Meister A (1979) Biochemistry of glutamate· glutamine and glutathione. In: Garattini et al. (eds) Adv Biochem Physiol. Raven Press, New York, pp 69–84

Enzymes of Renal Glutamine Metabolism

N. P. CURTHOYS, R. A. SHAPIRO, and W. G. HASER[1]

Introduction

Renal extraction and metabolism of plasma glutamine is an important component of the homeostatic adaptation that occurs in response to metabolic acidosis (Tannen 1978, Kovacevic and McGivan 1983). The amide and amine nitrogens of glutamine are released as ammonium ions. The increased ammoniagenesis provides an expendable cation which facilitates the excretion of anions and titratable acids without depleting the body's supply of sodium and potassium. Renal metabolism of glutamine also provides the precursor needed for the increased gluconeogenesis that is associated with metabolic acidosis. It has been estimated that following adaptation to a prolonged fast, approximately 50% of the total gluconeogenesis within man occurs in the kidney (Cahill 1970). Thus, the enzymes involved in the primary pathway of renal glutamine metabolism should exhibit properties consistant with a role in this process. The enzymology of renal glutamine metabolism has been most extensively characterized in the rat. Therefore, this review will be limited largely to this species.

During normal acid-base balance, rat renal extraction of glutamine is not significant (Squires et al. 1976, Hughey et al. 1980). However, during onset of acute acidosis, renal extraction of plasma glutamine is rapidly increased (Fine et al. 1978, Hughey et al. 1980). Simultaneously, the arterial plasma glutamine concentration is increased two-fold. Thus, increased renal ammoniagenesis initially requires acute adaptations in both renal and nonrenal metabolism of glutamine. In chronic metabolic acidosis, the arterial plasma glutamine concentration is decreased one-third compared to normal (Squires et al. 1976). Under these conditions, the increased renal glutamine utilization is partially compensated by a decreased splanchnic extraction (Heitmann and Bergman 1978).

Increased rat renal ammonia synthesis and excretion is evident within 30 min after onset of acute acidosis. Renal levels of glutamate, α-ketoglutarate and malate are decreased within 1 h (Boyd and Goldstein 1979). Within 2–3 h, isolated kidney slices exhibit increased ammoniagenesis from glutamine (Alleyne 1970). Increased synthesis of phosphoenolpyruvate carboxykinase, the rate limiting activity in renal gluconeogenesis, is evident by 2 h (Iynedjian and

1 Department of Biochemistry, University of Pittsburgh School of Medicine, Pittsburgh, PA 15261, USA

Glutamine Metabolism in Mammalian Tissues
Edited by D. Häussinger and H. Sies

Hanson 1977). The carboxykinase activity measured in crude homogenates is increased by 5–6 h and reaches a plateau three fold higher than normal within 2 days (Alleyne 1970). Glutaminase activity is also increased three-fold, but this adaptation requires 7 days (Leonard and Orloff 1955). Feeding NH_4Cl is the usual method of experimentally inducing acidosis in rats. However, a similar sequence of adaptations have been observed in response to feeding a high protein diet (Brosnan et al. 1978) or during onset of diabetic ketoacidosis (Goldstein et al. 1980). This sequence of adaptations also illustrates the changes which occur during the transition from acute to chronic acidosis. In chronic acidosis, the initial decreases observed in plasma pH and bicarbonate concentration are frequently compensated, whereas urinary acidification and increased ammoniagenesis are maintained (Parry and Brosnan 1978). Therefore, the factors that regulate renal ammonia synthesis may be different in the chronic and fully adapted state.

Renal metabolism of glutamine can be initiated by either transamination or deamidation. Rat kidney contains two isoenzymes of glutamine transaminase (Cooper and Meister 1977). The α-ketoglutaramate produced in this reaction is then deamidated by ω-deamidase.

$$\text{glutamine} + \alpha\text{-keto acid} \rightleftarrows \alpha\text{-ketoglutaramate} + \alpha\text{-amino acid}$$

$$\alpha\text{-ketoglutaramate} + H_2O \rightarrow \alpha\text{-ketoglutarate} + NH_4^+$$

The combined reactions were initially referred to as the glutaminase II pathway. However, more recent characterization of the kinetics and specificity of the glutamine transaminase suggest that it functions primarily to provide an exergonic salvage pathway for the reamination of α-keto acids derived from various essential amino acids. As such these enzymes play a relatively minor role in renal glutamine metabolism.

Katunuma et al. (1966) reported that rat kidney contains two distinct activities that deamidate glutamine; a phosphate-dependent glutaminase and a phosphate-independent, but maleate-stimulated glutaminase. The former glutaminase is immunologically identical to the glutaminase contained in rat brain and intestine, but distinct from the liver glutaminase (Curthoys et al. 1976a). The phosphate-independent glutaminase activity is a partial reaction of γ-glutamyltranspeptidase (Curthoys and Kuhlenschmidt 1975). Maleate increases the apparent glutaminase activity by inhibiting transpeptidation and by stimulating the hydrolytic activity. When assayed in crude homogenates of rat kidney tissue, the two enzymes catalyze a similar rate of glutamine hydrolysis. Thus, it is important to assess the relative contribution of the two enzymes to renal metabolism of glutamine.

Properties of γ-Glutamyltranspeptidase

γ-Glutamyltranspeptidase was originally described as an activity which catalyzes the hydrolysis of glutathione to glutamate and cysteinylglycine. It was subsequently shown that this enzyme can also catalyze the transfer of the γ-glutamyl group to a variety of amino acids and dipeptides to form γ-glutamyl peptides and cysteinylglycine. Some dipeptides, neutral amino acids, and cystine are the best acceptor substrates, whereas branched chain, acidic, and basic amino acids are poor acceptors, and D-amino acids and L-proline are inactive (Tate and Meister 1974). The enzyme also exhibits a broad specificity towards the γ-glutamyl donor. As a result, it will utilize various S-substituted glutathione derivatives as substrates. In addition, γ-glutamyltranspeptidase will utilize glutamine as a substrate for either a glutaminase or γ-glutamyltransferase reaction (Curthoys and Kuhlenschmidt 1975). Given the broad specificity of γ-glutamyltranspeptidase, the determination of its physiological function requires a characterization of its localization within the kidney and the identification of its primary physiological substrate.

Within kidney, the γ-glutamyltranspeptidase is primarily associated with the brush border membrane of the proximal tubules (Glossman and Neville 1972). Histochemical analysis has indicated that within rat kidney, the transpeptidase is distributed along the entire length of the proximal tubule (Glenner et al. 1962, Albert et al. 1964). However, analysis of individually dissected segments of renal tubules indicates that transpeptidase activity is three to ten times greater in the proximal straight tubule than in the proximal convoluted portion of the nephron (Curthoys and Lowry 1973, Heinle et al. 1977). Small amounts of γ-glutamyltranspeptidase may also be associated with the glomeruli (Marathe et al. 1979), the renal capillaries (Dass et al. 1981), and the basolateral membrane of the proximal tubule cells (Spater et al. 1982).

γ-Glutamyltranspeptidase is an amphipathic integral membrane glycoprotein (Curthoys and Hughey 1979). The enzyme possesses a limited hydrophobic domain that is responsible for its membrane association. The catalytic activity is contained within a separate hydrophilic domain that is highly glycosylated and positioned away from the membrane surface by a sequence of amino acids that is sensitive to papain. As a result, the γ-glutamyltranspeptidase purified following solubilization by limited papain proteolysis is soluble in aqueous buffers and has a mol. wt. of 69,000 (Hughey and Curthoys 1976). In contrast, purification following solubilization with Triton X-100 yields an amphipathic enzyme that is soluble only at concentrations of detergent above the critical micelle concentration and has a mol. wt. of 87,000. Only the amphipathic form of the enzyme can be reconstituted into unilamellar phosphatidylcholine vesicles (Hughey et al. 1979). The transpeptidase is composed of two nonidentical subunits. The molecular weight of the smaller subunit which contains the γ-glutamyl binding site (Tate and Meister 1977) is unaltered by papain treatment. However, the large subunit of the Triton purified enzyme is 21,000 daltons greater than that of the papain purified enzyme (Horiuchi et al. 1978). The isolated small subunits of the two forms of enzyme both contain NH_2-terminal threonine residues, whereas the NH_2-terminal residues of the large subunits are nonidentical (Tsuji et al. 1980). Furthermore, the reaction of the reconstituted enzyme with the membrane soluble, photoactivatable reagent, 3-trifluoromethyl-3-(m-[^{125}I]iodophenyl)-diazerine results in the labeling of only a limited segment of the large subunit (Frielle et al. 1982). Thus, the NH_2-terminal segment of the large subunit apparently contains the hydrophobic membrane binding domain of γ-glutamyltranspeptidase.

Various experiments have been carried out to determine the membrane orientation of γ-glutamyltranspeptidase. The transpeptidase activity associated with rat renal brush border membrane vesicles is inhibited by glutathione bound to impermeable dextran polymers (Horiuchi et al. 1978). In addition, electron microscopic analysis has established that ferritin conjugated antibodies specific for γ-glutamyltranspeptidase can interact with the luminal surface of the proximal tubule cells (Silbernagl et al. 1978) and with the external surface of isolated brush border membrane vesicles (Marathe et al. 1979). These observations indicate that the active site of the transpeptidase is at least in part localized on the external surface of the brush border membrane. The absolute asymmetry of this orientation was established by characterizing the proteolytic sensitivity and immunological reactivity of γ-glutamyltranspeptidase associated with isolated brush border membrane vesicles (Tsao and Curthoys 1980). Due to this asymmetry of orientation, the primary substrates available to γ-glutamyltranspeptidase in vivo are the metabolites contained within the lumen of the proximal tubule. The small amount of γ-glutamyltranspeptidase that may be associated with the peritubular membranes may react directly with metabolites contained in the plasma as it passes through the kidney.

Glutamine filtered by the renal glomeruli is rapidly and nearly quantitatively reabsorbed from the tubular lumen (Silbernagl 1980). Thus, the concentration of glutamine within the luminal fluid is unlikely to exceed the plasma concentration of glutamine. The concentration of glutathione within the luminal fluid is uncertain. The large pool of glutathione contained within the proximal tubular cells has a very rapid turnover (Sekura and Meister 1974, Brehe et al. 1976). An integral step in the catabolism of tubular glutathione is its translocation across the brush border membrane to the tubular lumen (Griffith and Meister 1979a, b). Thus, the concentration of glutathione within the luminal fluid may exceed its concentration within plasma.

The rate at which γ-glutamyltranspeptidase will utilize glutamine as a substrate is only 1%–2% of its rate of utilization of glutathione (Tate and Meister 1975). However, the glutaminase activity of this enzyme is stimulated seven to ten fold by the presence of 10–50 m*M* concentrations of maleate (Curthoys and Kuhlenschmidt 1975) or hippurate (Thompson and Meister 1980). It has been reported that metabolic acidosis results in a doubling of hippurate excretion (Welbourne and Dass 1981b). However, it is unlikely that the hippurate concentration in the lumen of the proximal tubule reaches the m*M* concentrations necessary to produce the stimulatory effect observed in vitro. The K_M of the transpeptidase for glutamine is difficult to determine due to stimulation of the γ-glutamyltransferase reaction which occurs at high concentrations of glutamine. However, the approximate K_M value determined by quantitating glutamate formation is greater than the plasma concentration of glutamine (Curthoys and Godfrey 1976). In contrast, the K_M of γ-glutamyltranspeptidase for glutathione hydrolysis is 5 µ*M* (McIntyre and Curthoys 1979), a value which is significantly less than the plasma concentration of glutathione. The difference in kinetic constants suggest that in vivo the γ-glutamyltranspeptidase is unlikely to significantly utilize glutamine as a substrate.

The observation that γ-glutamyltranspeptidase activity is increased in response to chronic metabolic acidosis has been offered as an indication that this enzyme may contribute to renal ammonia synthesis (McFarlane-Anderson and Alleyne 1977, Welbourne and Dass 1981a). Shapiro and Curthoys (1981) have confirmed that the specific activity of γ-glutamyltranspeptidase assayed with γ-glutamyl-p-nitroanilide is slightly, but significantly increased (1.3-fold) during prolonged acidosis. This increased activity is retained throughout a ten fold purification of brush border membrane vesicles. However, the specific activity of alkaline phosphatase is also increased slightly in both crude homogenates and in isolated brush border membranes derived from acidotic animals. Therefore, this adaptation may result from the generalized renal hypertrophy which occurs in response to chronic acidosis (Lotspeich 1967). When the isolated kidney is perfused in the absence of glutathione, or its amino acid precursors, the renal tissue is rapidly and significantly depleted of glutathione (Ormstad et al. 1980). In the absence of normal levels of glutathione, perfusion with glutamine may make an alternative substrate for the γ-glutamyltranspeptidase available. This nonphysiological situation may account for much of the data observed with isolated perfused kidney which suggests that γ-glutamyltranspeptidase significantly contributes to the metabolism of glutamine (Welbourne 1974, 1978).

We have recently utilized a selective and irreversible inhibitor of γ-glutamyltranspeptidase to determine whether this enzyme participates in renal metabolism of glutamine (Shapiro and Curthoys 1981). Between 1 and 7 h after injection of L-(αS,5S)-α-amino-3-chloro-4,5-dihydro-5-isoxazole acetic acid (AT-125), more than 95% of renal γ-glutamyltranspeptidase is inactivated. Even after 24 h, the transpeptidase activity returns to only 15%–20% of normal. In contrast, injection of AT-125 had no effect on the total renal phosphate-dependent glutaminase activity. When a set of AT-125 treated rats were made acutely acidotic, the observed increase in renal ammonia excretion was not significantly different from that observed in noninhibited, acutely acidotic control animals. In addition, injection of AT-125 had no effect on the increase in

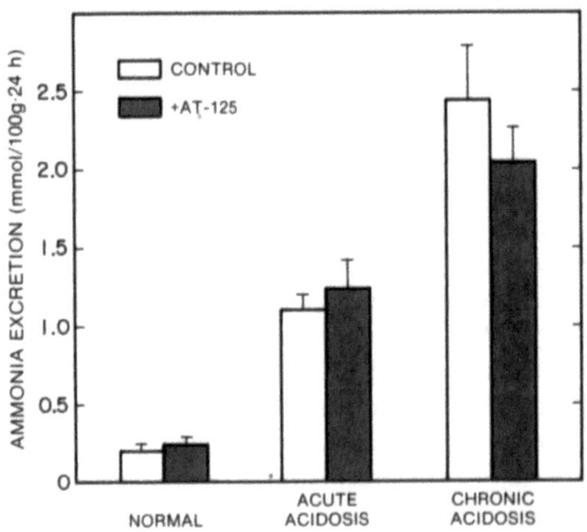

Fig. 1. Urinary ammonia excretion from control and *AT-125* treated rats. Acute acidosis was induced by stomach loading rats with 2 mmol NH_4Cl/100 g, whereas chronic acidosis was induced by providing rats with a 0.28 *M* NH_4Cl solution as their sole source of drinking water for 7 days

plasma glutamine concentration and the acidification of urine that are associated with onset of acute acidosis. Furthermore, the renal ammonia excretion observed in normal and chronic acidotic animals was unaffected by the injection of AT-125 (Fig. 1). In contrast, AT-125 inactivation of γ-glutamyltranspeptidase causes a mild glutathionemia (2.5-fold) and a pronounced glutathionuria (3,000-fold). These results indicate that γ-glutamyltranspeptidase activity is unlikely to participate in renal glutamine metabolism. Instead, this enzyme appears to function primarily in the extracellular catabolism of glutathione (McIntyre and Curthoys 1980, Meister 1981).

Properties of Phosphate-Dependent Glutaminase

The phosphate-dependent glutaminase is a mitochondrial enzyme that requires the presence of a polyvalent anion for activity (Curthoys et al. 1976b). Phosphate, the most commonly used activator, produces half-maximal activation, at 20–30 m*M* concentrations. However, the activation process exhibits a broad specificity and the physiological effector is unknown. Some monovalent anions also effect glutaminase activity, i.e., chloride, but not acetate, reverse the phosphate activation. The glutaminase activity is also strongly inhibited by renal tissue concentrations of glutamate (Goldstein 1966).

The phosphate-dependent glutaminase can be solubilized by resuspending renal mitochondria that have been lyophilized in the presence of a borate phosphate buffer (Curthoys et al. 1976b). Under these conditions, the solubilized enzyme undergoes extensive polymerization to yield an aggregate of sufficient molecular weight that is excluded during gel filtration on Sepharose 4B. Dialysis into a Tris buffer causes the selective dissociation of the glutaminase and permits its purification by rechromatography on Sepharose 4B. The Tris-form of the enzyme is labile and exhibits an absolute dependence upon the addition of polyvalent anions for activity.

Gel filtration and velocity sedimentation analysis were used to determine that the purified glutaminase has a mol. wt. of 160,000 in Tris or barbitol buffers (Godfrey et al. 1977). In the presence of saturating concentrations of phosphate, the protomeric enzyme forms dimers of 330,000 mol. wt. Increasing concentrations of phosphate cause an increase in the sedimentation coefficient that correlates well with the phosphate activation profile. Correlation between activation and dimer formation was also observed with other polyvalent anion activators. Moreover, the activation and dimer formation was found to be reversed to a similar extent by increasing concentrations of NaCl or of glutamate. These results suggest that the phosphate-dependent glutaminase is active only as a dimer or larger aggregate. However, these studies could not exclude the possibility that binding of phosphate changes the monomer conformation sufficiently to produce activation and that the new conformation leads to self-association.

In order to establish whether dimerization is essential for the phosphate-induced activation of the glutaminase, Morehouse and Curthoys (1981) characterized the properties of the protomeric and dimeric forms of the enzyme immobilized on CNBr-activated Sepharose. Covalent attachment of the dimeric form of the glutaminase was achieved with approximately 85% retention of activity. At least 70% of the bound glutaminase activity was found to be independent of phosphate. In addition, 6-diazo-5-oxo-L-norleucine, which interacts only with the catalytically active form of the glutaminase (Shapiro et al. 1979), inactivates the bound dimeric form of glutaminase at the same rate in either the absence or presence of added phosphate. Therefore, retention of dimeric structure is sufficient to maintain glutaminase activity even in the absence of phosphate. In contrast, the coupling of the protomeric form of the enzyme to Sepharose resulted in retention of only 3% of the phosphate-induced glutaminase activity. However, up to 48% of this activity could be reconstituted by the addition of soluble glutaminase under conditions that promote dimerization. These results indicate that the monomeric form of the glutaminase has minimal inherent activity and that dimerization is an essential step in the phosphate-induced activation of the glutaminase.

The protomeric and dimeric forms of the glutaminase also exhibit different sensitivities towards inactivation by two different glutamine affinity labels. The phosphate-dependent glutaminase is rapidly inactivated by the stoichiometric binding of L-2-amino-4-oxo-5-chloropentanoic acid (Shapiro et al. 1978). However, concentrations of phosphate which increase the glutaminase activity decrease the rate of inactivation by the chloroketone. Furthermore, inactivation is blocked by glutamate, but not by glutamine. These data indicate that the chloroketone reacts preferentially with the protomeric form of the glutaminase and that the enzyme may have a specific binding site for glutamate.

6-Diazo-5-oxo-L-norleucine is both an effective affinity label and a substrate for the phosphate-dependent glutaminase (Clark et al. 1982). Both reactions exhibit a similar phosphate-dependent activation profile. However, the hydrolysis of the diazoketone to L-glutamate occurs at a rate approximately 1,000-fold greater than the rate of enzyme inactivation. Glutamate protects against inactivation and proportionally reduces the extent of [6-^{14}C]diazo-ketone binding. However, the ability of glutamate to protect the glutaminase against inactivation is reversed by increasing concentrations of phosphate (Shapiro et al. 1979). Furthermore, the addition of o-carbamoyl-L-serine, a glutamine analog that promotes dimer formation, was found to stimulate inactivation by the diazoketone but to protect the glutaminase against inactivation by the chloroketone. These results indicate that the diazoketone interacts specifically with a catalytically active group that is located at the glutamine binding site of the glutaminase.

The results of the studies using the two affinity labels were initially interpreted to indicate that glutamine and glutamate interact with separate sites on the glutaminase. To further investigate this possibility, a sensitive radioactive assay was used to determine the kinetics of a particulate preparation of the mitochondrial glutaminase purified essentially free of γ-glutamyltranspeptidase (Shapiro et al. 1982). Increasing concentrations of phosphate cause a sigmoidal activation of the membrane-bound glutaminase. The K_M for glutamine was found to decrease from 36 to 4 mM as the phosphate concentration was increased from 5 to 100 mM. Glutamate was found to be a competitive inhibitor

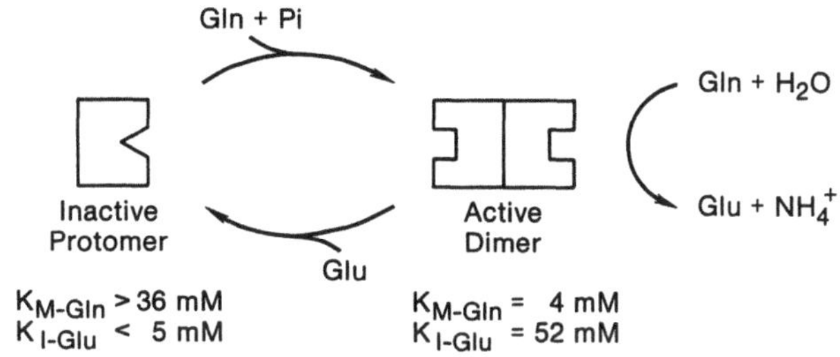

Fig. 2. Effect of phosphate and glutamate on the structure and activity of the phosphate-dependent glutaminase. Increasing concentrations of phosphate promote formation of the catalytically active dimer that preferentially binds glutamine. Increasing concentrations of glutamate reverse this process and favor the formation of the catalytically inactive protomer that preferentially binds glutamate

with respect to glutamine at both high and low concentrations of phosphate. However, the K_I for glutamate is increased from 5 to 52 m*M* with increasing phosphate concentration. Therefore, glutamine and glutamate interact with the same site on the glutaminase, but the conformation and specificity of this site is determined by the concentration of phosphate and the extent of dimerization (Fig. 2).

Native Form of Phosphate-Dependent Glutaminase

Purified rat renal phosphate dependent glutaminase migrates on SDS gels as a series of bands which range in mol. wt. from 48,000–59,000 (Fig. 3). The individual peptides found in the glutaminase preparation were isolated by eluting them from a sliced SDS gel. When subjected to partial hydrolysis with chymotrypsin and resubjected to electrophoresis, each of the peptides yielded the same pattern of proteolytic fragments (Clark and Curthoys 1979). Furthermore, when the purified glutaminase was stoichiometrically inactivated with either L-2-amino-4-oxo-5-chloro[5-^{14}C]pentanoic acid or 6-diazo-5-oxo-L-[6-^{14}C])-norleucine, fluorographic analysis indicated that all of the peptides bound the two affinity labels. Therefore, all of the peptides were structurally related and each peptide retained the active site of the glutaminase. Immunoprecipitates of freshly solubilized renal mitochondria contain only a 65,000 dalton form of the glutaminase (Fig. 3). If the Triton solubilized mitochondria are stored for 1–3 days at 4° C, the glutaminase is degraded to a series of peptides of molecular weight identical to those observed in purified enzyme. Therefore, the heterogeneity of subunit size found in the purified glutaminase results from partial proteolysis which occurs following solubilization of the mitochondrial glutaminase. These results also suggest that the native form of the glutaminase is a multimer of identical subunits that have a mol. wt. of 65,000.

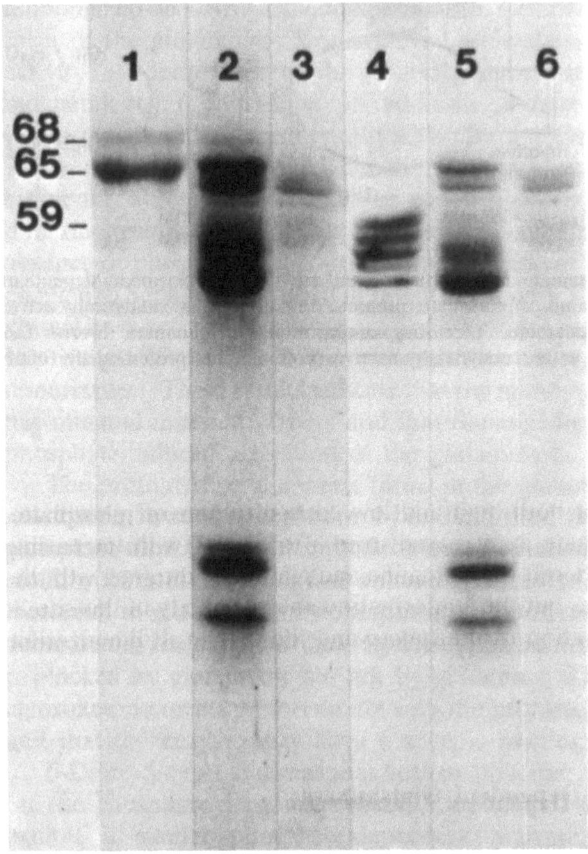

Fig. 3. Sodium dodecyl sulfate gel electrophoresis of purified or immunoprecipitated phosphate-dependent glutaminase. Glutaminase was purified from either rat brain or kidney (lanes *1* and *4*, respectively) or immunoprecipitated from crude homogenates of rat brain or kidney tissue solubilized with 1% Triton X-100 (lanes *2* and *5*, respectively). Nonspecific precipitation of contaminating proteins in the absence of antibody from crude homogenates using rat brain or kidney tissue solubilized with 1% Triton X-100 (lanes *3* and *6*, respectively)

Anti-glutaminase antibodies exhibit equivalent titer against the activities contained in Triton X-100 solubilized rat renal and brain mitochondria (Curthoys et al. 1976a). However, the purified brain glutaminase contains a predominant peptide of mol. wt. of 65,000 and a minor peptide of mol. wt. of 68,000 (Fig. 3). The purified brain glutaminase exhibits identical patterns of phosphate-induced dimerization and competitive inhibition by glutamate. Immunoprecipitates of freshly solubilized brain mitochondria contain both forms of glutaminase which remain stable during prolonged storage at 4° C. The presence of a kidney-specific protease was confirmed by incubating the supernatant of a renal mitochondrial extract depleted of glutaminase with solubilized brain mitochondria. The previously stable brain glutaminase was degraded to a pattern similar to that observed with the purified kidney enzyme.

Immunoblot analysis indicates that the glutaminase in intact rat brain and kidney mitochondria contains equal amounts of the 65,000 and 68,000 mol. wt. peptides. Cleveland digest analysis indicates that the two peptides are structurally related and may be derived from the same gene product. The functional significance of this observation remains to be determined.

Cellular Distribution of Phosphate-Dependent Glutaminase

The cellular distribution of the glutaminase has been determined by quantitative analysis of its activity in individually dissected segments of the renal nephron (Curthoys and Lowry 1973). In normal rat kidney, the glutaminase activity is greatest in the distal portions of the nephron, intermediate in the proximal convoluted tubules, and very low in glomeruli and proximal straight tubules. During chronic acidosis, glutaminase activity is increased only in the proximal convoluted tubules. As a result, the 3-fold increase in glutaminase activity observed in the whole organ is due to a 20-fold increase in activity in these tubules. Thus, in chronic acidosis, the glutaminase activity is greatest within the proximal convoluted tubule.

Immunological experiments have established that the increased glutaminase activity is due to an increased amount of the glutaminase enzyme (Cuthoys et al. 1976a). Experiments with actinomycin D indicate that the adaptation in the level of glutaminase is not required for increased ammoniagenesis during acute acidosis (Goldstein 1965), except in conditions where renal tissue mass is limiting (Benyajati and Goldstein 1975, 1978). However, after adaptation to chronic acidosis, the increase in glutaminase activity correlates well with the increased urinary excretion of ammonium ions (Rector et al. 1955). Mitochondria isolated from acidotic rats retain the ability to exhibit increased ammoniagenesis from glutamine. Therefore, the increased glutaminase activity may facilitate the maintenance of increased glutamine metabolism during chronic acidosis. However, following recovery from acidosis, the in vivo production of ammonium ions returns to normal within 1 day, whereas the increased glutaminase activity and ammonia synthesis by isolated mitochondria are retained for at least 3 days (Parry and Brosnan 1978). Therefore, in vivo glutamine metabolism is probably influenced by metabolism occurring in the cytoplasm.

Glutaminase activity is labile under the conditions generally used to carry out submitochondrial fractionation. However, when stabilized by the addition of borate, the glutaminase activity fractionates by both digitonin-lubrol, and swell-shrink sonication procedures as an enzyme associated with the inner mitochondrial membrane (Curthoys ana Weiss 1974). In addition, ten fold greater concentrations of various detergents are required to solubilize the glutaminase from kidney mitochondria than are required to release soluble matrix activities (Curthoys et al. 1976b). Furthermore, an Arrhenius plot of glutaminase activity in freeze-thawed mitochondria exhibits a discontinuity which disappears following treatment with Triton X-100 (Kovacevic 1976). When assayed at subsaturating concentrations of substrate and activator, the mitochondrial glutaminase exhibits a partial latency (Kalra and Brosnan 1974). Based on this evidence, it is generally accepted that the phosphate-dependent glutaminase is primarily associated with the internal or matrix surface of the inner mitochondrial membrane. However, none of the data currently available has quantitated the extent of asymmetry of this orientation.

Mitochondrial transport of glutamine occurs through a specific carrier and by a uniport mechanism (Adam and Simpson 1974, Brosnan and Hall 1977).

Considerable interest has been expressed in the possibility that regulation of mitochondrial transport could control the renal utilization of glutamine. Adam and Simpson (1974) studied the ability of rat renal mitochondria to transport and metabolize glutamine under conditions where glutamate oxidation was inhibited. They showed that the ability of mitochondria to accumulate radioactivity from [^{14}C]glutamine within the matrix increases in response to metabolic acidosis. This increase was proportional to the increase in glutaminase activity. Analysis of the radioactive material associated with mitochondria suggested that only glutamate and not glutamine could be detected within the matrix space. They therefore concluded that either transport was rate-limiting, or the glutaminase was a component of the transport system. In these studies, the amount of glutamine within the matrix space was calculated from the amount of glutamine associated with the mitochondria which was in excess of that required to equilibrate the space outside the inner mitochondrial membrane. The validity of this calculation is dependent upon the assumption that all of the glutaminase activity associated with the mitochondrial preparation is contained inside the inner mitochondrial membrane. The inability to detect glutamine in the matrix space of mitochondria isolated by differential centrifugation was found to be due to their contamination with brush border membranes containing γ-glutamyltranspeptidase (Curthoys and Shapiro 1978). The glutaminase activity of this enzyme was shown to be responsible for approximately half of the total activity measured under the conditions of the transport experiment. With mitochondria purified free of γ-glutamyltranspeptidase activity, glutamine was found to be present within the matrix space during glutamine transport. The apparent increase in glutamine transport observed in acidotic mitochondria was due to an accumulation of additional glutamate. Under these conditions, the concentration of glutamine in the matrix is unchanged. Therefore, glutamine transport and glutaminase activity are distinct processes. However, the inability to measure initial rates of transport in the absence of metabolism has made it impossible to firmly establish whether adaptive changes in mitochondrial transport of glutamine occurs during acidosis or if transport is rate-limiting for glutamine metabolism. In more recent experiments, Kovacevic and Bajin (1982) have used mitochondria preloaded with [^{14}C]glutamine to measure rates of efflux at 0° C. Under these conditions, the rate of glutamine transport was five times faster than the rate of glutaminase activity. These results suggest that the activity of the glutamine carrier is not the rate-limiting step in renal metabolism of glutamine.

Regulation of Renal Phosphate-Dependent Glutaminase

In mitochondria isolated from normal rats, glutamate produced from glutamine is metabolized primarily by transamination of oxalacetate to form aspartate (Schoolwerth et al. 1978). Extramitochondrial glutamate is also metabolized preferentially by this reaction. In contrast, during acidosis, the flux of mitochondrial glutamate through glutamate dehydrogenase is greatly increased (Kunin and Tannen 1979). As a result, the amino nitrogen of glutamine is also released as ammonium ions. The α-ketoglutarate produced from glutamine is then metabolized primarily to form glucose (Vinay et al. 1978). It has been proposed that increased renal gluconeogenesis during acidosis causes the stimulation of renal glutamine metabolism (Goodman et al. 1966). According to this theory, accelerated use of α-ketoglutarate for glucose synthesis causes an increased flux of glutamate through glutamate dehydrogenase and a net decrease in the intramitochondrial concentration of α-ketoglutarate and glutamate. A decreased concentration of glutamate would cause release of glutaminase from inhibition and would thus stimulate ammonia synthesis. In support of this theory, renal tissue levels of glutamine, glutamate, and α-ketoglutarate are decreased during acidosis (Hems and Brosnan 1971) and phosphoenolpyruvate carboxykinase activity is increased (Alleyne and Scullard 1969).

The decrease in plasma pH observed following onset of acute acidosis could also initiate the alterations in renal glutamine metabolism. However, kidney slices (Relman and Narins 1975) and isolated mitochondria (Tannen and Kunin 1976) when incubated in a slightly acidified medium (pH = 7.0) exhibit either a decreased or an unaltered level of ammonia production. In contrast, the isolated perfused kidney responds to acute acidification of the perfusate with a prompt increase in urine flow, a decrease in fractional sodium reabsorption, and acidification of the urine pH from 6.0 to 5.1 (Tannen and Ross 1979). These changes are accompanied by an increased ammonia production, the magnitude of which closely correlates with the observed decrease in urine pH. The increase in ammonia production is diminished in studies carried out with a nonfiltering perfused kidney. Following acidification of the perfusion medium, the renal concentrations of malate and α-ketoglutarate are decreased. Inhibition of gluconeogenesis by blocking phosphoenolpyruvate carboxykinase activity with mercaptopicolinic acid restores malate concentrations to normal and prevents the increase in ammonia production without affecting the acidification of the urine (Ross and Tannen 1979). These studies support the hypothesis that removal of ammonium ions by urinary acidification and the increased disposal of glutamine carbon, via gluconeogenesis are critical stimuli for increasing glutamine metabolism during acute acidosis. Both of these adaptations would remove products of the glutamate dehydrogenase reaction and would shift its equilibrium towards increased α-ketoglutarate formation, thus reducing mitochondrial glutamate levels. Addition of either fluorocitrate (Bourke et al. 1979) or fluoroacetate (Lemieux et al. 1979) to renal slices effectively blocks citrate metabolism and results in decreased tissue glutamate and α-ketoglutarate levels. This is accompanied by significant increases in glutamine metabolism. These results are also compatible with the hypothesis that glutamine deamidation is increased secondary to enhanced glutamate deamination.

In summary, these observations suggest that the sensitivity of the mitochondrial phosphate-dependent glutaminase activity to changes in effector metabolite concentrations may be the key regulator of renal glutamine metabolism. Only the dimeric form of the purified glutaminase is catalytically active. Various activators and inhibitors of the glutaminase affect its activity by altering its extent of dimerization. However, little is known about the biochemical mechanism by which the membrane associated form of the glutaminase responds to changes in effector molecules. In addition, the mechanism by which glutaminase is induced specifically in the proximal convoluted tubules during acidosis remains to be characterized.

References

Adam W, Simpson DP (1974) Glutamine transport in rat kidney mitochondria in metabolic acidosis. J Clin Invest 54: 165–174

Albert Z, Orlowski M, Szewczuk S (1964) Histochemical and biochemical investigations of γ-glutamyltranspeptidase in the tissues of man and laboratory rodents. Acta Histochem 18: 78–89

Alleyne GAO (1970) Renal metabolic response to acid base changes. II. The early effects of metabolic acidosis on renal metabolism in the rat. J Clin Invest 49: 943–951

Alleyne GAO, Scullard GA (1969) Renal metabolic response to acid base change. I. Enzymatic control of renal ammoniagenesis in the rat. J Clin Invest 48: 364–370

Benyajati S, Goldstein L (1975) Renal glutaminase adaptation and ammonia excretion in infant rats. Am J Physiol 228: 693–698

Benyajati S, Goldstein L (1978) Relation of ammonia excretion adaptation to glutaminase activity in acidotic, subtotalnephrectomized rats. Kidney Int 14: 50–57

Bourke E, Frindt G, Schreiner GE, Preuss HG (1979) Effects of fluorocitrate on renal ammoniagenesis and glutamine metabolism in the intact dog kidney. Kidney Int 15: 255–263

Boyd TA, Goldstein L (1979) Kidney metabolite level and ammonia production in acute acid-base alterations in the rat. Am J Physiol 236: E289–E295

Brehe JE, Chan AW, Alvey TR, Burch HB (1976) Effect of methionine sulfoximine on glutathione and amino acid levels in the nephron. Am J Physiol 231: 1536–1540

Brosnan JT, Hall B (1977) The transport and metabolism of glutamine by kidney-cortex mitochondria from normal and acidotic rats. Biochem J 164: 331–337

Brosnan JT, McPhee P, Hall B, Parry DM (1978) Renal glutamine metabolism in rats fed high-protein diets. Am J Physiol 235: E261–E265

Cahill GF Jr (1970) Starvation in man. New Engl J Med 282: 668–675

Clark VM, Curthoys NP (1979) Cause of subunit heterogeneity in purified rat renal phosphate-dependent glutaminase. J Biol Chem 254: 5939–5941

Clark VM, Shapiro RA, Curthoys NP (1982) Comparison of the hydrolysis and the covalent binding of 6-diazo-5-oxo-L-[6-^{14}C]norleucine by rat renal phosphate-dependent glutaminase. Arch Biochem Biophys 213: 232–239

Cooper AJL, Meister A (1977) The glutamine transaminase-ω-amidase pathway. CRC Crit Rev Biochem 4: 281–303

Curthoys NP, Godfrey SS (1976) Properties of rat kidney glutaminase enzymes and their role in renal ammoniagenesis. In: Schmidt U, Dubach UC (eds) Renal metabolism in relation to renal function. Huber, Bern, pp 346–356

Curthoys NP, Hughey RP (1979) Characterization and physiological function of rat renal γ-glutamyltranspeptidase. Enzyme 24: 383–403

Curthoys NP, Kuhlenschmidt T (1975) Phosphate-independent glutaminase from rat kidney. Partial purification and identity with γ-glutamyltranspeptidase. J Biol Chem 250: 2099–2105

Curthoys NP, Lowry OH (1973) The distribution of glutaminase isoenzymes in the various structures of the nephron in normal, acidotic and alkalotic rat kidney. J Biol Chem 248: 162–168

Curthoys NP, Shapiro RA (1978) Effect of metabolic acidosis and phosphate on the presence of glutamine within the matrix space of rat renal mitochondria during glutamine transport. J Biol Chem 253: 63–68

Curthoys NP, Weiss RF (1974) Regulation of renal ammoniagenesis. Subcellular localization of rat kidney glutaminase isoenzymes. J Biol Chem 249: 3261–3266

Curthoys NP, Kuhlenschmidt T, Godfrey SS, Weiss RF (1976a) Phosphate-dependent glutaminase from rad kidney. Cause of increased activity in response to acidosis and identity with glutaminase from other tissues. Arch Biochem Biophys 172: 162–167

Curthoys NP, Kuhlenschmidt T, Godfrey SS (1976b) Regulation of renal ammoniagenesis purification and characterization of phosphate-dependent glutaminase from rat kidney. Arch Biochem Biophys 174: 82–89

Dass PD, Misra RP, Welbourne TC (1981) Presence of γ-glutamyltranspeptidase in the renal microvascular compartment. Can J Biochem 59: 383–386

Fine A, Bennett FI, Alleyne GAO (1978) Effects of acute acid base alterations on glutamine metabolism and renal ammoniagenesis in the dog. Clin Sci 54: 503–508

Frielle T, Brunner J, Curthoys NP (1982) Isolation of the hydrophobic domain of rat renal γ-glutamyltranspeptidase selectively labeled with 3-trifluoromethyl-3-(m-[^{125}I]idophenyl)diazerine. J Biol Chem 257: 14979–15982

Glenner CG, Folk JE, McMillan PJ (1962) Histochemical demonstration of γ-glutamyltranspeptidase-like activity. J Histochem Cytochem 19: 481–489

Glossman H, Neville DM Jr (1972) γ-Glutamyltranspeptidase in kidney brush border membranes. FEBS Lett 19: 340–344

Godfrey SS, Kuhlenschmidt T, Curthoys NP (1977) Correlation between activation and dimer formation of rat renal phosphate-dependent glutaminase. J Biol Chem 252: 1927–1931

Goldstein L (1965) Actinomycin D inhibition of the adaptation of renal glutamine-deamidating enzymes in the rat. Nature (London) 205: 1330–1331

Goldstein L (1966) Relation of glutamate to ammonia production in the rat kidney. Am J Physiol 210: 661–666

Goldstein L, Boylan JM, Schröck H (1980) Adaptation of renal ammonia production in the diabetic ketoacidotic rat. Kidney Int 17: 57–65

Goodman AD, Fuisz RE, Cahill GF (1966) Renal gluconeogenesis in acidosis, alkalosis and potassium deficiency: its possible role in regulation of renal ammonia production. J Clin Invest 45: 612–619

Griffith OW, Meister A (1979a) Translocation of intracellular glutathione to membrane-bound γ-glutamyltranspeptidase as a discrete step in the γ-glutamyl cycle: glutathionuria after inhibition of transpeptidase. Proc Natl Acad Sci USA 76: 268–272

Griffith OW, Meister A (1979b) Glutathione: Inter-organ translocation, turnover and metabolism. Proc Natl Acad Sci USA 76: 5606–5610

Heinle H, Wendel A, Schmidt U (1977) The activities of the key enzymes of the γ-glutamyl cycle in microdissected segments of the rat nephron. FEBS Lett 73: 220–224

Heitmann RN, Bergman EN (1978) Glutamine metabolism, interorgan transport, and glucogenicity in the sheep. Am J Physiol 234: E197–E203

Hems DA, Brosnan JT (1971) Effect of metabolic acidosis and starvation on the content of intermediary metabolites in rat kidney. Biochem J 123: 391–397

Horiuchi S, Inoue M, Morino Y (1978) γ-Glutamyltranspeptidase: sidedness of its active site on renal brush border membrane. Eur J Biochem 87: 429–437

Hughey RP, Curthoys NP (1976) Comparison of the size and physical properties of γ-glutamyltranspetidase purified from rat kidney following solubilization with papain or Triton X-100. J Biol Chem 251: 7863–7870

Hughey RP, Coyle PJ, Curthoys NP (1979) Comparison of the association and orientation of γ-glutamyltranspeptidase in lecithin vesicles and in native membranes. J Biol Chem 254: 1124–1128

Hughey RP, Rankin BB, Curthoys NP (1980) Acute acidosis and renal arteriovenous differences of glutamine in normal and adrenalectomized rats. Am J Physiol 238: F199–F204

Iynedjian PB, Hanson RW (1977) Messenger RNA for renal phosphoenolpyruvate carboxykinase (GTP). Its translation in a heterologous cell-free system and its regulation by glucocorticoids and by changes in acid-base balance. J Biol Chem 252: 8398–8403

Kalra J, Brosnan JT (1974) The subcellular localization of glutaminase isoenzymes in rat kidney cortex. J Biol Chem 249: 3255–3260

Katunuma H, Tomino I, Nishino H (1966) Glutaminase isoenzymes in rat kidney. Biochem Biophys Res Commun 22: 321–328

Kovacevic Z (1976) Importance of the flux of phosphate across the inner membrane of kidney mitochondria for the activation of glutaminase and the transport of glutamine. Biochim Biophys Acta 430: 399–412

Kovacevic Z, Bajin K (1982) Kinetics of glutamine efflux from liver mitochondria loaded with the ^{14}C-labeled substrate. Biochim Biophys Acta 687: 291–295
Kovacevic Z, McGivan JD (1983) Mitochondrial metabolism of glutamine and glutamate and its physiological significance. Physiol Rev 63: 547–605
Kunin AS, Tannen RL (1979) Regulation of glutamate metabolism by renal cortical mitochondria. Am J Physiol 237: F55–F62
Lemieux G, Vinay P, Baverel G, Briere R, Gougoux A (1979) Relationship between lactate and glutamine metabolism in vitro by the kidney: differences between dog and rat and importance of alanine synthesis in the dog. Kidney Int 16: 451–458
Leonard E, Orloff J (1955) Regulation of ammonia excretion in the rat. Am J Physiol 182: 131–138
Lotspeich WD (1967) Metabolic aspects of acid-base change. Science 155: 1066–1075
Marathe GV, Nash B, Haschemeyer RH, Tate SS (1979) Ultrastructural localization of γ-glutamyltranspeptidase in rat kidney and jejunum. FEBS Lett 107: 436–440
McFarlane-Anderson N, Alleyne GAO (1977) The effect of metabolic acidosis on γ-glutamyltranspeptidase activity in rat kidney. FEBS Lett 79: 51–53
McIntyre TM, Curthoys NP (1979) Comparison of the hydrolytic and transfer activities of rat renal γ-glutamyltranspeptidase. J Biol Chem 254: 6499–6504
McIntyre TM, Curthoys NP (1980) The inter-organ metabolism of glutathione. Int J Biochem 12: 545–551
Meister A (1981) On the cycles of glutathione metabolism and transport. In: Estabrook R, Srere P (eds) Current topics in cellular regulation, vol 18. Academic Press, New York, pp 21–58
Morehouse RF, Curthoys NP (1981) Properties of rat renal phosphate-dependent glutaminase coupled to Sepharose. Evidence that dimerization is essential for activation. Biochem J 193: 709–716
Ormstad K, Lastbom T, Orrenius S (1980) Translocation of amino acids and glutathione studied with the perfused kidney and isolated renal cells. FEBS Lett 112: 55–59
Parry D, Brosnan JT (1978) Glutamine metabolism in the kidney during induction of, and recovery from, metabolic acidosis in the rat. Biochem J 174: 387–396
Rector FC Jr, Seldon DW, Copenhaver JH (1955) The mechanism of ammonia excretion during ammonium chloride acidosis. J Clin Invest 34: 20–26
Relman AS, Narins RG (1975) The control of ammonia production in the rat. Med Clin N Am 59: 583–593
Ross BD, Tannen RL (1979) Effect of decrease in bicarbonate concentration on metabolism of the isolated perfused rat kidney. Clin Sci 57: 103–111
Schoolwerth AC, Nazar BL, LaNoue KF (1978) Glutamate dehydrogenase activation and ammonia formation by rat kidney mitochondria. J Biol Chem 253: 6177–6183
Sekura R, Meister A (1974) Glutathione turnover in the kidney; considerations relating to the γ-glutamyl cycle and the transport of amino acids. Proc Natl Acad Sci USA 71: 2969–2972
Shapiro RA, Curthoys NP (1981) Differential effect of AT-125 on rat renal glutaminase activities. FEBS Lett 133: 131–134
Shapiro RA, Clark VM, Curthoys NP (1978) Covalent interaction of L-2-amino-4-oxo-5-chloropentanoic acid with rat renal phosphate-dependent glutaminase. J Biol Chem 253: 7086–7090
Shapiro RA, Clark VM, Curthoys NP (1979) Inactivation of rat renal phosphate-dependent glutaminase with 6-diazo-5-oxo-L-norleucine. Evidence for interaction at the glutamine binding site. J Biol Chem 254: 2835–2838
Shapiro RA, Morehouse RF, Curthoys NP (1982) Inhibition by glutamate of phosphate-dependent glutaminase of rat kidney. Biochem J 207: 561–566
Silbernagl S (1980) Tubular reabsorption of L-glutamine studied by free-flow micropuncture and microperfusion of rat kidney. Int J Biochem 12: 9–16
Silbernagl S, Pfaller W, Heinle H, Wendel A (1978) Topology and function of renal γ-glutamyltranspeptidase. In: Sies H, Wendel A (eds) Functions of glutathione in liver and kidney. Springer, Berlin Heidelberg New York, pp 60–69
Spater HW, Poruchynsky MS, Quintana N, Inoue M, Novikof AB (1982) Immunocytochemical localization of γ-glutamyltransferase in rat kidney with protein A-horseradish peroxidase. Proc Natl Acad Sci USA 79: 3547–3550

Squires EJ, Hall DE, Brosnan JT (1976) Arteriovenous differences for amino acids and lactate across kidney of normal and acidotic rats. Biochem J 160: 125–128
Tannen RL (1978) Ammonia metabolism. Am J Physiol 235: F265–F277
Tannen RL, Kunin AS (1976) Effect of pH on ammonia production by renal mitochondria. Am J Physiol 231: 1631–1637
Tannen RL, Ross BD (1979) Ammoniagenesis by isolated perfused rat kidney: the critical role of urinary acidification. Clin Sci 56: 353–364
Tate SS, Meister A (1974) Interaction of γ-glutamyltranspeptidase with amino acids, dipeptides, and derivatives and analogs of glutathione. J Biol Chem 249: 7593–7602
Tate SS, Meister A (1975) Identity of malate-stimulated glutaminase activity with γ-glutamyltranspeptidase in rat kidney. J Biol Chem 250: 4619–4627
Tate SS, Meister A (1977) Affinity labeling of γ-glutamyltranspeptidase and location of the γ-glutamyl binding site on the light subunit. Proc Natl Acad Sci USA 74: 931–935
Thompson GA, Meister A (1980) Modulation of γ-glutamyltranspeptidase activities by hippurate and related compounds. J Biol Chem 255: 2109–2113
Tsao B, Curthoys NP (1980) The absolute asymmetry of orientation of γ-glutamyltranspeptidase and aminopeptidase on the external surface of the rat renal brush border membrane. J Biol Chem 255: 7708–7711
Tsuji A, Matsuda Y, Katunuma (1980) Studies on the structure of γ-glutamyltranspeptidase. Location of the segment achoring γ-glutamyltranspeptidase to the membrane. J Biochem 87: 1567–1571
Vinay P, Mapes JP, Krebs HA (1978) Fate of glutamine carbon in renal metabolism. Am J Physiol 234: F123–F129
Welbourne TC (1974) Ammonia production and pathways of glutamine metabolism in the isolated perfused rat kidney. Am J Physiol 226: 544–548
Welbourne TC (1978) Cytoplasmic γ-glutamyltranspeptidase: isolation, product formation and physiological role. In: Guder WG, Schmidt U (eds) Biochemical nephrology. Hans Huber, Bern, pp 201–212
Welbourne TC, Dass PD (1981a) Mechanism of the acidosis induced adaptation in renal γ-glutamyltranspeptidase. Life Sci 28: 1219–1224
Welbourne TC, Dass PD (1981b) Role of hippurate in acidosis induced adaptation in renal γ-glutamyltranspeptidase. Life Sci 29: 253–258

Enzymes of Cerebral Glutamine Metabolism

E. KVAMME[1]

Introduction

Glutamine is a major amino acid in CNS. The glutamine family, which consists of glutamine (Gln), glutamate (Glu), γ-aminobutyrate (GABA), and aspartate (Asp), constitutes 70%–80% of the free amino acid nitrogen in mammalian brain (Timiras et al. 1973). Gln serves as metabolic precursor for the additional members of the Gln family, for other amino acids, amonia, amino sugars, proteins, purines, and pyrimidines (Review: Kvamme 1983a). Gln has important transport functions as a carrier of Glu, GABA, and ammonia; and may in this way contribute to energy metabolism and control the availability of amino acid transmitters. Evidence has thus been produced to show that Gln is most likely a precursor of transmitter Glu and probably also of GABA (Bradford et al. 1978, Hamberger et al. 1978, Reubi et al. 1978). Since brain lacks a complete urea cycle, ammonia is largely detoxified by formation of the nontoxic Gln, whereby nitrogen is also saved for synthesis of important compounds, such as proteins.

The following enzymes involving Gln metabolism are dealt with in this article, phosphate-activated glutaminase (EC 3.5.1.2), maleate-activated glutaminase (EC 2.3.2.2), transglutaminase (EC 2.3.2.13), and the Gln aminotransferase (EC 2.6.1.15), -ω-amidase (EC 3.3.1.3) pathway (Review: Kvamme 1983b). Glutamine synthetase (EC 6.3.1.2) which is the most important Gln synthesizing enzyme in brain, is considered to be outside the scope of this paper (Review: Kvamme 1983a). Phosphate activated glutaminase appears to be the major Gln metabolizing enzyme in brain.

1 Neurochemical Laboratory, Preclinical Medicine, University of Oslo, P.O. Box 1115 – Blindern, Oslo 3, Norway

Abbreviations. AOA: aminooxyacetic acid; Asp: aspartate; BHBD: β-hydroxybutyrate dehydrogenase; CNS: central nervous system; CSF: cerebrospinal fluid; cyclic-AMP: cyclic adenylic acid; cyclic GMP: cyclic guanylic acid; GABA: γ-aminobutyrate; γ-GT: γ-glutamyl transpeptidase; Gln: glutamine; Glu: glutamate; Hepes: N-2-hydroxy-ethylpiperazine-N′-2-ethanesulphonic acid; Lys: lysine; MAG: maleate-activated glutaminase; Mers: mersalyl; Met: methionine; NEM: N-ethylmaleimide; PAG: phosphate-activated glutaminase; PMB: p-mercuribenzoate; SEM: standard error of means

Glutamine Metabolism in Mammalian Tissues
Edited by D. Häussinger and H. Sies

Phosphate-Activated Glutaminase (PAG) [phosphate-dependent glutaminase, "brain type" of glutaminase (Krebs 1935) EC 3.5.1.2]

$$Gln + H_2O \rightarrow Glu + NH_4^+$$

The old term phosphate-activated glutaminase (PAG) is preferred to phosphate-dependent glutaminase, since the enyzme is activated by a variety of other compounds and is thus not dependent on phosphate.

The PAG reaction, which catalyzes the hydrolytic cleavage of Gln to yield Glu and ammonia, is for practical purposes irreversible. It is fast, energy-independent, and independent of any known cofactor or substrate other than Gln.

Molecular Properties

PAG was first described by Krebs in 1935; Errera and Greenstein (1949) discovered that the enzyme is activated by phosphate. Pig renal PAG was solubilized in 1954 (Otey et al.) and partially purified (ca. 300-fold) in 1958 (Klingman and Handler, Sayre and Roberts). Pig renal PAG and pig brain PAG were purified by our group to apparent homogeneity in 1970 and 1973, respectively (Kvamme et al. 1970, Svenneby et al. 1973, Kvamme and Svenneby 1975), making use of the property of the enzyme to solubilize and polymerize in a reversible manner. Later rat kidney PAG (Curthoys et al. 1976) and pig brain (Nimmo and Tipton 1980) have been purified using modifications of our method.

Purified pig kidney and brain PAG have mol. wts. of 130,000–150,000 and 120,000–135,000, respectively, when solubilized in Tris-HCl buffer and keeping the concentration of enzyme protein below 0.1 mg/ml. At higher protein concentrations PAG will dimerize in the presence of phosphate, but polymers are also found (Kvamme et al. 1970, Kvamme and Svenneby 1975, Curthoys et al. 1976). On addition of borate to the phosphate buffer, insoluble polymers with mol. wts. of $1.5-2.0 \times 10^6$ are formed. Three forms of the enzyme are known: the Tris-HCl enzyme (monomeric form), the phosphate enzyme (dimeric form), and phosphate-borate enzyme (polymeric form). In addition, a membrane-bound insoluble form of pig brain PAG has been suggested (Nimmo and Tipton 1981). Rat kidney PAG has a mol. wt. of 160,000 (monomeric form) (Curthoys et al. 1976); it dimerizes in the presence of phosphate and polymeric forms are formed in the presence of phosphate and borate (Godfrey et al. 1977). We found that pig kidney PAG is composed of two subunits with mol. wts. of 64,000 and 57,000, respectively, and that pig brain PAG has only one subunit with a mol. wt. of 64,000 (Kvamme and Svenneby 1975), but others report that the subunit of pig brain PAG has a mol. wt. of 73,000 (Nimmo and Tipton 1980). Rat kidney PAG has been reported to contain subunits with mol. wts. of 83,000. However, subunits with lower mol. wts. are formed, probably due to proteolysis (Clark and Curthoys 1979).

Electron micrographs of pig brain PAG appear to be similar to those for pig kidney PAG (Olsen et al. 1970, 1973, Svenneby 1970). After negative staining, the Tris-HCl enzyme consists of small, rounded, square, or rectangular particles. In high magnification micrographs a number of molecules show the presence of a stain-filled cleft which appear to divide each molecule into two parts. After addition of phosphate to the Tris-HCl enzyme, an elongated dimer composed of two Tris-HCl enzyme molecules becomes a major component of this preparation. On addition of phosphate-borate buffer, large helical polymers are formed.

Activators and Inhibitors of Purified PAG

Activators

PAG is subject to a very complex regulation. Phosphate is a potent activator of PAG, whereas members of the citric acid cycle, such as succinate and citrate, are less potent (Svenneby 1971, 1972, Svenneby et al. 1973, Kvamme and Svenneby 1975). Maleate activates the purified enzyme to the same extent as do intermediates of the citric acid cycle. Since the effect of maleate disappears following inactivation of PAG by heat treatment, it is not caused by contamination of PAG with maleate activated glutaminase (γ-glutamyl transpeptidase) which is resistant to this treatment (Kvamme and Olsen 1981). Also PAG in nonpurified tissue preparations, e.g., in homogenates, mitochondria (Greenstein and Leuthardt 1948, O'Donovan and Lotspeich 1966, Katunuma et al. 1966, Weil-Malherbe 1969, 1972), synaptosomes, and astrocytes (see below) is affected by the activators mentioned above. Other general activators of PAG are phosphorylated compounds, such as riboflavin phosphate (Weil-Malherbe and Beall 1970), trinucleotides (Weil-Malherbe 1972), and acyl-CoA derivatives (Kvamme and Torgner 1974a, b, 1975). The dye Bromothymol Blue exerts an interesting effect on pig kidney and brain PAG. In low concentrations the dye strongly increases the activation by phosphate, but at higher concentrations it inhibits the enzyme (Kvamme et al. 1965, 1966, 1970, Kvamme and Torgner 1974a). However, phosphate protects against the inhibition. The effects of Bromothymol Blue mimic similar effects by physiological compounds, such as acyl-CoA derivatives (Kvamme and Torgner 1974a, b, 1975), thyroxine (Hovhannissian et al. 1970, Badalian et al. 1975), and calcium (see below). Furthermore, it has been reported that PAG in intact brain mitochondria is activated by increased oxygen pressure (Kritschevskaya et al. 1959), by acetyl-Asp, and acetyl-Glu (Hovhannissian et al. 1969, 1970).

Activation by anions differs for the monomeric and polymeric enzyme forms, because the latter form is more susceptible to activation by low concentrations of anions than the former (Tris-HCl enzyme) (Kvamme 1979).

Since the citric acid cycle intermediates and acetyl-CoA produce no additional activation to that of phosphate in high concentrations, they appear to compete for the same site (Weil-Malherbe 1969, Svenneby et al. 1970, Svenneby 1971, Kvamme and Torgner 1974b). The specific activity of the dimeric and polymeric forms of PAG is three times that of the monomeric Tris-HCl form (Svenneby et al. 1970, Kvamme and Svenneby 1975). For that reason, in addition to the described activation of the various enzyme forms, the pig kidney and brain Tris-HCl enzymes are characterized by time-dependent activation. This occurs when the protein concentration is sufficiently high to permit dimerization or polymerization on incubation with anionic activators, such as phosphate, phosphate-borate (Svenneby et al. 1970, Svenneby 1972, Nimmo and Tipton 1981), or acyl-CoA derivatives (Kvamme and Torgner 1974b, 1975). Half time ($T^1/_2$) of the time-dependent activation at room temperature with phosphate or phosphate and borate has been found to be about 53 s for both the pig kidney and pig brain enzyme (Svenneby et al. 1970, Svenneby 1972). Time-dependent activation seems to be a hysteretic effect and may initially be accompanied by a conformational change of PAG, followed by polymerization (Frieden 1970). However, dimerization or polymerization does not appear to be a prerequisite for activation of the Tris-HCl enzyme, because its sedimentation coefficient does not change following activation with malonate or citrate (Kvamme et al. 1970).

Pig kidney and brain PAG are allosteric enzymes (Kvamme et al. 1965, 1970). Double-reciprocal plots of activity against Gln give straight lines only at a pH of about 8, whereas the plots become increasingly concave-upwards at higher pH values and convex-upwards at decreasing pH values, indicating a switch from positive to negative cooperativity (Kvamme and Svenneby 1975). In addition, double-inverse plots of activity against phosphate concentration are concave-upwards (Svenneby 1971). Moreover, such plots are generally convex-upwards against acyl-CoA derivatives with relatively short carbon chains and concave-upwards for dervatives with long carbon chain (Kvamme and Torgner 1974 a, b, 1975).

Inhibitors

The reaction product Glu is a potent inhibitor of both purified and nonpurified PAG. Phosphate protects particularly the polymeric form of pig kidney and brain PAG against Glu inhibition. It is noteworthy that the other reaction product of the PAG reaction, ammonia, exerts differential effects on purified and structural-bound PAG (except PAG in cultured astrocytes), because only the structural-bound PAG is inhibited (see below). However, 10–20 m*M* ammonia may activate the purified enzyme (Kvamme et al. 1970, Svenneby 1971). Similarly, N-ethylmaleimide, which inhibits structural-bound PAG, has no effect on the purified enzyme (Kvamme and Olsen 1981).

In addition, PAG is inhibited by cyclic-AMP and cyclic-GMP (Weil-Malherbe 1972), by protons (Kvamme et al. 1970), and by long-chain acyl-CoA derivatives, the latter being required in higher concentrations than those which activate the enzyme (Kvamme and Torgner 1974a, 1975). Moreover, diethylstilbestrol and testosterone have been reported to counteract the activation produced by thyroxine, although these compounds appear to have little effect on the activation by phosphate, citrate, or other anions (Mikirtumova et al. 1976).

PAG of Synaptosomal-Enriched Preparation and Astrocytes

Effects of Ammonia, Glu, and 2-Oxoglutarate

Although synaptosomal preparations are never pure, some important results have emerged using such preparations. Thus, the PAG content of synaptosomes prepared from rat brain tissue is very high (Bradford and Ward 1976, Kvamme and Lenda 1981). These organelles contain about 40% of the total tissue PAG as compared to only 10% of the protein (Bradford and Ward 1976). Moreover, as mentioned above, stimulus-released Glu and GABA appear to a large extent to be derived from Gln, suggesting that PAG may be involved in production of these transmitters.

Synaptosomal PAG is principally subject to regulation by similar activators and inhibitors as the purified enzyme. However, there are some important distinctions. In contrast to purified PAG, synaptosomal PAG is remarkably sensitive to inhibition by low concentrations of Glu and ammonia, as shown in Fig. 1. The inhibition by Glu and ammonia increases markedly on raising the concentration of these compounds from 0–0.4 m*M*, whereby the PAG activity is reduced by 40%–50%, but the activity remains almost constant when higher inhibitor concentrations are added. It is thus striking that Glu appears to penetrate the membranes separating mitochondrial-bound PAG from its surroundings just as easily as ammonia, indicating that synaptosomal PAG may be regulated by the extramitochondrial variation in the concentration of these compounds. Moreover, the inhibition of PAG by its reaction products is affected by the extracellular pH. When pH in the medium is lowered from 7.6 to 7.0, the

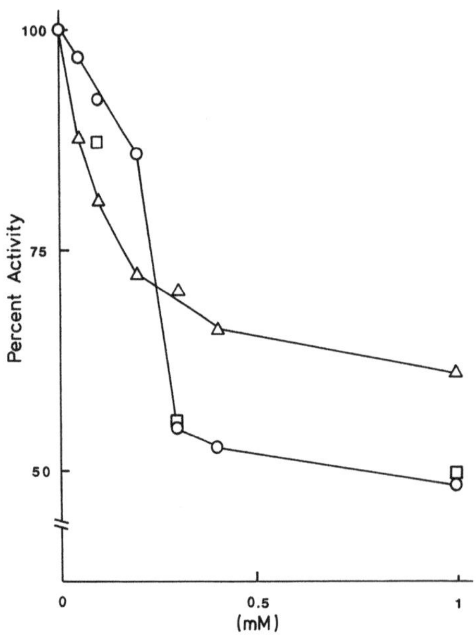

Fig. 1. Percent inhibition of PAG in rat brain synaptosomes following incubation for 2 min, 25° C, pH 7.4 with glutamate ○, 2-oxoglutarate □ (Na-salts) or ammonium chloride △ in concentrations as indicated. Other additions: 2 m*M* L-[(U)-^{14}C] glutamine, 5 m*M* Na-phosphate, 90 m*M* NaCl, 56 m*M* KCl, 4 m*M* Hepes, 5 m*M* $MgCl_2$, 10 mg/l oligomycin, and 0.6 mg/liter antimycin A. Mean of two experiments run in duplicate. Experimental procedure as described by Kvamme and Lenda (1982)

Table 1. The effect of pH on the inhibition of synaptosomal phosphate-activated glutaminase by ammonia or glutamate[a]

Additions	Synaptosomal phosphate-activated glutaminase (percent activity)	
	pH 7.6	pH 7.0
None	100	42 ± 5
Glutamate (1 m*M*)	47 ± 5	35 ± 5
NH_4Cl (1 m*M*)	52 ± 8	30 ± 4

[a] Otherwise conditions as in Fig. 1
Data are given as means ± SEM ($n = 6$)

PAG activity is inhibited by 50%–60% and the effects of Glu and ammonia are markedly reduced (Table 1).

Figure 1 also demonstrates that 2-oxoglutarate inhibits PAG similarly to Glu in comparable concentrations. Aspartate (or alanine) alone has no effect on PAG, but may produce some additional inhibition to that of 2-oxoglutarate. However, on incubating the synaptosomes with 2-oxoglutarate in the presence of the aminotransferase inhibitor aminooxyacetic acid (AOA), the inhibition by 2-oxoglutarate is abolished (Table 2). Therefore, PAG is apparently not inhibited by 2-oxoglutarate as such, but by Glu formed from the oxoacid by aminotransferase reactions. This has been confirmed by the finding of a concentration dependent relationship between 2-oxoglutarate added to the incubation medium, and Glu found at the end of the incubation (Kvamme and Lenda 1982). The PAG activity may thus be controlled by the operation of the citric acid cycle in vivo.

Ammonia at 1–2 m*M* does not stimulate Glu formation from 2-oxoglutarate, even when adding glucose and malate and omitting the metabolic inhibitors antimycin A and oligomycin. This is in accordance with low Glu dehydrogenase activity (Dienel et al. 1977, Kvamme and Lenda 1982). We have also been unable to detect any maleate activated glutaminase in synaptosomes (Kvamme and Olsen 1981), indicating that PAG is a dominant Gln metabolizing enzyme in this structure.

The inhibition by Glu of PAG in astrocytes cultured from mouse brain appears to follow the same pattern as that of synaptosomes, but interestingly, ammonia has no inhibitory effect, similar to the case when purified enzyme preparations are used (Kvamme et al. 1982).

Synaptosomal enriched preparations contain a high concentration of endogenous Glu (4 m*M*) (Kvamme and Lenda 1981). If not compartmentalized, this Glu would inhibit PAG considerably. In that case, PAG of intact synaptosomes should be much less susceptible to activation by phosphate and inhibition by Glu in low concentrations than PAG in disrupted synaptosomes where endogenous Glu presumably has leaked out. We found that intact and disrupted synaptomes are equally susceptible to activation by phosphate as well as to inhibition by Glu, indicating that the endogenous Glu in intact

Table 2. The effect of aminooxyacetate on the inhibition of synaptosomal phosphate-activated glutaminase by 2-oxoglutarate (m*M*)[a]

Additions	Synaptosomal phosphate-activated glutaminase (percent of control)
None	56 ± 6
Aspartate (2 m*M*)	50 ± 6
Aminooxyacetate (5 m*M*)	99
Aspartate (2 m*M*) plus aminooxyacetate (5 m*M*)	110 ± 9

[a] Further additions see Fig. 1. Control activity in the absence of 2-oxoglutarate was set to 100%. Mean ± SEM of five experiments or of two experiments run in duplicate

synaptosomes is compartmentalized (Kvamme and Lenda 1981). This is in accord with recent findings by Storm-Mathisen et al. (1983). They provide evidence by using immunocytochemical methods that the transmitter amino acids are concentrated in synaptic vesicles.

Structural Localization of PAG

PAG has been known for a long time to be mitochondrial-bound (Errera and Greenstein 1949), but the localization inside the mitochondria has been a matter of discussion. Thus, PAG has been reported to be localized to the matrix region (Kalra and Brosnan 1974), whereas others have provided strong evidence that the enzyme is bound to the inner mitochondrial membrane (Curthoys and Weiss 1974, Kovácević 1976, Kvamme 1982).

This problem has been approached using compounds which are known either to be impermeable or permeable to the inner mitochondrial membrane. Since mersalyl (Mers) and p-mercuribenzoate (PMB) in contrast to N-ethylmaleimide (NEM) do not bind reduced glutathione in intact liver mitochondria, and glutathione is localized to the matrix region, Mers and PMB are considered to be unable to penetrate the inner mitochondrial membrane, in contrast to NEM (Tietze 1969). The use of the three inhibitors permitted to localize rat liver mitochondrial β-hydroxybutyrate dehydrogenase (BHBD) to the inner face of the inner membrane (Gaudemer and Latruffe 1975, McIntyre et al. 1978) and the enzyme can be considered a marker for this region. When incubating the inhibitors with a suspension of intact synaptosomes or with mitochondria prepared from synaptosomes, BHBD is strongly inhibited by NEM (80%–90%) and relatively little affected by Mers and PMB, but following sonication, the inhibition by Mers and PMB increases, and no appreciable difference is found between the three compounds with regard to inhibition of β-hydroxybutyrate dehydrogenase (Table 3, top). Hence, similarly to rat liver mitochondria, Mers and PMB do not seem to penetrate the inner membrane of synaptosomal mitochondria, whereas NEM penetrates the membrane, and these inhibitors can therefore be used to localize PAG.

We have found that NEM as well as Mers and PMB inhibit PAG in intact synaptosomes and synaptosomal mitochondria 70%–80% (Table 3, bottom).

Table 3. β-Hydroxybutyrate and phosphate-activated glutaminase in synaptosomal preparations

Additions	β-Hydroxybutyrate activity (percent of control)		
	Synaptosomes	Synaptosomal mitochondria[a]	Sonicated synaptosomal mitochondria
NEM (1 m*M*)	18 ± 9	6 ± 4	37 ± 4
Mersalyl (0.1 m*M*)	87 ± 8	75 ± 11	20 ± 2
p-Mercuribenzoate (0.5 m*M*)	–	78 ± 3	20 ± 2
	Phosphate-activated glutaminase (percent of control)		
NEM (1 m*M*)	50 ± 4		40 ± 13
Mersalyl (0.1 m*M*)	33 ± 3		–
p-Mercuribenzoate (0.5 m*M*)	29 ± 2		–

[a] Prepared according to Lai and Clark (1976)

This may be explained by assuming either that PAG is localized to the outer face of the inner mitochondrial membrane or by an effect on the carriers of Gln (Simpson and Adams 1975) or phosphate (Meijer et al. 1970) which are known to be sensitive to sulfhydryl group reagents. However, preloading the synaptosomes or synaptosomal mitochondria with phosphate does not affect the extent of inhibition. Moreover, sonication of the synaptosomes or synaptosomal mitochondria to an extent which makes them leaky to the matrix enzyme malate dehydrogenase and permits Mers and PMB to inhibit BHBD is also without appreciable effect on the inhibition of PAG by Mers, PMB, and NEM. Therefore, a direct inhibition of PAG appears to be likely, similar to that found using rat liver mitochondria (Josef and Meijer 1981). These findings support the conclusion that PAG is localized to the outer face of the inner membrane. However, other experiments indicate that PAG is also localized to the inner face of the inner mitochondrial membrane and that the enzyme may be approached by Gln from both sides.

The inhibition by NEM is only partial. On increasing the NEM concentration from 0–0.5 m*M*, the inhibition of PAG increases gradually to reach a maximal level of 50% at 0.5 m*M*, and the percentage inhibition remains constant at higher NEM concentrations (Kvamme and Olsen 1979, 1981). Thus, some Mers- and PMB-sensitive sulfhydryl groups on PAG may not be accessible to NEM. Moreover, only the NEM sensitive fraction of PAG is sensitive to changes in pH and the phosphate concentration of the incubation medium (Kvamme and Olsen 1981); these parameters may, therefore, be important for regulation of the PAG activity in vivo.

It is of interest that the sulfhydryl group reagents produce a similar pattern of inhibition of PAG in astrocytes cultured from mouse brain as in synaptosomes (Kvamme et al. 1982).

Using pig kidney mitochondria we found that neither NEM nor Mers nor PMB inhibits β-hydroxybutyrate dehydrogenase in intact mitochondria appreciably, but the extent of inhibition increases markedly following sonication. These results are somewhat at variance with those obtained with liver mitochondria and synaptosomes, and indicate that neither NEM nor Mers nor PMB penetrate the inner membrane of pig kidney mitochondria. However, PAG in the intact mitochondria is inhibited 60%–70% by these inhibitors. Slightly higher inhibition of PAG has been obtained using pig kidney mitoplasts, demonstrating that the outer membrane is of little importance for the inhibition. Moreover, PAG in sonicated mitochondria is also inhibited to the same extent by NEM, Mers, and PMB, excluding an indirect effect on the Gln or phosphate carrier. This gives further support to the view that PAG itself is inhibited and partially localized externally in the inner mitochondrial membrane. Experiments using inside-out vesicles prepared from the inner mitochondrial membrane suggest that PAG also is localized to the inner face of the inner mitochondrial membrane. Whether two distinct species of PAG exist or if the same enzyme species penetrates the entire inner membrane with binding sites for Gln on both sides, is at present unclear, but the experiments suggest that PAG in renal and synaptosomal mitochondria have the same localization.

Activation by Calcium

Calcium (1 m*M*) activates PAG in rat brain synaptosomes (Kvamme 1979, Kvamme and Olsen 1979, Kvamme et al. 1983), brain slices, brain homogenates (Benjamin 1981), pig renal mitochondria (Kvamme and Olsen 1979, Kvamme 1982), and astrocytes cultured from mouse brain (Kvamme et al. 1982). Characteristically, the activation is dependent on phosphate (Table 4). Without phosphate added, calcium inhibits the enzyme. However, purified preparations of PAG are not activated by calcium, and the effect appears, thus, to be a property of structurally-bound PAG. As mentioned above, the dye Bromothymol Blue (Kvamme et al. 1970, Kvamme and Torgner 1974a) and several acyl-CoA derivatives (Kvamme and Torgner 1974a, 1975) activate purified kidney and brain PAG in cooperation with phosphate, but to our knowledge calcium is the only cationic compound which has this effect.

The calcium activation of PAG in synaptosomes and synaptosomal mitochondria is unaffected by the uncoupler 2,4-dinitrophenol which inhibits the mitochondrial proton pump (Mitchell 1967), and can therefore not be ascribed to energy requiring transport mechanisms. It is, however, abolished by the product inhibitors of PAG, ammonia, and Glu (Kvamme et al. 1983 (Table 5).

Potassium in depolarizing concentrations (56 m*M*) does not by itself affect calcium activation (Kvamme et al. 1983). Moreover, preincubating the synaptosomes with NEM before the addition of calcium also has no effect on the calcium activation, indicating that the action of calcium, similar to that of NEM, is not caused by an effect on phosphate transport mechanisms. This notion is supported by the finding that the calcium activation can be detected in intact synaptosomes as well as in synaptosomes disrupted by freezing and thawing or sonication, and in disrupted synaptosomal mitochondria (Table 6). Further-

Table 4. Calcium activation of phosphate-activated glutaminase in synaptosomes as affected by phosphate[a]

Additions	Percent activity		
	No P_i	P_i (5 mM)	P_i (10 mM)
None	100 ± 4	178 ± 22	265 ± 17
$CaCl_2$ (1 mM)	52 ± 4	217 ± 9	465 ± 30

[a] Conditions as in Fig. 1; mean of five experiments ± SEM

Table 5. Effect of glutamate and ammonia on calcium activation of synaptosomal phosphate-activated glutaminase[a]

Additions	Phosphate-activated glutaminase (percent of control)	
	No calcium	$CaCl_2$ (0.5 mM)
None	100	163 ± 10
Glutamate (0.4 mM)	56 ± 5	38 ± 4
Amonia (0.4 mM)	53 ± 2	53 ± 2

[a] Conditions as in Fig. 1, except that phosphate was 10 mM. Mean ± SEM of six experiments

Table 6. Counteraction by calcium of the NEM inhibition of phosphate-activated glutaminase[a]

Additions[a]	Percent activity of phosphate-activated glutaminase			
	Intact synaptosomes		Sonicated synaptosomal mitochondria	
	No NEM	NEM	No NEM	NEM
None	100 ± 13	57 ± 12	100 ± 14	40 ± 13
$CaCl_2$ (1 mM)	172 ± 14	154 ± 8	206 ± 35	184 ± 37

[a] Additions: 1 mM N-ethylmaleimide (NEM) and 10 mM phosphate. The preparations were preincubated (5 min 25° C) with NEM (wi.en added) before the addition of glutamine or glutamine + calcium. Otherwise, conditions as in Fig. 1. Mean ± SEM of five experiments

more, calcium counteracts the inhibition by Mers, PMB, and NEM, both in intact synaptosomes, disrupted synaptosomes, and sonicated synaptosomal mitochondria, as shown with NEM in Table 6.

Calcium may possibly act by liberating an activator of PAG, e.g., an acyl-CoA derivative, which makes PAG more sensitive to phosphate. Thus, calcium has been reported to inhibit the hydrolysis of acyl-CoA by acyl-CoA hydrolase (Strosznajder et al. 1981). However, the most likely explanation appears to be that calcium affects PAG by causing a conformational change which makes the enzyme more sensitive to phosphate, e.g., as visualized in Fig. 2, by moving the phosphate-binding activating site towards the Gln-binding catalytic site on the enzyme.

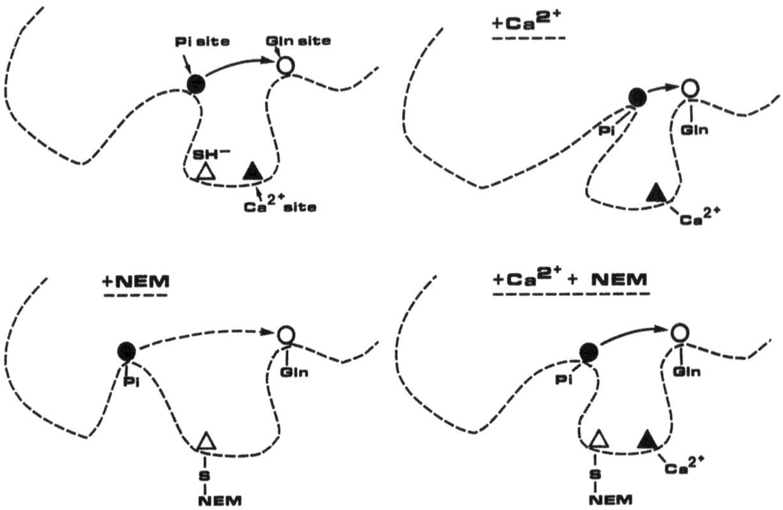

Fig. 2. A hypothetical scheme showing the action of calcium and N-ethylmaleimide (NEM) on PAG localized to the outer face of the inner mitochondrial membrane. The effect of calcium and NEM is vsiualized by conformational changes of the phosphate activating site, whereby the sensitivity to phosphate activation of Gln hydrolysis is affected

Since NEM is known to bind sulfhydryl groups covalently, calcium cannot counteract the NEM inhibition by displacement of NEM. The NEM-binding sulfhydryl groups are, therefore, not essential for the enzyme activity, but may be important for the phosphate activation of PAG. It is unlikely that NEM binds to the phosphate activating site as such or prevents the accessibility of phosphate to this site, because saturating the phosphate site with phosphate before the NEM addition does not abolish the inhibition by NEM. Thus, calcium and NEM may both affect the phosphate activation of PAG indirectly, e.g., as visualized in Fig. 2 by hypothetical conformational changes.

As shown in Table 5 calcium does not counteract the inhibition by the reaction products Glu and NH_4^+. This may be explained if these product inhibitors affect the catalytic site, an assumption which is supported by kinetic data.

Functional Properties

PAG is mitochondrial-bound and appears to be present in all brain regions. Following intrastriatal injection of kainic acid, a correlation between the PAG and Glu decarboxylase activities have been reported, suggesting that 60% of PAG in neostriatum is located in GABA-ergic structures (McGeer and McGeer 1979).

PAG is regulated by a great variety of activators and inhibitors. Some of these compounds may exist in too low concentrations in vivo to have any regulatory function. However, the concentration itself may not be the only important factor. More important may be the range of physiological variation of a regulatory ligand and its interaction with other ligands affecting the enzyme activity. Since PAG appears to be partially localized to the outer face of the inner mitochondrial membrane, it may be regulated by compounds in the extramitochondrial and extracellular space. Compounds, such as Glu, ammonia, calcium, and acyl-CoA derivatives, may under special conditions undergo great variations in their extramitochondrial concentration, permitting the PAG activity to be adjusted according to the physiological needs. It is possible, however, that the outer face PAG is normally inhibited in renal mitochondria due to the high extramitochondrial Glu concentration. In brain synaptosomes, however, this Glu appears to be compartmentalized and unavailable to PAG (Kvamme and Lenda 1981).

Since the calcium activation is unaffected by potassium in depolarizing concentrations, it may not be directly involved in neurotransmission processes, but possibly in replenishing depleted stores of transmitter Glu (Kvamme 1979).

The concept of the Gln cycle has been suggested, based on the well-documented compartmentalization of cerebral Gln-Glu metabolism (Shank and Aprison 1981). Thus, high Gln synthetase activity has been found in glial cells and little or no activity in neurons. The PAG activity has also been reported to be higher in nerve endings than in glial cells (Salganicoff and De Robertis 1965, Bradford and Ward 1976). However, since the extracellular Gln concentration is assumed to be 0.5 m*M* (Gjessing et al. 1972, Johnson 1978) and the mitochondrial phosphate concentrations is 5–8 m*M*, the activity of pig brain PAG is at most 5%–10% of that obtained by maximal activation in the presence of 20 m*M* Gln and 100 m*M* phosphate. Therefore, the functional activity of PAG may be unrelated to its maximal capacity. It is also possible that the nerve endings may be supplied with the necessary Gln, e.g., for transmitter synthesis, from the extracellular space instead of directly from glial cells by the operation of the Gln cycle.

Maleate-Activated Glutaminase (MAG) (Phosphate-independent glutaminase, γ-glutamyl transpeptidase, γ-glutamyl transferase, EC 2.3.2.2)

MAG was discovered by Katunuma et al. (1966) in rat kidney, liver, and brain. Rat kidney MAG has been partially purified and shown to be identical with γ-glutamyl transpeptidase (γ-GT) (Curthoys and Kuhlenschmidt 1975, Tate and Meister 1975, Allison and Meister 1981). γ-GT is localized in choroid plexus, ciliary body, retina, visual receptor cells or their capillaries, Purkinje cells, and anterior horn cells (Meister and Tate 1976, Prusiner and Prusiner 1978).

The physiological role of MAG remains uncertain, as maleate does not occur in vivo. In contrast to γ-GT MAG has not been demonstrated in brain with certainty. Most workers have found that brains from rat, pig, and guinea pig (Weil-Malherbe 1972, Huang and Knox 1976, Nimmo and Tipton 1979) as well as synaptosomes (Kvamme and Olsen 1981) contain only PAG and not MAG. It is, thus, possible that γ-GT in brain has different properties than in kidney.

Transglutaminase (Glutaminyl-peptide γ-glutamyl transferase, EC 2.3.2.13)

N^2-R-glutaminylpeptide + R′-NH_2
→ NH_4^+ + N^2-R-R′-glutaminylpeptide

Transglutaminase catalyzes a calcium-dependent acyl transfer reaction in which γ-carboxamide groups of peptide-bound Gln residues are the acyl donors and primary amino groups may function as acyl acceptors. The enzyme has been purified from guinea pig liver and plasma, and the enzyme exists in several forms (Folk and Finlayson 1977). Hemoglobin (Pincus and Waelsch 1968), chromogranins (Wajda et al. 1971), and polyamines (Folk et al. 1980) may serve as substrates. Hair follicles, wool, and epidermis contain high concentrations of transglutaminase. Moreover, factor XIII a (thrombin-activated blood coagulation factor XIII) which catalyzes the last step in hemostasis, the covalent cross linking of fibrin through ε-(y-Glu) Lys bonds, is a transglutaminase (Folk and Finlayson 1977, Gorman and Folk 1981). The enzyme may also participate in receptor-mediated endocytosis in cultured normal kidney cells (Levitzki et al. 1980, Davies et al. 1980) and in human erythrocytes, whereby proteins, drugs, and peptide hormones (e.g., insulin, epidermal growth factor) are taken up. Human lens appears to contain a transglutaminase which forms a protein polymer present in human cataract (Lorand et al. 1981) and it has been proposed that a transglutaminase is responsible for essential pathological findings in the brains of patients with Alzheimer's disease (Selkoe et al. 1982).

Glutamine Aminotransferase [Glutaminase II (EC 2.6.1.15) ω-amidase (EC 3.3.1.3) pathway]

Glutamine may be converted to 2-oxoglutarate and ammonia by successive actions of these two enzymes:

1) L-Gln + 2-oxoacid ⇔ 2-oxoglutaramic acid + L-α-amino acid

2) 2-oxoglutaramic acid + H_2O → 2-oxoglutarate + NH_4^+

Glutamine aminotransferase has been purified to apparent homogeneity (1,300-fold) (Van Leuven 1975) from rat brain. It has a mol. wt. of 98,000 and is probably composed of two subunits. Pyridoxal phosphate is likely to be a cofactor; glyoxylate and phenylpyruvate are found to be the best amino acceptors, but there is a broad oxoacid substrate specificity. The most active amino donors are Gln and Met (review Benuck and Lajtha 1975). ω-Amidase has been purified from rat liver (Hersh 1972). The brain mitochondrial fraction contains Gln aminotransferase, whereas ω-amidase is found in the soluble fraction. Both enzyme activities have been assayed in the brains of mouse, rabbit guinea pig, and man. Glutamine aminotransferase appears to be higher in rat brain than in human brain in contrast to ω-amidase (Cooper and Gross 1977, Lockwood and Duffy 1977). Since elevated concentrations of 2-oxoglutaramic acid are found in the CSF of patients in hepatic coma (Vergara et al. 1974, Duffy et al. 1974), it has been suggested that the two enzymes are compartmentalized or that ω-amidase is inhibited in these patients (Cooper and Gross 1977). High concentrations of ammonia inhibit purified ω-amidase (Hersh 1972, Lockwood and Duffy 1977), but this probably does not explain the raised level of 2-oxoglutaramic acid, because the measured increase in ammonia is not sufficient to inhibit ω-amidase.

References

Allison RD, Meister A (1981) Evidence that transpeptidation is a significant function of gamma-glutamyl transpeptidase. J Biol Chem 256: 2988–2992

Badalian LL, Buniatian HC, Hovhannissian VS (1975) Effect of glutamic acid in the interaction of various activators of brain glutaminase. Vopr Biolchim Mozga Akad Nauk Armjan SSR 10: 40–54

Benjamin AM (1981) Control of glutaminase activity in rat brain cortex in vitro: Influence of glutamate, phosphate, ammonium, calcium and hydrogen ions. Brain Res 208: 363–377

Benuck M, Lajtha A (1975) Aminotransferase activity in brain. Int Rev Neurobiol 17: 85–129

Bradford HF, Ward HK (1976) On glutaminase activity in mammalian synaptosomes. Brain Res 110: 115–125

Bradford HF, Ward KH, Thomas AJ (1978) Glutamine – a major substrate for nerve endings. J Neurochem 30: 1453–1459

Clark VM, Curthoys NP (1979) Cause of subunit heterogeneity in purified rat renal phosphate-dependent glutaminase. J Biol Chem 254: 4939–4941

Cooper AJL, Gross M (1977) The glutamine transaminase-omega-amidase system in rat and human brain. J Neurochem 28: 771–778

Curthoys NP, Kuhlenschmidt T (1975) Phosphate independent glutaminase from rat kidney. Partial purification and identity with gamma-glutamyltranspeptidase. J Biol Chem 250: 2099–2105

Curthoys NP, Weiss R (1974) Regulation of renal ammoniagenesis. Subcellular localization of rat kidney glutaminase isoenzymes. J Biol Chem 249: 3261–3266

Curthoys NP, Kuhlenschmidt T, Godfrey SS (1976) Regulation of renal ammoniagenesis. Purification and characterization of phosphate-dependent glutaminase from rat kidney. Arch Biochem Biophys 174: 82–89

Davies PJA, Davies DR, Levitzki A, Maxfield FR, Milhaud P, Willingham MC, Pastan IH (1980) Transglutaminase is essential in receptor-mediated endocytosis of α_2-macroglobulin and polypeptide hormones. Nature (London) 283: 162–167

Dienel G, Ryder E, Greengard O (1977) Distribution of mitochondrial enzymes between the perikaryal and synaptic fractions of immature and adult rat brain. Biochim Biophys Acta 496: 484–494

Duffy TE, Cooper AJL, Meister A (1974) Identification of alpha-ketoglutarate in rat liver, kidney and brain. Relationship to glutamine transaminase and omega-amidase activities. J Biol Chem 249: 7603–7606

Errera M, Greenstein JP (1949) Phosphate-activated glutaminase in kidney and other tissues. J Biol Chem 178: 495–502

Folk JE, Finlayson S (1977) The epsilon(gamma-glutamyl)lysine crosslink and the catalytic role of transglutaminase. Adv Protein Chem 31: 1–133

Folk JE, Park MH, Chung SI, Schrode J, Lester EP, Cooper HL (1980) Polyamines as physiological substrates for transglutaminases. J Biol Chem 255: 3695–3700

Frieden C (1970) Kinetic aspects of regulation of metabolic processes. The hysteretic concept. J Biol Chem 245: 5788–5799

Gaudemer Y, Latruffe N (1975) Evidence for penetrant and non-penetrant thiol reagents and their use in the location of rat liver mitochondrial D(-)-beta-hydroxybutyrate dehydrogenase. FEBS Lett 54: 30–34

Gjessing LR, Gjesdahl P, Sjaastad O (1972) The free amino acid in human cerebrospinal fluid. J Neurochem 19: 1807–1808

Godfrey S, Kuhlenschmidt T, Curthoys NP (1977) Correlation between activation and dimer formation of rat renal phosphate-dependent glutaminase. J Biol Chem 252: 1927–1931

Gorman JJ, Folk JE (1981) Structural features of glutamine substrates for transglutaminases. J Biol Chem 256: 2712–2715

Greenstein JP, Leuthardt FM (1948) Effect of phosphate and other anions on the enzymatic desamidation of various amides. Arch Biochem Biophys 17: 105–114

Hamberger A, Chiang G, Nylén ES, Scheff SW, Cotman CW (1978) Stimulus evoked increase in the biosynthesis of the putative neurotransmitter glutamate in the hippocampus. Brain Res 143: 549–555

Hersh LB (1972) Rat liver omega-amidase. Kinetic evidence for an acylenzyme intermediate. Biochemistry 11: 2251–2255

Hovhannissian VS, Buniatian HC, Badalian LL, Mikirtumova GS (1969) On the isoenzymes of the glutaminase system of brain mitochondrial fraction. Vopr Biolchim Mozga Adad Nauk Armjan SSR 5: 5–16

Hovhannissian VS, Buniatian HC, Ukrdumova GS, Badalian LL (1970) The participation of thyroxine in the interaction of the isoenzymes of brain glutaminase and certain features of its action. Vopr Biokhim Mozga Akad Nauk Armjan SSR 6: 5–13

Huang Y-Z, Knox WE (1976) A comparative study of glutaminase isoenzymes in rat tissues. Enzyme 21: 408–426

Johnson JL (1978) The excitant amino acids glutamic and aspartic acid as transmitter candidates in the vertebrate central nervous system. Prog Neurobiol 10: 155–202

Josef SK, Meijer AJ (1981) The inhibitory effects of sulphydryl reagents on the transport and hydrolysis of glutamine in rat-liver mitochondria. Eur J Biochem 119: 523–529

Kalra J, Brosnan JT (1974) The subcellular localization of glutaminase isoenzymes in rat kidney cortex. J Biol Chem 249: 3255–3260

Katunuma N, Tomino L, Nishino H (1966) Glutaminase isoenzymes in rat kidney. Biochem Biophys Res Commun 22: 321–328

Klingman JD, Handler P (1958) Partial purification and properties of renal glutaminase. J Biol Chem 232: 369–380

Kovaćević Z (1976) Importance of the flux of phosphate across the inner membrane of kidney mitochondria for the activation of glutaminase and the transport of glutamine. Biochim Biophys Acta 430: 399–412

Krebs HA (1935) Metabolism of amino-acids. IV. The synthesis of glutamine from glutamic acid and ammonia, and the enzymic hydrolysis of glutamine in animal tissue. Biochem J 29: 1951–1969

Kritschevskaya AA, Gerschenovitsch Z, Scherbatzker S (1959) Ammonia formation from amides by the brain and liver homogenates under increased oxygen pressure. Biochimija 64: 459–464

Kvamme E (1979) Regulation of glutaminase and its possible implication for GABA metabolism. In: Mandel P, DeFeudis FV (eds) GABA-biochemistry and CNS functions. Plenum, New York, pp 111–138

Kvamme E (1982) Regulation of pig kidney phosphate-activated glutaminase: In: Tannen RL, Goldstein L, Lemieus G, Simpson D, Vinay P (eds) Renal ammonia metabolism. In: Berlyne GM, Giovannetti S, Thomas S (eds) Contributions to nephrology, vol 31. Karger, Basel, pp 60–70

Kvamme E (1983a) Glutamine. In: Lajtha A (ed) Handbook of neurochemistry, vol III. Plenum Press, New York, pp 405–422

Kvamme E (1983b) Deaminases and amidases. In: Lajtha A (ed) Handbook of neurochemistry, vol IV. Plenum Press, New York, pp 85–110

Kvamme E, Lenda K (1981) Evidence for compartmentalization of glutamate in rat brain synaptosomes using the glutamate sensitivity of phosphate activated glutaminase as a functional test. Neurosci Lett 25: 193–198

Kvamme E, Lenda K (1982) Regulation of glutaminase by exogenous glutamate, ammonia and 2-oxoglutarate in synaptosomal enriched preparation from rat brain. Neurochem Res 7: 667–678

Kvamme E, Olsen BE (1979) Evidence for two species of mammalian phosphate-activated glutaminase having different regulatory properties. FEBS Lett 107: 33–36

Kvamme E, Olsen BE (1981) Evidence for compartmentation of synaptosomal phosphate-activated glutaminase. J Neurochem 36: 1916–1923

Kvamme E, Svenneby G (1975) Phosphate activated glutaminase in brain. In: Marks N, Rodnight T (eds) Research methods in neurochemistry, vol III. Plenum Press, New York, pp 277–290

Kvamme E, Torgner I (1974a) Phosphate-dependent effect of palmitoyl-CoA and stearyl-CoA on phosphate-activated pig brain and pig kidney glutaminase. FEBS Lett 47: 244–247

Kvamme E, Torgner I (1974b) The effect of acetyl-coenzyme A on phosphate-activated glutaminase from pig kidney and brain. Biochem J 137: 525–530

Kvamme E, Torgner I (1975) Regulatory effect of fatty acyl-coenzyme A derivatives on phosphate-activated pig brain and kidney glutaminase in vitro. Biochem J 149: 83–91

Kvamme E, Tveit B, Svenneby G (1965) Glutaminase from pig kidney, an allosteric protein. Biochem Biophys Res Commun 20: 566–572

Kvamme E, Svenneby G, Tveit B (1966) Studies of mammalian glutaminase. In: Walaas O (ed) Molecular basis of some aspects of mental activity, vol I. Academic Press, London, New York, pp 211–219

Kvamme E, Tveit B, Svenneby G (1970) Glutaminase from pig renal cortex. I. Purification and general properties. J Biol Chem 245: 1871–1877

Kvamme E, Svenneby G, Hertz L, Schousboe A (1982) Properties of phosphate activated glutaminase in astrocytes cultured from mouse brain. Neurochem Res 7: 761–770

Kvamme E, Svenneby G, Torgner I (1983) Calcium stimulation of glutamine hydrolysis in synaptosomes from rat brain. Neurochem Res 8: 23–36

Lai JCK, Clark JB (1976) Preparation and properties of mitochondria derived from synaptosomes. Biochem J 154: 423–432

Levitzki A, Willingham M, Pastan I (1980) Evidence for participation of transglutaminase in receptor-mediated endocytosis. Proc Natl Acad Sci USA 77: 2706–2710

Lockwood AH, Duffy TE (1977) Glutamine transaminase and omega-amidase species variations in brain activity and effect of portacaval shunting. J Neurochem 28: 673–675

Lorand L, Hsu LKH, Siefring GE Jr, Rafferty NS (1981) Lens transglutaminase and cataract formation. Proc Natl Acad Sci USA 78: 1356–1360

McGeer EC, McGeer PL (1979) Localization of glutaminase in the rat neostriatum. J Neurochem 32: 1071–1075

McIntyre JO, Bock H-GO, Fleischer S (1978) The orientation of D-beta-hydroxybutyrate dehydrogenase in the mitochondrial inner membrane. Biochim Biophys Acta 513: 255–267

Meijer AJ, Groot GSP, Tager JM (1970) Effect of sulfhydryl-blocking reagents in mitochondrial anion-exchange reactions involving phosphate. FEBS Lett 8: 41–44

Meister A, Tate SS (1976) Glutathione and related gamma-glutamyl compounds. Biosynthesis and utilization. Ann Rev Biochem 45: 559–604

Mikirtumova KS, Hairapetian HL, Hovhannissian VS, Buniatian HS (1976) The effect of diethylstilboestrol and steroid hormones on glutaminase activity on rat brain mitochondrial fractions. Vopr Biokhim Mozga Akad Nauk Armjan SSR 11: 17–28

Mitchell P (1967) Proton-translocation phosphorylation in mitochondria, chloroplasts and bacteria: natural fuel cells and solar cells. Fed Proc 26: 1370–1379

Nimmo GA, Tipton KF (1979) The distribution of soluble and membrane-bound forms of glutaminase in pig brain. J Neurochem 33: 1083–1094

Nimmo GA, Tipton KF (1980) Purification of soluble glutaminase from pig brain. Biochem Pharmacol 29: 359–367

Nimmo GA, Tipton KF (1981) Kinetic comparisons between soluble and membrane-bound glutaminase preparations from pig brain. Eur J Biochem 11: 57–64

O'Donovan J, Lotspeich WD (1966) Activation of kidney mitochondrial glutaminase by inorganic phosphate and organic acids. Nature (London) 212: 930–932

Olsen BR, Svenneby G, Kvamme E, Tveit B, Eskeland T (1970) Formation and ultrastructure of enzymically active polymers of pig renal glutaminase. J Mol Biol 52: 239–245

Olsen BR, Torgner I, Christensen TB, Kvamme E (1973) Ultrastructure of pig renal glutaminase. Evidence for conformational changes during polymer formation. J Mol Biol 74: 239–251

Otey MC, Birnbaum SM, Greenstein JP (1954) Solubilized kidney glutaminase. I. Arch Biochem Biophys 49: 245–247

Pincus JH, Waelsch H (1968) The specificity of transglutaminase. I. Human hemoglobin as a substrate for the enzyme. Arch Biochem Biophys 126: 34–43

Prusiner PE, Prusiner SB (1978) Partial purification and kinetics of gamma-glutamyl transpeptidase from bovine choroid plexus. J Neurochem 30: 1253–1259

Reubi JC, Van den Berg CJ, Cuénod M (1978) Glutamine as precursor for the GABA and glutamate transmitter pools. Neurosci Lett 10: 171–174

Salganicoff L, De Robertis E (1965) Subcellular distribution of the enzymes of the glutamic acid, glutamine and gamma-aminobutyric acid cycles in rat brain. J Neurochem 12: 287–309

Sayre FW, Roberts E (1968) Preparation and some properties of a phosphate-activated glutaminase from kidneys. J Biol Chem 233: 1128–1134

Selkoe DJ, Ihara Y, Salazar FJ (1982) Alzheimer's disease: Insolubility of partially purified paired helical filaments in sodium dodecyl sulfate and urea. Science 215: 1243–1245

Shank RP, Aprison MH (1981) Minireview: Present status and significance of the glutamine cycle in neural tissues. Life Sci 28: 837–842

Simpson and Adams (1975) Glutamine transport and metabolism by mitochondria from dog renal cortex. General properties and response to acidosis and alkalosis. J Biol Chem 250: 8148–8158

Storm-Mathisen J, Leknes AK, Bore AT, Vaaland JL, Edminson P, Haug F-MS, Ottersen OP (1983) First visualization of glutamate and GABA in neurones by immunocytochemistry. Nature (London) 301: 517–520

Strosznajder J, Tang W, Manning R, Lin AY-T, MacQuarrie R, Sun GY (1981) Metabolism of oleoyl-CoA in rat brain synaptosomes: Effect of calcium and post-decapitative ischemia. Neurochem Res 6: 1231–1240

Svenneby G (1970) Pig brain glutaminase. Purification and identification of different enzyme forms. J Neurochem 19: 1591–1599

Svenneby G (1971) Activation of pig brain glutaminase. J Neurochem 18: 2201–2208

Svenneby G (1972) Time- and temperature-dependent activation of pig brain glutaminase. J Neurochem 19: 165–174

Svenneby G, Torgner I, Kvamme E (1973) Purification of phosphate-dependent pig brain glutaminase. J Neurochem 20: 1217–1224

Svenneby G, Tveit B, Kvamme E (1970) Glutaminase from pig renal cortex. II. Activation by inorganic and organic anions. J Biol Chem 245: 1878–1882

Tate SS, Meister A (1975) Identity of maleate-stimulated glutaminase with gamma-glutamyl transpeptidase in rat kidney. J Biol Chem 250: 4619–4627

Timiras PS, Hudson DB, Oklund S (1973) Changes in central nervous system. Free amino acids with development and aging. Prog Brain Res 40: 267–275

Tietze F (1969) Enzymic method for quantitative determinations of nanogram amounts of total and oxidized glutathione: Applications to mammalian blood and other tissues. Anal Biochem 27: 502–522

Van Leuven F (1975) Highly purified glutamine transaminase from rat brain. Physical and kinetic properties. Eur J Biochem 58: 153–158

Vergara F, Plum F, Duffy TE (1974) alpha-Ketoglutarate: Increased concentrations in the cerebrospinal fluid of patients in hepatic coma. Science 183: 81–83

Wajda IJ, Hanbauer I, Manigault I, Lajtha A (1971) Chromogranins as substrate for transglutaminase. Biochem Pharmacol 20: 3197–3206

Weil-Malherbe H (1969) Activators and inhibitors of brain glutaminase. J Neurochem 16: 855–864

Weil-Malherbe H (1972) Modulators of glutaminase activity. J Neurochem 19: 2257–2267

Weil-Malherbe H, Beall GD (1970) Riboflavin 5'phosphate: a potent activator of brain glutaminase. J Neurochem 17: 1101–1103

Glutamine Transport Across Biological Membranes

Z. Kovačević[1] and J. D. McGivan[2]

Transport of Glutamine Across the Plasma Membrane of Mammalian Cells

The transport of neutral amino acids across cell membranes has been studied in great detail over the past 20 years. Oxender and Christensen (1963) originally proposed the existence of two distinct transport systems with overlapping specificity in the Ehrlich ascites cell. These were termed System A and System L. System A catalysed the Na^+-dependent accumulation of amino acids with short, unbranched side-chains, e.g., alanine, glycine and serine. System L was Na^+-independent and catalysed the equilibration across the cell membrane of branched-chain and some aromatic amino acids. A further system (System ASC) was subsequently characterised in the Ehrlich cell which was Na^+-dependent and catalysed the transport of alanine, serine and cysteine (Christensen et al. 1967).

The route by which a particular amino acid is transported was defined by the use of non-metabolisable amino acid analogues. The analogue N-methylaminoisobutyrate is transported only by System A and competition for Na^+-dependent transport between a given amino acid and N-methylaminoisobutyrate serves to distinguish Transport via System A from that via System ASC. System L may be distinguished from the other two systems by its lack of Na^+-dependence. The analogue 2-aminobicyclo-[2,2,1]-heptane-2-carboxylate (BCH) is a specific substrate for System L.

Extension of this approach has shown that amino acid transporting systems similar to those defined for the Ehrlich cell occur in the plasma membranes of a great variety of different mammalian cell types (Guidotti et al. 1978), although the exact specificity of the transport systems towards particular amino acids is somewhat variable. In spite of the very large volume of published work on the transport of neutral amino acids across cell membranes, relatively little attention has been paid to the transport of glutamine. This amino acid is transported via System A in Ehrlich cells (Oxender and Christensen 1963), via System ASC in pigeon erythrocytes (Thomas and Christensen 1971). However, in several comprehensive papers on amino acid transport in Chinese hamster ovary cells (Bass et al. 1981), chick embryo heart cells (Gazzola et al. 1972) and on regulation of amino acid transport in a variety of cell types (Guidotti ct al. 1975), the transport of glutamine is not discussed.

1 Department of Biochemistry, Medical Faculty, 21000 Novi Sad, Yugoslavia
2 Department of Biochemistry, Medical School, University of Bristol, Bristol BS8 1TD, England

Glutamine Metabolism in Mammalian Tissues
Edited by D. Häussinger and H. Sies

It is of particular importance to define the transport of glutamine in tissues such as brain, kidney, liver and intestinal epithelium where glutamine is of primary importance in metabolism (for a review see Kovačević and McGivan 1983). The mechanism and regulation of transport of glutamine have been studied mainly in hepatocytes. More work is required to elucidate the control of glutamine transport into kidney and gut.

Glutamine Transport Across the Plasma Membrane of Rat Hepatocytes

Joseph et al. (1978) showed that glutamine was accumulated into isolated rat hepatocytes. The transport was largely Na^+-dependent and was inhibited by ouabain. Glutamine inhibited the transport of alanine and serine non-competitively. On the basis of these and other observations, it was proposed that glutamine was accumulated in hepatocytes by a transport system distinct from that which catalysed the transport of alanine.

Kilberg et al. (1980) subsequently showed that glutamine transport was not inhibited by N-methylaminoisobutyrate and was hence not transported by System A. Glutamine also failed to inhibit the transport of cysteine, a specific substrate for System ASC in hepatocytes. It was, therefore, proposed that glutamine is transported in hepatocytes by a Na^+-dependent system distinct from both System A and System ASC. This newly discovered transport system was termed System N and proved to catalyse also the transport of asparagine and histidine (Kilberg et al. 1980). Glutamine transport via System N was not stimulated by insulin or glucagon, but the rate of transport of glutamine was stimulated twofold in hepatocytes cultured for 8 h in the absence of amino acids. Under these conditions the transport of N-methylaminoisobutyrate increased 16-fold.

Further evidence for the existence of separate transporting systems in hepatocytes for glutamine and alanine was provided by Hayes and McGivan (1982) who showed that the rate of glutamine transport in hepatocytes from starved rats was not increased over that in hepatocytes from normally fed animals, while the transport of alanine was significantly stimulated. Thus, the transport of glutamine into hepatocytes is regulated independently of that of alanine. System N has been so far found only in hepatocytes and certain hepatoma cells and may be unique to the liver. The physiological importance of the presence of a specific glutamine transporting system in hepatocytes distinct from those in other cells is not yet clear. It is unlikely that the transport of glutamine across the cell plasma membrane is an important regulatory site for glutamine metabolism in the liver (see McGivan et al. this volume).

Transport of Glutamine in Various Preparations of Brain Cells

Elucidation of the characteristics of glutamine transport in brain cell is of particular significance in view of the role of glutamine as a precursor of the neurotransmitters glutamate and γ-aminobutyrate. In slices of rat brain, glutamine was found to be accumulated by a Na^+-dependent process having an overall K_m for glutamine of 47 μM (Balcar and Johnston 1975); this reflected the overall uptake of glutamine by a number of different cell types. A similar "high affinity" glutamine uptake had been previously demonstrated in rat dorsal roots (Roberts and Keen 1974).

In both isolated glial cells and isolated synaptosomes, glutamine transport occurs by systems having a much higher K_m for glutamine ("low affinity" system). Glutamine transport in each case was shown to be Na^+-independent and insensitive to ouabain (Weiler et al. 1979). Glutamine transport showed similar characteristics in preparations of isolated astrocytes (Schousboe et al. 1979). In cultured neuronal cells, glutamine transport is mediated by a Na^+-dependent system showing a high glutamine specificity. In contrast, glutamine transport into glial cells was Na^+-independent and had a lower specificity for glutamine. These findings are consistent with the concept that glutamine is taken up and hydrolysed mainly by the neurons and is resynthesised from glutamate and ammonia in the glial cells (Weiler et al. 1979).

Transport of Glutamine in Kidney and Intestine

In neither the kidney nor the intestinal epithelium is the measurement of glutamine transport straightforward. The major problems in each case have been the lack of a satisfactory isolated cell preparation and the rapidity with which glutamine is hydrolysed by various preparations of these tissues. Measurement of glutamine "uptake" by preparations such as kidney slices, perfused kidney and loops of intestine does not give unambiguous information about the properties of the transport processes involved and will not be considered here.

The transport of glutamine into rat kidney cortex brush border and basolateral membrane vesicles has been studied by Foreman et al. (1982). Both membrane types possess saturable high and low affinity systems mediating glutamine uptake. The high affinity system of brush border membrane vesicles had a K_m for glutamine of 0.26 m*M*, whereas the same kinetic parameter of the low affinity system was 1.05 m*M*. The transport kinetic parameters of the high and low affinity systems of basolateral membrane vesicles were 0.04 m*M* and 3.18 m*M*, respectively. It was found that acidosis leads to an increase in glutamine uptake by the brush border membrane vesicles, but had no effect on glutamine uptake by basloateral membrane vesicles, although the former membrane preparation was much more efficient at accumulating glutamine.

Although intestinal brush border membrane vesicles have been used to study the transport of alanine (Sigrist-Nelson et al. 1975) and proline (Hayashi et al. 1982), there appear to be no reports of glutamine transport in this system.

Recently, Bradford and McGivan (1980) have shown that the isolated rat enterocyte preparation of Watford et al. (1979) can be used to study some aspects of plasma membrane glutamine transport. Glutamine transport in this system was Na^+-dependent and inhibited by ouabain. Analysis of the inhibition by N-methylaminoisobutyrate and alanine indicated that the major part of glutamine transport was catalysed by System A, in contrast to the situation in liver. However, in this preparation it was not possible to discriminate between transport across the mucosal and serosal membranes.

The picture which emerges from these studies is that, in general, glutamine is accumulated across cell membrane by a Na^+-dependent active transport process. Exceptions to this generalisation are encountered in certain brain cells. Hepatocytes appear to possess a glutamine-transporting system which is distinct from that found elsewhere. Further work is required to investigate the regulation of glutamine metabolism in various tissues and to assess the possible regulatory role of glutamine transport in glutamine metabolism.

Glutamine Transport Across the Inner Mitochondrial Membrane

There are mainly two reasons why the study of glutamine transport in mitochondria attracted attention during the last 10 years. The first is related to the investigation of regulation of glutaminase (EC 3.5.1.2), the enzyme which has exclusively an intramitochondrial location, and which, especially in quantitative terms, plays a key role in glutamine metabolism (Kovačević and McGivan 1983). The second reason is related to the assumption suggested by Pitts (1973) and some other investigators (Simpson and Adam 1975; Goldstein 1975) that stimulation of renal ammoniagenesis during acute metabolic acidosis results primarily from the stimulation of glutamine transport across the inner mitochondrial membrane. The problem of glutamine transport in mitochondria was partially reviewed by Kovačević and McGivan (1983) and La Noue and Schoolwerth (1979).

Evidence for the Existence of Carrier-Mediated Transport of Glutamine in Mitochondria

There is now no doubt that glutamine transport in mitochondria is a carrier-mediated process. This has been clearly demonstrated in the case of kidney and liver mitochondria which possess high activity of glutaminase. One of the first pieces of evidence which suggested the existence of glutamine carrier in mitochondria was the finding that they swell in an isoosmotic solution of L-glutamine, but not in the isoosmotic solution of D-glutamine (Kovačević et al. 1970; Brosnan and Hall 1977).

Demonstration of the latency of glutaminase suggested the assumption that the transport could be the rate-limiting step in the deamidation of glutamine. Disruption of mitochondria or their swelling increases activity of glutaminase two to three times, which can be explained by the removal of the permeability barrier between the enzyme and the substrate. However, the latency of an enzyme may result from a number of other factors and it is known that, generally, the activity of occluded enzyme increases with decreasing osmotic pressure and the swelling of organelles.

Strong evidence for the existence of carrier-mediated transport of glutamine in mitochondria is the inhibition of the transport by a -SH blocking reagent mersalyl. This was shown by Simpson and Adam (1975), Goldstein (1975), and Kovačević (1976) for kidney mitochondria, and Joseph and Meijer (1981) for liver mitochondria. Approximately 20 nmol mersalyl/mg protein is required for the complete inhibition of glutamine influx into the mitochondria. In the presence of mersalyl ^{14}C-glutamine space is the same as the ^{14}C-sucrose space, whereas in the absence of the inhibitor ^{14}C-glutamine space is equal to the ^{3}H-H_2O space. It is assumed that ^{14}C-sucrose does not penetrate the inner mitochondrial membrane, whereas the membrane is freely permeable for tritiated water. It should be noted that the same amount of mersalyl which inhibits the glutamine carrier inhibits the inorganic phosphate (Pi) carrier as well, whereas glutamate transport is not inhibited at all (Kovačević 1976). The glutamine carrier was found to be one of the most sensitive mitochondrial carriers to inhibition by this organomercurial compound.

Once the existence of the carrier was demonstrated the investigation was directed toward the elucidation of its kinetics, mechanism and regulation. This was stimulated especially by the important pathophysiological problem regarding regulation of renal ammoniagenesis during metabolic acidosis since it was suggested that the transport of glutamine is the rate-limiting step in glutaminase activity (Pitts 1973, Simpson and Adam 1975, Goldstein 1975). It was found that under no experimental condition could ^{14}C-glutamine be detected in the mitochondrial matrix, because almost all the ^{14}C-label was recovered as ^{14}C-glutamate. This was explained as evidence that the influx of glutamine into the mitochondria was the rate-limiting step in glutamine deamidation (Simpson and Adam 1975). Simpson and collaborators assumed that a close functional relation between the glutamine transport system and glutaminase exists, so that during the transport process glutamine is hydrolysed and glutamate is released into the matrix space. It is now clear that ^{14}C-glutamine was not found in mitochondria for the following reasons: (1) Separation of the mitochondria from incubation medium was not rapid enough to prevent hydrolysis of glutamine by the very active glutaminase; (2) the mitochondrial preparation was contaminated with brush border membrane vesicles which contain phosphate-independent glutaminase, so that hydrolysis of glutamine appeared even outside of the mitochondria (Curthoys and Shapiro 1978). Besides, recently Kovačević and Bajin (1982) were able to load the mitochondria with ^{14}C-glutamine which proves that glutamine does appear in the mitochondrial matrix.

In relation to the study of the kinetics of glutamine transport an important advance was achieved by Goldstein and Boylan (1978), who applied automated rapid mixing and rapid filtration techniques. They obtained data in the one-second range and were able to measure the initial rate kinetics. They found that at pH 7.4 and 23° C the K_m of the carrier is 2.7 m*M* and V_{max} 150–300 nmol glutamine/mg protein. The same investigators also reported that 2-oxoglutarate at 0.3 m*M* is a competitive inhibitor of the carrier. It must be noted, however, that these experiments were carried out under the conditions of intensive metabolism of glutamine, so that metabolic process interfered with the transport.

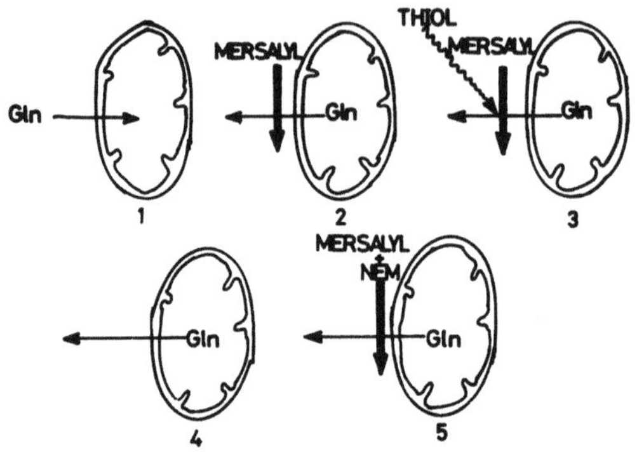

Scheme 1. Loading of mitochondria with ^{14}C-glutamine and measurement of its efflux from the organelles: *1* Loading of mitochondria with ^{14}C-glutamine; *2* inhibition of ^{14}C-glutamine efflux with mersalyl; *3* removal of mersalyl inhibition with a thiol reagent; *4* efflux of ^{14}C-glutamine; and *5* stop of the efflux with mersalyl plus N-ethylmaleimide

Loading of Mitochondria with ^{14}C-Glutamine and Measurement of Its Efflux

Because of the importance of the measurement of the initial rate of glutamine transport, not only for the study of the kinetics, mechanism and regulation of the carrier, but also due to difficulties which appear in the measurement of the influx of glutamine into the mitochondria, Kovačević and Bajin (1982) introduced a new technique in order to solve this problem. They loaded rat liver and kidney mitochondria with ^{14}C-glutamine and measured the rate of its efflux from the organelles. In order to avoid interference with metabolism of glutamine, the experiments were carried out at 0° C using the Pi-free medium, so that the activity of glutaminase was negligible compared to the rate of the transport. After loading of the mitochondria with ^{14}C-glutamine mersalyl was added (approx. 100 nmol/mg protein) in order to prevent glutamine efflux during the washing of the organelles (Scheme 1). In the experiments where the kinetic was measured the efflux was started by the addition of glutathione. The efflux was rapidly inhibited by the addition of mersalyl plus N-ethylmaleimide (inhibitor stop technique). Results of these investigations allowed the following conclusions: (1) glutamine appears in the matrix space; (2) the rate of the transport of glutamine is much faster than the rate of its hydrolysis. The same results were obtained with both rat liver and rat kidney mitochondria. In accordance with Goldstein (1975, 1976) we also found that glutamate, 2-oxoglutarate and some other anions inhibit glutamine efflux, especially in the case of liver mitochondria.

The technique of loading mitochondria with a substrate and measurement of the rate of its efflux has several advantages compared to the measurement of the influx. In the first case it is not necessary to measure ^{3}H-water and ^{14}C-sucrose space which is connected with several technical and theoretical difficulties. For that reason the technique of loading of mitochondria with substrate and the measurement of its efflux is applied whenever it is possible. However, in the case of our technique an important question appeared: Is the rate of the removal of mersalyl inhibition of the carrier by the addition of GSH fast enough not to limit the rate of carrier-mediated glutamine efflux. The initial experimental data already indicated that on the molecular basis much more GSH was needed for the fast removal of mersalyl inhibition (approx. 3,000 nmol/mg protein)

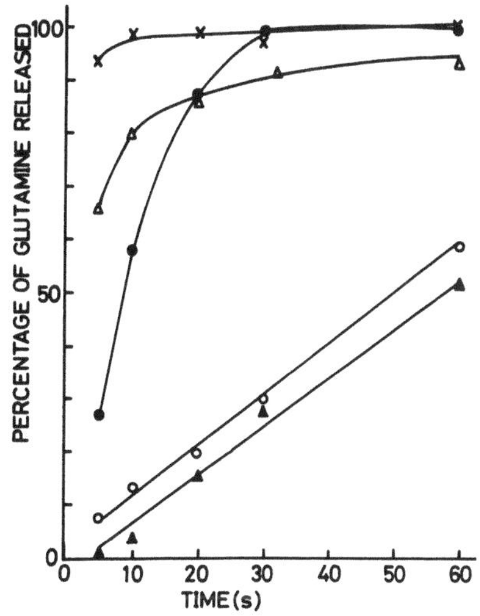

Fig. 1. Kinetics of the release of ^{14}C-glutamine from rat kidney mitochondria (1.5 mg protein) after the addition of various thiol reagents: cysteine (×), thioglycolate (△), GSH (●), dithiothreitol (○) and 2-mercaptoethanol (▲). Incubation medium: KCl (100 m*M*), Tris-HCl (10 m*M*) and rotenone (1 μg); pH 7.0, temperature 0° C

compared to the amount of mersalyl added. This suggested that the inhibitor-carrier complex is not easily accessible to GSH whereas, on the other hand, the carrier is very accessible to mersalyl which is a hydrophilic and non-penetrant -SH blocking reagent.

In order to answer the question, "Is the removal of the inhibition or the activity of the carrier itself the rate-limiting step in the efflux of ^{14}C-glutamine from the mitochondria?", several thiol reagents were tested. The fastest rate was achieved by the addition of cystein (Fig. 1). GSH and thioglycolate were less efficient, whereas dithiothreitol and 2-mercaptoethanol were slow in the removal of the inhibition, especially in the case of liver mitochondria. Thus, the question arose why there are differences between the thiols tested. It is known from the work of Neubert and Lehninger that thiol reagents induce swelling of mitochondria. They probably react with specific, structure-controlling thiol or disulfide groups in the mitochondrial membrane. We tested the effect of the mentioned thiols on the volume change of the mitochondria under identical experimental conditions. A direct correlation between the effect of the thiols on the removal of the carrier inhibition by mersalyl and their ability to induce swelling of the mitochondria was found. The thiols which were fastest in the removal of the inhibition of glutamine efflux were also fastest in the induction of the swelling (Fig. 2). Dithiothreitol and 2-mercaptoethanol do not cause swelling and release only 25%–30% of ^{14}C-glutamine within 5 min.

It is interesting that some anions like glutamate, succinate, malate and Pi prevent GSH and cysteine-induced swelling and under the same conditions strongly inhibit glutamine efflux induced by the thiols (Kovačević and Bajin 1982). Besides, addition of the anions induces contraction of the mitochondria which were already swollen in the presence of the thiols (Fig. 3). It is concluded that the rate of glutamine efflux initiated by the addition of GSH is limited not by activity of the carrier itself but by the rate of the removal of mersalyl inhibition. Compared to GSH, cysteine is much more efficient in the removal of the

2
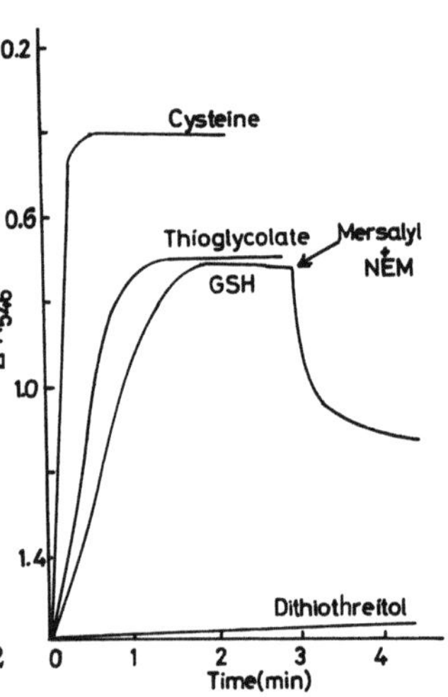

3
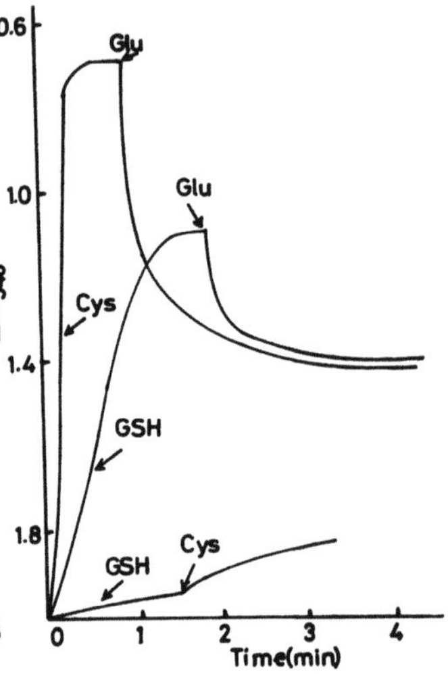

Fig. 2. Swelling of rat liver mitochondria (1.2 mg protein) in the KCl medium after the addition of cysteine (2 m*M*), thioglycolate (2 m*M*), GSH (2 m*M*) and dithiothreitol (6 m*M*) and the effect of mersalyl (6 m*M*) plus N-ethylmaleimide (6 m*M*). The final volume was 2 ml

Fig. 3. Swelling of rat liver mitochondria (1.2 mg protein) in the KCl medium (2 ml) after the addition of cysteine (2 m*M*) or GSH (2 m*M*) and reversal of the swelling by glutamate (10 m*M*). When glutamate was added before GSH and cysteine, subsequent addition of the thiols did not induce swelling of the mitochondria

inhibition, and it is possible that the removal is not the rate-limiting step in the glutamine efflux. Inhibition of glutamine efflux by glutamate and some other anions is the result of the inhibition of swelling of the mitochondria caused by the thiols.

The described phenomena complicate the use of the technique for the study of the kinetics of glutamine transport in mitochondria. However, they suggest at the same time several interesting assumptions in relation to the mechanism and regulation of the carrier. The fact that mersalyl as a hydrophilic, non-penetrant reactant is easily accessible to the carrier, and the finding that the swelling of the mitochondria is necessary for the removal of mersalyl inhibition by thiols suggests that the inhibitor induces a strong conformational change of the membrane and/or the carrier. This results in the reorientation of the inhibitor-carrier complex in the membrane. For that reason the complex becomes much less accessible to the thiol reagents. It is known, for instance, that the binding of carboxyatractylate or bongkrekate to the ATP/ADP carrier results in its reorientation with carrier sites facing only the matrix or cytosolic side of the membrane (Klingenberg et al. 1976). After the binding of mersalyl to the glutamine carrier the inhibitor-carrier complex probably diffuses or rotates in the membrane. The same could happen after the binding of glutamine to the carrier. Conformational change of the membrane either due to the binding of a ligand or change of the energy state of the mitochondria may have a strong influence on the conformation of the carrier, resulting in the change of orientation of its binding sites for the substrate. Also, the carrier may be more or less surrounded by phospholipids. All these factors may have a great influence on the accessibility of the carrier to the substrate and, consequently, on the regulation of its activity. It was shown that the glutamine carrier is less sensitive to mersalyl inhibition in energized than in deenergized mitochondria. It is also known that energization stimulates glutaminase activity in intact mitochondria, both of kidney and liver (Kovačević and McGivan 1983). The mechanism of this phenomenon is not elucidated, and it is not ruled out that there is stimulation of glutamine transport rather than glutaminase directly.

The results of Goldstein and Boylan (1978) and Kovačević and Bajin (1982) suggest that activity of the glutamine carrier as such is not the rate-limiting factor in glutamine hydrolysis. However, the described experiments with thiols indicate that the conformation of the mitochondrial membrane and the carrier itself, its orientation and position in the membrane may have a profound influence on the accessibility of glutamine to the binding site of the carrier and in that way regulate the rate of glutamine transport, possibly via a physiological compound with an influence on membrane conformation and accessibility of glutamine to the carrier. It is known that some hormones like, for instance, glucagon, anti-diuretic hormon, ACTH, are good swelling agents like thiol compounds. There is now a considerable body of evidence which indicates that the volume change of mitochondria and therefore, conformational change of the membrane and its proteins, may regulate numerous processes in these organelles (Halestrap et al. 1983).

References

Balcar VJ, Johnston GAR (1975) High-affinity of uptake of L-glutamine in rat brain slices. J Neurochem 24: 875–879

Bass R, Hedegaard HB, Dillehay L, Moffet J, Engelsberg E (1981) The A, ASC and L system for the transport of amino acids in chinese hamster ovary cells (CHO-Kl). J Biol Chem 256: 10259–10266

Bradford NM, McGivan JD (1982) The transport of alanine and glutamine into isolated rat intestinal epithelial cells. Biochem Biophys Acta 589: 55–62

Brosnan JT, Hall B (1977) The transport and metabolism of glutamine by kidney cortex mitochondria from normal and acidotic rats. Biochem J 164: 331–337

Christensen HN, Liang M, Archer EG (1967) A distinct Na^+-requiring transport system for alanine, serine, cysteine and similar amino acids. J Biol Chem 242: 5237–5246

Curthoys NP, Shapiro RA (1978) Effect of metabolic acidosis and phosphate on the presence of glutamine within the matrix space of rat renal mitochondria during glutamine transport. J Biol Chem 253: 63–68

Foreman JW, Reynolds RA, Pepe LM, Segal S (1982) Glutamine transport into isolated renal membrane vesicles from normal and acidotic rats. In: Tannen RL, Goldstein L, Lemieux G, Simpson D, Vinay P (eds) Renal ammonia metabolism. Karger, Basel, pp 101–104

Gazzola GC, Franchi R, Saibene V, Ronchi P, Guidotti GG (1972) Regulation of amino acid transport in chick embryo heart cells l. Adaptive system of mediation for neutral amino acids. Biochem Biophys Acta 266: 407–421

Goldstein L (1975) Glutamine transport by mitochondria isolated from normal and acidotic rats. Am J Physiol 229: 1027–1033

Goldstein L (1976) α-ketoglutarate regulation of glutamine transport and deamidation by renal mitochondria. Biochem Biophys Res Commun 70: 1136–1141

Goldstein L, Boylan JM (1978) Renal mitochondrial glutamine transport and metabolism: Studies with rapid-mixing, rapid filtration technique. Am J Physiol 234: F514–521

Guha SR (1962) Glutaminase of guinea-pig liver mitochondria. Enzymologia 24: 310–326
Guidotti GG, Gazzola GC, Borghetti AF, Franchi-Gazzola R (1975) Adaptive regulation of amino acid transport across the cell membrane in avian and mammalian tissues. Biochim Biophys Acta 406: 264–279
Guidotti GG, Borghetti AF, Gazzola GC (1978) The regulation of amino acid transport in animal cells. Biochim Biophys Acta 515: 329–366
Halestrap AP, Armston AE, Quinlan PT, Whipps DE (1983) Hormonal control of liver mitochondrial metabolism. In: Dumont JE, Nunez J, Denton RM (eds) Hormones and cell regulation. Elsevier/North Holland Biomedical Press, Amsteram, pp 186–200
Hayashi K, Yamamoto S, Ohe K, Miyoshi A, Kawasaki T (1980) Na^+-gradient dependent transport of L-proline and analysis of its carrier system in brush border membrane vesicles of the guinea-pig ileum. Biochim Biophys Acta 601: 654–663
Hayes MR, McGivan JD (1982) Differential effects of starvation on alanine and glutamine transport in isolated rat hepatocytes. Biochem J 204: 365–368
Joseph SK, Bradford NM, McGivan JD (1978) Characteristics of the transport of alanine, serine and glutamine across the plasma membrane of isolated rat liver cells. Biochem J 176: 827–836
Joseph SK, Meijer AJ (1981) The inhibitory effect of sulphydryl reagents on the transport and hydrolysis of glutamine in rat liver mitochondria. Eur J Biochem 119: 523–529
Kilberg MS, Handlogten ME, Christensen HN (1980) Characteristics of an amino acid transport system in rat liver for glutamine, asparagine, histidine and closely related analogs. J Biol Chem 225: 4011–4019
Klingenberg M, Riccio P, Aquila H, Buchanan BB, Grebe K (1976) Mechanism of carrier transport and ADP, ATP carrier. In: Hatefi Y, Djavadi-Ohaniance L (eds) The structural basic of membrane frunction. Academic Press, London New York, pp 293–311
Kovačević Z (1976) Importance of the flux of phosphate across the inner membrane of kidney mitochondria for the activation of glutaminase and the transport of glutamine. Biochim Biophys Acta 430: 399–412
Kovačević Z, Bajin K (1982) Kinetics of glutamine efflux from liver mitochondria loaded with the ^{14}C-labeled substrate. Biochim Biophys Acta 687: 291–295
Kovačević Z, McGivan JD (1983) Mitochondrial metabolism of glutamine and glutamate and its physiological significance. Physiol Rev 63: 547–605
Kovačević Z, McGivan JD, Chappell JB (1970) Condition for activity of glutaminase in kidney mitochondria. Biochem J 118: 265–274
La Noue KF, Schoolwerth AC (1979) Metabolite transport in mitochondria. Annu Rev Biochem 48: 871–922
Oxender DL, Christensen HN (1963) Distinct mediating system for the transport of neutral amino acids by the Ehrlich cell. J Biol Chem 238: 3686–3699
Pitts RF (1973) Production and excretion of ammonia in relation to acid-base regulation. In: Orloff J, Berliner RW (eds) Handbook of physiology. Renal physiology. Am Physiol Soc, Washington DC, pp 455–496
Neubert D, Lehninger AL (1962) The effect of thiols and disulfides on water uptake and extrusion by rat liver mitochondria. J Biol Chem 273: 952–958
Roberts PJ, Keen P (1974) High-affinity uptake system for glutamine in rat dorsal roots but not in nerve endings. Brain Res 67: 352–357
Schousboe A, Hertz L, Svenneby G, Kvamme E (1979) Phosphate-activated glutaminase activity and glutamine uptake in primary cultures of astrocytes. J Neurochem 32: 943–950
Sigrist-Nelson K, Murer H, Hopfer U (1975) Active alanine transport in isolated brush border membranes. J Biol Chem 250: 5674–5680
Simpson DP, Adam W (1975) Glutamine transport and metabolism by mitochondria from dog renal cortex. J Biol Chem 250: 8148–8158
Thomas EC, Christensen HN (1971) Nature of the cosubstrate action of Na^+- and neutral amino acids in a transport system. J Biol Chem 246: 1682–1688
Watford M, Lund P, Krebs HA (1979) Isolation and metabolic characteristics of rat and chicken enterocytes. Biochem J 178: 589–596
Weiler CT, Nystrom B, Hamberger A (1979) Characteristics of glutamine vs. glutamate transport in isolated glia and synaptosomes. J Neurochem 32: 559–565

Intestine and Liver

Metabolism of Vascular and Luminal Glutamine by Intestinal Mucosa in Vivo

H. G. WINDMUELLER[1]

Introduction

Studies over the last 15 years have revealed that the small intestine is a major site for the metabolism of circulating glutamine in most animal species examined, including man. Because of the quantitative importance of glutamine as a vehicle for transporting nitrogen among tissues, the small intestine is now seen to play a significant role in processing the waste nitrogen from other organs. Meanwhile, oxidation of the glutamine carbon provides an important source of energy for the epithelial cells of the intestinal mucosa, where glutamine utilization is largely localized. Glutamine has emerged as a quantitatively more important respiratory fuel than glucose in this tissue. Also, it is now apparent that the small intestine is more than a site for digestion and absorption of dietary ingredients; it also serves some vital metabolic functions for the whole organism. Neptune's work in 1965 [23], showing that glutamine was vigorously oxidized to CO_2 by incubated ileal tissue from several laboratory animal species, appears to be the first reported indication that glutamine may be a preferred substrate for intestine. These studies seem to have attracted little attention, however, perhaps because the conditions employed in vitro were not easily related to conditions in vivo. More direct clues emerged from arteriovenous difference measurements across the nonhepatic splanchnic organs of dogs [1, 7, 14], sheep [49], rats [2], and man [9], studies that revealed an uptake of circulating glutamine by the combined organs drained by the hepatic portal vein – the stomach, small and large intestine, cecum, spleen, and pancreas. The quantitative importance of this process, the specific role of the small intestine, and the metabolic fate of the glutamine were the subject of additional studies that will be described.

The topic of intestinal glutamine utilization was reviewed in considerable detail in 1981 [38], including discussions of glutamine metabolism by disrupted preparations of intestine in vitro, glutamine utilization by extraintestinal tissues, and the sources of circulating glutamine. Since the most numerous cells in the intestine, the epithelial cells of the mucosa, have a dual source of nutrients – the intestinal lumen and arterial blood – preparations of intact intestine in situ that allow separate access to both sources are particularly well suited for metabolic studies. In this chapter, therefore, I will review primarily work from our laboratory with whole animals and with two different preparations of intact rat

1 Laboratory of Cellular and Developmental Biology, National Institute of Arthritis, Diabetes, and Digestive and Kidney Diseases, National Institutes of Health, Bethesda, MD 20205, USA

Glutamine Metabolism in Mammalian Tissues
Edited by D. Häussinger and H. Sies

intestine. Also included will be relevant recent work from other laboratories. The following topics will be considered: (a) a description of the animal preparations; (b) metabolism of arterial glutamine, including localization, quantification, and the metabolic fate of glutamine carbon and nitrogen; (c) metabolism of dietary glutamine from the lumen; (d) relevant enzymology; and (e) the apparent physiologic significance of intestinal glutamine metabolism.

Utilization of Arterial Glutamine

Whole Animal Studies and Localization

Arteriovenous difference measurements for glutamine in vivo across the tissues drained by the superior mesenteric vein were found to be even larger than those measured across all portal-drained organs, localizing much of the uptake in the small intestine and proximal colon [40]. A large net glutamine uptake was observed, equaling 20%–30% of the total plasma glutamine in each blood pass through the intestines in rats, dogs, cats, hamsters, and monkeys. Net uptake was less in rabbits and no net glutamine uptake was found in guinea pigs and chickens, animals that also have substantially lower arterial blood glutamine concentrations. Glutamine utilization was as great in germfree as in conventional rats, indicating that intestinal tissue and not the intestinal microflora was responsible. The glutamine uptake was entirely from the plasma fraction of the blood, since the glutamine concentration of erythrocytes passing through intestine did not change [40]. Further localization was determined by measuring the tissue and product distribution of ^{14}C at 20 s following the infusion of a tracer dose of L-[U-^{14}C]glutamine into the superior mesenteric artery of anesthetized rats [40]. The highest uptake and most extensive metabolism occurred in the mucosal epithelial cells of the small intestine where at 20 s 74% of the glutamine carbon was already present in metabolic products. Subsequent studies have confirmed that isolated mucosal epithelial cells in vitro are capable of metabolizing glutamine [38] and that they contain a group of enzymes that can account for the metabolic products of glutamine observed in vivo (see Sect. Enzymology). It was surprising to find that despite the continuous high uptake of glutamine by the mucosal cells, their steady-state glutamine content was low, only 0.2 μmol/g tissue, or 3% the concentration found in liver and 6% that of skeletal muscle [40]. This is related apparently to the high substrate affinity of the intestinal phosphate-dependent glutaminase (see Sect. Phosphate-Dependent Glutaminase). Villus columnar cells, or enterocytes, which constitute more than 80% of the cells of mucosal epithelium, must participate in metabolizing the glutamine. Whether the other cell types – Paneth cells, mucous-secreting cells, and enteroendocrine cells – also participate is not known. All four cell types are derived from a common stem cell in the mucosal crypts [6].

Animal Preparations and Quantification

We have used two isolated rat intestinal preparations to quantify the rate of intestinal glutamine utilization and to identify and quantify the metabolic products.

Isolated, Blood-Perfused Rat Gut

Initially we used an isolated, perfused gut preparation that includes nearly the entire small intestine, the cecum, the proximal portion of the colon and associated mesentery. The tissue is excised en bloc after beginning vascular perfusion with approximately 100 ml of heparinized rat blood in a recycling system that includes a membrane oxygenator, a valve for regulating venous pressure, and an oximeter for monitoring arterial and venous blood oxygen content (Fig. 1). Blood is pumped into the superior mesenteric artery of the perfused tissue and collected from the superior mesenteric vein. Lymph produced by the perfused tissue is also collected.

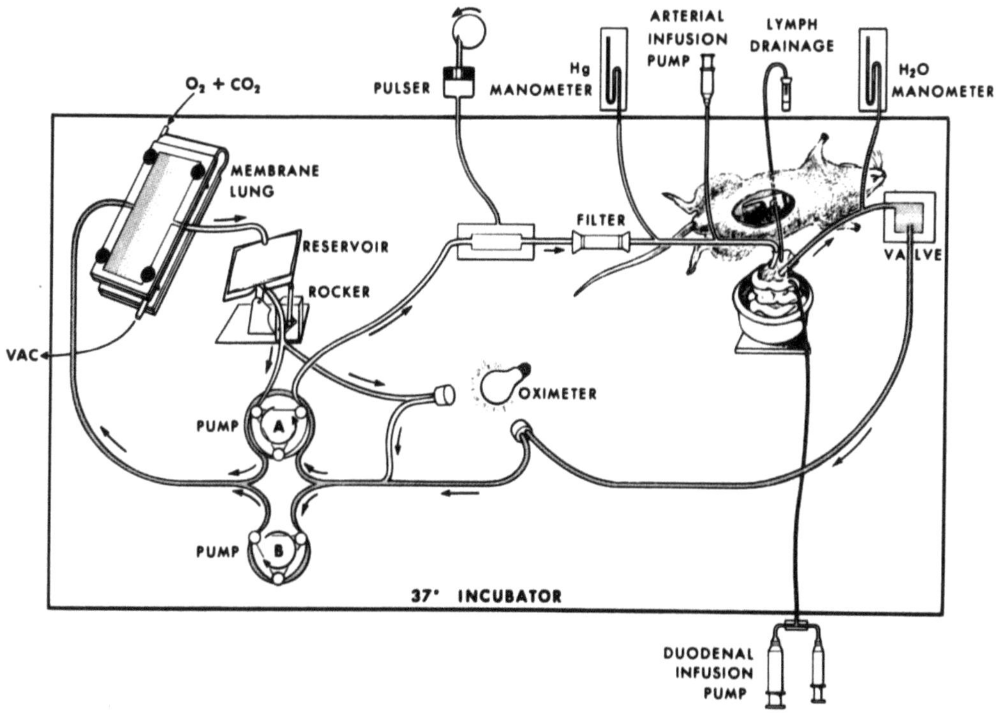

Fig. 1. Schematic diagram of isolated, vascularly perfused rat gut preparation. In the perfusion circuit, indicated by the *arrows*, *stippling* indicates venous blood from the intestine and *no stippling* indicates fully oxygenated blood. Pump *A* has two pumping chambers, one in the arterial circuit, the other in the venous circuit. *VAC*, vacuum. [39]

The isolated small intestine has been historically a very difficult organ to maintain by the perfusion technique. Reported problems, which we also initially encountered, included tissue edema, hyperemia, abnormal vascular resistance, profuse hypersecretion of fluid into the lumen, severe hypermotility, and progressive tissue necrosis. From the response of the preparation to a series of drugs and anesthetic agents, we concluded that these abnormalities were responses to the loss of central sympathetic innervation resulting from isolating the tissue [48]. We were successful in preventing all of these deleterious effects by continuously infusing small amounts of norepinephrine (approx. 5×10^{-8} *M*) into the arterial cannula carrying blood to the tissue (Fig. 1, arterial infusion pump). The further addition to the perfusate of dexamethasone (9α-fluoro-16α-methylprednisolone, 6×10^{-7} *M*), a synthetic glucocorticoid, was found to be required to maintain the responsiveness of the preparation to norepinephrine [48]. Also important, a surgical protocol was developed to eliminate tissue anoxia at the onset of extracorporeal tissue perfusion. This precaution plus use of the two hormones resulted in a preparation that remained quite normal functionally and histologically through 5 h of perfusion. The method has been described in detail [47].

It was during unrelated studies with this preparation that we observed a rapid disappearance of glutamine, but no other amino acid, from the recycling blood [40], a striking and unexpected finding that first stimulated our interest in this phenomenon. Gut preparations from 280 g rats, fasted overnight, showed a net glutamine uptake of 74 µmol/h when sufficient glutamine was continuously infused into the perfusate to maintain the normal blood concentration at 0.6 m*M*. The following considerations pointed to the quantitative importance of this process. Glutamine is the most abundant amino acid in rat plasma, and 30% was removed in each blood pass through the preparation. Meanwhile, there was no net uptake of any other amino acid. This utilization rate for glutamine would produce a turnover of the rat's entire plasma glutamine pool every 4.6 min. Furthermore, it could account for approx. 20% of the turnover rate of the entire body glutamine pool [31]. Finally, the net uptake rate for glutamine was nearly as large as that for glucose, generally considered to be the primary energy source for small intestine [18]. As further evidence for quantitative importance, recent arteriovenous difference measurements by Schröck and Goldstein [29] across major organs of the rat indicate that glutamine uptake by the nonhepatic splanchnic vascular bed is sufficient to account for the entire glutamine output by skeletal muscle, the major site that releases glutamine into the circulation.

The net rate of glutamine utilization by the perfused gut was concentration dependent, approaching a maximum near the normal blood concentration of 0.6 m*M*. The actual rate of glutamine utilization was apparently higher than the measured net uptake rate, because evidence showed that some cells in the preparation synthesize glutamine and release it into the blood. Thus, without glutamine addition, the glutamine concentration in the recycling perfusate fell quickly from 0.6 m*M* to 0.2 m*M*, but no lower. However, when 4 m*M* L-methionine sulfoximine, an inhibitor of glutamine synthetase, was added to the perfusate, the glutamine concentration fell to near zero [40].

Autoperfused Intestinal Segments In Vivo

We further quantified the process with autoperfused segments of rat small intestine in vivo, a preparation established to avoid the use of hormones and to permit single-pass perfusions of both the vasculature and lumen of a defined region of intestine for an hour or more. This allowed us to determine quantitatively the metabolic fate of luminal as well as arterial glutamine and other substrates.

In this method (Fig. 2), a 3–5 cm jejunal or ileal segment of an anesthetized 300 g rat is exteriorized into a tissue bath. The ends of the segment are ligated, forming a closed sac and including an in-flow cannula. After administering heparin to the rat intravenously, the single vein draining the segment is cannulated and the entire blood outflow serially collected. The arterial blood supply to the segment and neural and lymphatic connections remain intact. To replace the blood collected, the rat is transfused continuously via a saphenous vein from a pool of heparinized rat blood. By way of another cannula in the aorta, arterial pressure is monitored and arterial blood can be sampled, allowing arteriovenous difference measurements to be made across the segment. Since the blood flow rate is also measured, the net rate of metabolite exchange across the segment can be determined. The metabolic fate of radioactively labeled luminal and arterial substrates can be studied quantitatively under physiological conditions by infusing them into the segment lumen or into the artery supplying

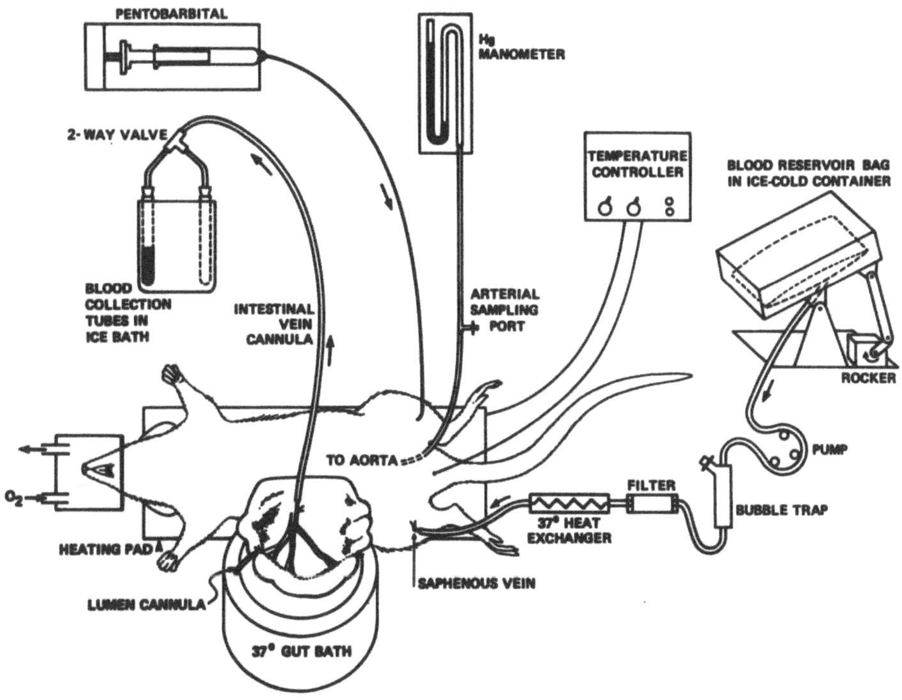

Fig. 2. Schematic diagram of autoperfused intestinal segment preparation in vivo. [45]

the segment, respectively, and then identifying and quantifying the metabolic products in the collected venous blood [46, 47].

Jejunal or proximal ileal segments from rats fasted for 16 h extracted 30%–35% of the total plasma glutamine from blood perfusing the tissue [41, 43]. The net glutamine utilization rate was 220 nmol/min g intestine, approximately equal to the net glucose utilization rate, and equivalent to 59.4 µmol glutamine per h by the entire small intestine, which in a 300 g rat weighs about 4.5 g. This rate is 80% of that observed in the isolated, perfused rat gut [40], where the pancreas [24] and colon [36] apparently also utilize some glutamine. A similar rate of glutamine utilization was found by Hanson and Parsons [11] with an isolated rat intestinal preparation vascularly perfused with a high glutamine concentration (1.5 m*M*). In rats not fasted, intestinal segments, rinsed free of diet residue, utilized arterial glutamine and glucose at about twice the rates, per unit weight of tissue, as observed in fasted animals [44]. The net uptake rate for glutamine by these intestinal segments was more than 20 times the net uptake rate observed for any other individual plasma amino acid. Glutamine and glucose utilization were not increased further when the rinsed lumen of the segment was perfused continuously with a solution of 3-0-methyl-D-glucose and several amino acids, all of which undergo carrier-mediated transport into the mucosal cells, but none of which is metabolized [44]. The rate of glutamine metabolism was also unaffected by perfusing the lumen with 70 m*M* glucose, which was readily transported and only 3% of which was metabolized during absorption [44]. To summarize, the utilization rate of arterial glutamine is concentration dependent, is approximately equal to the rate for glucose, is reduced by fasting, and is not increased during carrier-mediated transport of nonmetabolized luminal substrates.

Metabolic Fate

Glutamine Carbon

The radioactive products of L-[U-^{14}C]glutamine metabolism were identified with both isolated intestinal preparations. After recycling labeled glutamine through the isolated gut for 3 h, labeled CO_2, lactate, citrate, citrulline, proline, and smaller amounts of several other amino acids were recovered from the perfusate and also from the perfused tissue [40]. The production rate of each metabolite was then determined with the perfused gut [40] and also with the autoperfused segment preparation [43] under single-pass perfusion conditions where product recycling through the intestine was avoided. The product distribution of the arterial glutamine carbons was similar with the two preparations and also similar whether conventional or germfree rats were used, precluding the intestinal microflora as a source for any of the metabolic products (Table 1, arterial administration). The large recovery of glutamine carbon in CO_2 and organic acids indicates extensive glutamine oxidation. Incorporation into protein and other acid-insoluble tissue components accounted for less than 15% of the glutamine carbon. Also noteworthy were a small production of glucose from

Table 1. Metabolic fate of the carbon of arterial and luminal glutamine utilized by rat small intestine [38]

Metabolic product	Percent of total glutamine carbon		
	Isolated vascularly perfused gut, arterial administration[a]	Autoperfused jejunal segment in vivo	
		Arterial administration[a]	Luminal administration[b]
Released to intestinal venous blood			
CO_2	57.2	55.5	53.4
Lactate	8.1	8.5	14.8
Citrate	5.4	1.6	1.7
Other organic acids[c]	4.0	1.1	2.1
Citrulline	5.5	4.7	4.2
Proline	4.6	5.2	5.5
Alanine	0.7	3.3	3.8
Other amino acids[d]	0.1	2.0	6.6
Glucose	–	4.7	2.4
Incorporated into tissue	14.4	13.4	5.5
Total	100.0	100.0	100.0

[a] With each preparation, a tracer dose of L-[U-^{14}C]glutamine was infused continuously into the arterial blood flowing into the intestine. The ^{14}C-labelled metabolic products were quantified in the intestinal venous blood, which was totally collected in a series of sequential samples. The rate of appearance of all products reached a steady state within 10 min. This steady state rate and the specific radioactivity of the perfusing glutamine were used to calculate the conversion rate of glutamine carbon to each product. With both preparations, the total recovery rate of glutamine carbon in the combined products, determined radiochemically, agreed closely with the net rate of glutamine uptake by the preparation, determined chemically. Results were similar with preparations from conventional and germfree rats. Every value is the mean from four to eight steady state determinations from each of two to three experiments [40, 43]

[b] L-[U-^{14}C]glutamine, 2 µmol in 0.3 ml of a balanced salt solution, was introduced into the lumen of a closed jejunal segment from which all venous blood was collected for 60 min, by which time 97% of the radioactive dose had been absorbed and either translocated into the blood (92%) or incorporated into the acid-insoluble fraction of the tissue (5%). Of the translocated radioactivity, 37% remained in glutamine and 63% was distributed among the indicated metabolic products as shown. Values are means from three experiments [41]

[c] Malate, succinate, pyruvate, α-ketoglutarate, and 5-oxoproline

[d] Aspartate, ornithine, glutamate, serine, and glycine

glutamine and a very small release of labeled glutamate. The latter is in contrast to findings with intestinal slices [22] and isolated enterocytes incubated in vitro [35], where much of the glutamate produced from glutamine remains unmetabolized and is released into the medium.

Glutamine Nitrogen

In studies with autoperfused intestinal segments, the quantities of ammonia, citrulline, alanine, and proline released into intestinal venous blood from rinsed

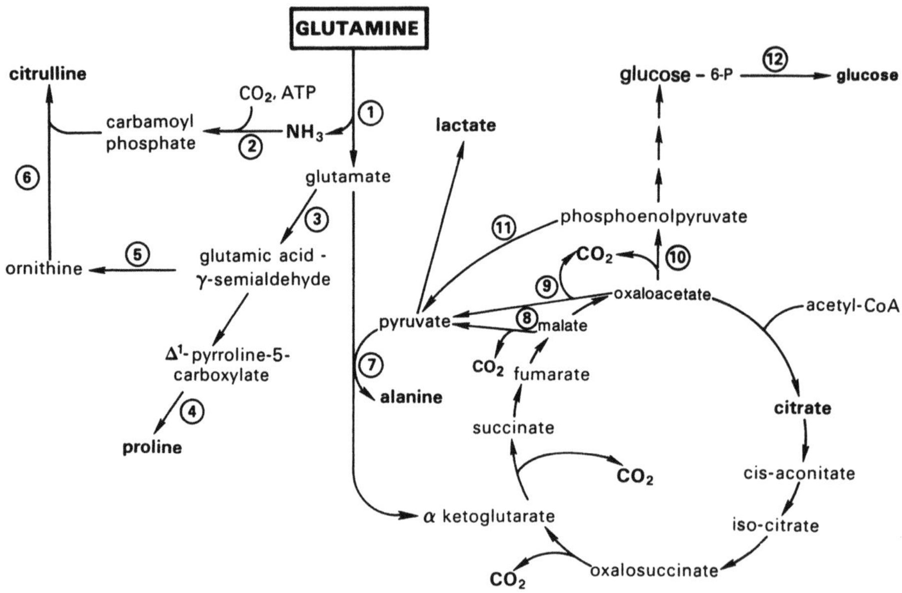

Fig. 3. Proposed pathways of glutamine metabolism in rat intestinal mucosa. *Bold print* indicates metabolic products released into intestinal venous blood. *Numbered reactions* have all been demonstrated individually in vitro with mucosal preparations from rat small intestine [38] and are discussed under Enzymology [37] and [38]

segments could fully account for the amounts of vascular glutamine nitrogen taken up [44]. Approximately 38% of the glutamine nitrogen was recovered as ammonia, 28% as citrulline, 24% as alanine, and 7% as proline. A very small net release of glutamate and ornithine accounted for the remaining 3%. The same products accumulated in the recycling perfusate of isolated rat gut preparations during metabolism of vascular glutamine [11, 40] and were released into the circulation in vivo by the nonhepatic splanchnic organs in dogs, sheep, and man [38].

The proposed metabolic pathways shown in Fig. 3 can account for the metabolic products of glutamine carbon and nitrogen in rat small intestine. The necessary enzymes are discussed further under Enzymology.

Utilization of Luminal Glutamine

Absorption and Metabolic Fate

Glutamine was absorbed as rapidly from the lumen of autoperfused jejunal segments as was asparagine, asparate, arginine, and several other amino acids and more rapidly than glutamate [41, 42]. When L-[U-^{14}C]glutamine was given intraluminally at a concentration of 6 mM, approximating postprandial luminal levels, more than 94% of the radioactivity was recovered in the collected venous blood in 30 min, mostly during the first 10 min. One-third of the ^{14}C remained in glutamine and the remainder was distributed among a group of metabolic products closely resembling those observed following the metabolism of arterial glutamine (Table 1, luminal administration). Also analogous with arterial glutamine, luminal doses of glutamine or glutamate increased the venous output of alanine, citrulline, and proline; with glutamine, the output of ammonia was also incresed [41]. These results indicate that glutamine was metabolized similarly whether it entered the mucosal cells across the brush border from the lumen or across the basolateral membrane from the arterial blood.

The proportion of absorbed glutamine that was metabolized by intestine was a function of the luminal concentration administered. At 6 m*M* only 34% was absorbed intact, but at 45 m*M* this increased to 70% [41]. The rate limiting step for glutamine catabolism is apparently its deamidation (see Sect. Phosphate-Dependent Glutaminase) and not metabolism of the resulting glutamate. Thus, no glutamate was released during luminal perfusion with either 6 m*M* or 45 m*M* glutamine. Furthermore, when 6 m*M* L-[U-^{14}C]glutamate was administered luminally, none of it escaped metabolism during its translocation across the intestine. The ^{14}C-labeled products of glutamate catabolism were the same as those arising from L-[U-^{14}C]glutamine [41]. The tissue's capacity to metabolize glutamate is large, but it was exceeded when the luminal concentration was increased to 30 m*M*. At that concentration some intact glutamate was translocated into the blood.

Relationship Between Luminal and Arterial Glutamine

Mucosal cells of the intestine have access to large amounts of glutamine from two sources during the digestion of a protein meal. As might be expected, the two sources of glutamine compete for the same enzymes. Thus, we found that the presence of glutamine or glutamate in the lumen reduced the rate of glutamine utilization from the blood. Luminal administration of 30 m*M* glutamate in autoperfused jejunal segments reduced the uptake and metabolism of arterial glutamine by 25%, and 6 m*M* luminal glutamine reduced it by 40%. However, since most of the luminal glutamine is also metabolized, the total rate of glutamine metabolized was increased by 70% when both sources of glutamine were available [41]. This was confirmed in an isolated rat intestinal preparation by Hanson and Parsons [11]. These studies underscore the large capacity of the small intestine for catabolizing glutamine. Calculations indicate that the small intestine of a 280 g rat ingesting a diet with 20% protein would metabolize a total of 4 mmol or more of glutamine daily.

Enzymology

Phosphate-Dependent Glutaminase

Several lines of evidence indicate that the initial reaction of glutamine catabolism in intestinal mucosal cells is catalyzed by a phosphate-dependent glutaminase [25], hereafter referred to as glutaminase (Fig. 3, reaction 1).

(a) Mucosal scrapings of rat small intestine contain glutaminase activity equivalent to 3–6 μmol/h mg protein, sufficient to account for the rates of glutamine breakdown observed. Meanwhile, there is only low and insufficient activity for several other glutamine-degrading enzymes (carbamoyl phosphate synthetase, phosphoribosylpyrophosphate amidotransferase, fructose-6-phosphate aminotransferase, cytidine triphosphate synthetase, glutamine transaminase, and γ-glutamyltranspeptidase) [38]. (b) The distribution of glutaminase is consistent with its proposed role. High activity was found in duodenal, jejunal, and ileal mucosa of the rat, dog, cat, hamster, rabbit, and monkey, all species in which small intestine utilizes arterial glutamine. Much lower activity was found in the chicken and guinea pig, where the intestine did not exhibit net glutamine utilization. Glutaminase was found in mature mucosal cells from the villi and also in the rapidly dividing mucosal cells from the crypts, but it is virtually absent from intestinal muscularis, consistent with the identification of mucosa as the region of highest glutamine metabolism (see Sect. Whole Animal Studies and Localization). (c) Glutaminase and glutamine catabolism are sensitive to the same inhibitors. Glutaminase is inhibited irreversibly by two glutamine analogs, 6-diazo-5-oxo-L-norleucine (DON) and L-2-amino-4-oxo-5-chloropentanoic acid (chloroketone). Chloroketone administered luminally into an autoperfused jejunal segment preparation inhibited the whole tissue glutaminase by 50% and inhibited totally the hydrolysis of glutamine subsequently absorbed from the lumen. DON administered arterially in this preparation inhibited the tissue glutaminase almost completely and inhibited the hydrolysis of arterial as well as luminal glutamine. (d) The high substrate affinity of the glutaminase is consistent with the low tissue levels of glutamine. The apparent K_m of the glutaminase for glutamine at pH 8.1 is about 2.2 m*M*, a concentration lower by a factor of 10 than the K_m reported for glutaminase from liver [38]. This could explain the much lower tissue levels of glutamine in intestinal mucosa (0.2 μmol/g) than in liver (5.9 μmol/g) [40]. The tissue/plasma concentration ratio for glutamine in intestine is approximately 0.3, while in liver it is near 10.

Compared with the activity found in small intestinal mucosa, relatively low glutaminase activity was found in mucosa from rat stomach, cecum, and colon [25]. Consistent with this, rat stomach utilizes relatively little vascular glutamine [3]. Nearly all of the intestinal glutaminase is membrane-bound, and, as is also true for other tissues [38], it is associated with the mitochondria, from which it can be partially extracted with hypotonic phosphate-borate [25]. The enzyme in isolated mitochondria from rat intestine has a rather broad pH optimum between pH 7.6 and pH 9.2. The apparent $K_{0.5}$ for activation by phosphate is 22 m*M*. However, with zero or suboptimal phosphate, a number of other compounds will acitvate the enzyme, including ammonium sulfate, hexose phosphate, maleate, ammonium bicarbonate, pyrophosphate, and arsenate. The enzyme will not hydrolyze D-glutamine or L-asparagine, consistent with the translocation of luminal asparagine across rat intestine without hydrolysis [42] and the failure of intestine to utilize arterial asparagine [40]. The crude intestinal glutaminase resembles the glutaminase of kidney, brain, and ascites cells in many of its properties, but it is different from the liver enzyme [19, 25], a relationship supported by immunological studies [38].

Intestinal glutaminase activity in rats was not increased by feeding large amounts of glutamine [25], nor was it significantly decreased by fasting [5, 25]. Adult levels of glutaminase are present in small intestine of fetal rats immediately before birth and up to 5 days of age. Activity increases about two fold by days 12–15 before returning to the adult level at 20 days of age [25].

Other Relevant Enzymes

In addition to glutaminase, all of the other enzymes needed to support the proposed metabolic pathways in Fig. 3 have been individually demonstrated in cell-free mucosal preparations of small intestine by various laboratories [38]. Approximately 80% of the ammonia generated by the glutaminase (Fig. 3, reaction 1) is released into the portal vein, contributing importantly to the large ammonia flux between intestine and liver that also includes ammonia generated by the intestinal microflora [36, 40]. The remaining ammonia is apparently converted to carbamoyl phosphate (reaction 2). In addition to carbamoyl phosphate synthetase, rat intestinal mucosa contains N-acetylglutamate, a cofactor for the enzyme, and N-acetylglutamate synthase [32]. Part of the glutamate produced in reaction 1 is reduced to glutamate-γ-semialdehyde (reaction 3), which can be further reduced to proline (reaction 4) or transaminated to ornithine (reaction 5), a reaction in which glutamate also serves as the amino donor. Intestine appears to be the rare mammalian tissue that can effect a net synthesis of ornithine and proline from glutamate [12, 13, 34]. Glutamate-γ-semialdehyde is converted nonenzymatically to pyrroline-5-carboxylate (P5C). The protein(s) that convert glutamate to P5C (P5C synthase) has been characterized in mitochondria from rat intestinal mucosa by Jones and co-workers [33, 34]. Ornithine carbamoyltransferase catalyzes the conversion of ornithine and carbamoylphosphate to citrulline (reaction 6). Citrulline is not further metabolized [45] and is released into intestinal venous blood. Intestine contains an arginase active in vivo [42], but apparently it does not have the other cytosolic urea cycle enzymes needed to convert citrulline to arginine.

The glutamate not utilized in reactions 3 and 5 can be transaminated with pyruvate to yield alanine and α-ketoglutarate (reaction 7). Glycolysis is a source for much of the pyruvate (see Sect. Glutamate as Intestinal Energy Source). Additional enzymes present to metabolize glutamate include asparate aminotransferase and glutamate dehydrogenase. The reaction catalyzed by the latter enzyme does not appear to be of quantitative importance, however, considering the failure of glutamate metabolism in intestine in vivo to increase ammonia output [38].

The following pathways are present to explain the appearance of glutamine carbon in the 3-carbon metabolites, pyruvate, and lactate: "malic enzyme" (reaction 8), oxaloacetate decarboxylase (reaction 9), and phosphoenolpyruvate carboxykinase (PEPCK) plus pyruvate kinase (reactions 10 and 11). The relative importance of the three pathways remains uncertain [38]. PEPCK plus the glycolytic pathway and glucose-6-phosphatase (reaction 12) can explain the conversion of glutamine carbon to glucose.

Physiologic Implications

Glutamine as Intestinal Energy Source

The importance of glutamine as a respiratory fuel became apparent from measurements of its contribution to the total CO_2 produced by intestine. With the isolated, perfused gut preparation, total CO_2 output was 9,750 nmol/min and the steady-state rate of CO_2 production from arterial glutamine was 3,130 nmol/min, or 32% of the total [40]. A more detailed analysis in vivo of glutamine and other respiratory fuels for intestine was made using autoperfused jejunal segments [43, 44]. Potential substrate fuels were first identified chemically from arteriovenous difference measurements. Then in subsequent studies, the CO_2 contribution from each substrate utilized was quantified radiochemically. In individual experiments, a tracer dose of each substrate, labeled with ^{14}C, was infused directly into the arterial supply to the segment, and the rate of product formation was calculated from the steady state rate of appearance of each labeled metabolic product in the venous blood (see Table 1, footnote a), The results for conversion of the substrates to CO_2 are summarized in Table 2. More than one-half of the CO_2 produced by intestine of fasted rats was derived from the metabolism of circulating ketone bodies and 36% was derived from glutamine. Although glucose and glutamine were utilized in approximately equal amounts, 75% of the glucose carbon appeared in alanine and lactate, and only 14% in CO_2. Thus, glucose contributed only 7% of the total CO_2. Additional small amounts of CO_2 were contributed by lactate and unesterified fatty acids. These arterial substrates together accounted fully for all the released CO_2, measured manometrically.

In rinsed intestinal segments from nonfasted rats, the CO_2 contribution from glutamine increased to nearly 50% of the total and that from ketone bodies was reduced (Table 2). The contribution from glucose remained small even though the animals had been fed a diet containing 50% glucose and 15% corn starch for 12 days or more in an attempt to induce a greater capacity for glucose utilization. The role of other potential substrates was not evaluated under these conditions and only 70% of the CO_2 produced was accounted for. In additional segments from the nonfasted rats, when the lumen was perfused with a glucose-amino acid mixture to simulate conditions after a meal, a total of 77% of the CO_2 was derived from the metabolism of arterial glutamine and luminal glutamine, glutamate, and asparate. Only 3% of the absorbed luminal glucose was metabolized and contributed only 6% of the CO_2 (Table 2).

These studies reveal the importance of amino acids, particularly glutamine, as energy sources for small intestine. The quantitative importance of glutamine in mucosal epithelial cells must be even greater than indicated by the values in Table 2, which reflect the combined metabolic activity of the full thickness of the tissue, including the muscularis, which is glutaminase-free. The intestine is the

Table 2. Contribution of arterial and luminal substrates to respiratory CO_2 in rat jejunum in vivo[a]

	Animal treatment		
	Fasted 16 h	Nonfasted, lumen emptied	Nonfasted, lumen perfused with a "meal"[b]
1. Total CO_2 produced (nmol/min g intestine)[c]	1,650	2,670	2,780
2. CO_2 yield from substrates (% of total produced)			
Arterial:[d]			
Glutamine	36.2	46.3	38.3
3-Hydroxybutyrate	26.9	2.9	0.9
Acetoacetate	24.8	10.7	5.2
Glucose	7.0	10.3	
Lactate	4.8		
Unesterified fatty acids	3.5		
Luminal:			
Glutamine + glutamate + aspartate (5 m*M* each)[e]			39.0
Glucose (70 m*M*)[f]			5.8
Total	103.3	70.1	89.2

[a] Experiments were performed with autoperfused jejunal segments of 280 g animals. Results are taken from [43] and [44]

[b] Earle's balanced salt solution plus glucose (70 m*M*) and 19 L-amino acids (0.5–6.5 m*M* each), to simulate a meal

[c] Determined from arteriovenous difference measurements for CO_2 across the segments using the Van Slyke manometric procedure

[d] The rate of CO_2 production from each arterial substrate was determined in separate experiments. As described for glutamine in Table 1, footnote a, the ^{14}C-labeled substrate was infused at a constant rate into the arterial supply to the segment, the steady state rate of $^{14}CO_2$ appearing in the venous blood was determined, and the rate of CO_2 production was calculated

[e] These three amino acids in the luminal perfusate were uniformly labeled with ^{14}C; the specific radioactivity of the carbon atoms of all three compounds was the same. The CO_2 yield was calculated from this specific radioactivity and the steady state rate of appearance of $^{14}CO_2$ in the collected jejunal venous blood

[f] The glucose in the luminal perfusate was uniformly labeled with ^{14}C and the CO_2 yield was determined as in footnote [e]

only organ known with such a glutamine preference, with the possible exception of the kidney during metabolic acidosis [26]. The low glucose utilization rate in vivo is in sharp contrast to the much higher, apparently artifactual rates observed in vitro, which result perhaps from the relative tissue anoxia under such conditions [43, 44].

There is little evidence that, aside from its role as a respiratory fuel, glutamine serves a more specific function in the physiology of intestine. Glutamine has been reported to stimulate the transport of glucose [27], water

[20], and sodium ions [20] by intestine incubated in vitro. This may be a reflection of its energy-producing capacity, however.

Relevance for the Whole Animal

As a primary site for metabolizing plasma glutamine, intestine plays an important role in processing the waste nitrogen produced by other tissues, particularly skeletal muscle, where glutamine and alanine are the main nitrogenous end-products of amino acid catabolism and ammonia fixation. The alanine nitrogen is further processed to urea in the liver and the glutamine nitrogen is converted to ammonia, alanine, proline, and citrulline and released into the portal blood by the intestine (Fig. 4). The ammonia, alanine, and proline are efficiently taken up by the liver where the nitrogen is mostly converted to urea. The citrulline is not taken up by the liver but reaches the general circulation and is largely taken up by the kidneys, where it is converted to arginine and released back into the blood (see below). Some of this arginine can then be also degraded in the liver. Thus, all of the nitrogenous end products of intestinal glutamine metabolism can serve ultimately as urea precursors [21]. The nitrogen flux along this pathway is large. From the measured rates of urinary urea output and intestinal glutamine metabolism, we estimate that 30% of the nitrogen utilized for hepatic ureagenesis in nonfed rats is derived from the nitrogenous products of intestinal glutamine breakdown [44]. The carbon atoms of the alanine and proline, as well as those of the lactate produced, are available for gluconeogenesis in the liver [28].

The citrulline from intestine serves more immediately as an arginine precursor in the brain and, to a greater extent, in the kidneys, which are an important endogenous arginine source for protein biosynthesis in muscle [8]. The only other tissue known to make citrulline is the liver, where it is an

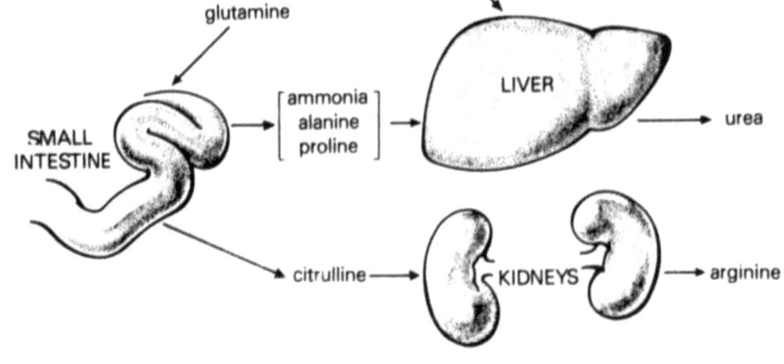

Fig. 4. Fate of nitrogenous end-products from skeletal muscle

intermediate in the urea cycle. Although it was previously believed that the liver also releases significant amounts of citrulline into the circulation, the results of liver perfusions performed under physiological conditions have recently indicated that this is not the case [45]. Rat livers were perfused with whole blood supplemented with sufficient ammonia, ornithine, and 15 other amino acids to generate 20 µmol/min of urea. This is about 3.5 times the average output observed over 24 h in vivo in rats fed a diet with 20% protein. Despite the large quantities of citrulline obviously generated to account for this high rate of ureagenesis, less than 40 nmol/min of citrulline was released into the perfusate, a small fraction of the quantities released continuously by the small intestine. Arteriovenous difference measurements in rats in vivo confirmed the failure of liver to release citrulline; they also showed that the kidneys take up circulating citrulline at about the same rate as its release by intestine and that they release arginine equivalent to 80% of the citrulline taken up. No additional significant sites for citrulline release or uptake were revealed by arteriovenous measurements across other major organs [45]. The available data indicate that the intestines and kidneys cooperate to provide an important endogenous source of arginine for the animal. This would explain why arginine is an essential dietary amino acid only when the rates of body protein biosynthesis are high, e.g., during periods of rapid growth, pregnancy, and lactation.

Metabolic Acidosis

The kidneys, which utilize little or no circulating glutamine normally, become an important site of glutamine utilization for urinary ammonia production during metabolic acidosis [29, 38]. Skeletal muscle and the liver provide the increased amounts of glutamine required under these conditions, while the glutamine utilization rate by intestine remains largely unchanged [10, 29, 38]. Although the phosphate-dependent glutaminases of kidney and intestine share several other characteristics (see Sect. Phosphate-Dependent Glutaminase), only the kidney enzyme shows a large adaptive increase in activity during metabolic acidosis [4, 25].

References

1. Addae SK, Lotspeich WD (1968) Relation between glutamine utilization and production in metabolic acidosis. Am J Physiol 215: 269–277
2. Aikawa T, Matsutaka H, Yamamoto H, Okuda T, Ishikawa E, Kawano T, Matsumura E (1973) Gluconeogenesis and amino acid metabolism. II. Interorganal relations and roles of glutamine and alanine in the amino acid metabolism of fasted rats. J Biochem (Tokyo) 74: 1003–1017
3. Anderson NG, Hanson PJ (1983) Arteriovenous differences for amino acids across control and acid-secreting rat stomach in vivo. Biochem J 210: 451–455
4. Anderson NM, Bennett FI, Alleyne GAO (1976) Ammonia production by the small intestine of the rat. Biochim Biophys Acta 437: 238–243
5. Budohoski L, Challis RAJ, Newsholme EA (1982) Effects of starvation on the maximal activities of some glycolytic and citric acid-cycle enzymes and glutaminase in mucosa of the small intestine of the rat. Biochem J 206: 169–172

6. Cheng H, Leblond CP (1974) Origin, differentiation and renewal of the four main epithelial cell types in the mouse small intestine: V. Unitarian theory of the origin of the four epithelial cell types. Am J Anat 141: 537–562
7. Elwyn DH, Parikh HC, Shoemaker WC (1968) Amino acid movements between gut, liver, and periphery in unanesthetized dogs. Am J Physiol 215: 1260–1275
8. Featherston WR, Rogers QR, Freedland RA (1973) Relative importance of kidney and liver in synthesis of arginine by the rat. Am J Physiol 224: 127–129
9. Felig P, Wahren J, Karl I, Cerasi E, Luft R, Kipnis DM (1973) Glutamine and glutamate metabolism in normal and diabetic subjects. Diabetes 22: 573–576
10. Fine A (1982) The effects of chronic metabolic acidosis on liver and muscle glutamine metabolism in the dog in vivo. Biochem J 202: 271–273
11. Hanson PJ, Parsons DS (1977) Metabolism and transport of glutamine and glucose in vascularly perfused small intestine of rat. Biochem J 166: 509–519
12. Henslee JG, Jones ME (1982) Ornithine synthesis from glutamate in rat small intestinal mucosa. Arch Biochem Biophys 219: 186–197
13. Herzfeld A, Raper SM (1976) Enzymes of ornithine metabolism in adult and developing rat intestine. Biochim Biophys Acta 428: 600–610
14. Hills AG, Reid EL, Kerr WD (1972) Circulatory transport of L-glutamine in fasted mammals: cellular sources of urine ammonia. Am J Physiol 223: 1470–1476
15. Ishikawa E (1976) The regulation of uptake and output of amino acids by rat tissues. Adv Enzyme Regul 14: 117–136
16. Knox WE, Horowitz ML, Friedell GH (1969) The proportionality of glutaminase content to growth rate and morphology of rat neoplasms. Cancer Res 29: 669–680
17. Kovačević Z, Morris HP (1972) The role of glutamine in the oxidative metabolism of malignant cells. Cancer Res 32: 326–333
18. Krebs HA (1972) Some aspects of the regulation of fuel supply in omniverous animals. Adv. Enzyme Regul 10: 397–420
19. Krebs HA, Baverel G, Lund P (1980) Effect of bicarbonate on glutamine metabolism. Int J Biochem 12: 69–73
20. Love AHG, Mitchell TG, Neptune EM Jr. (1965) Transport of sodium and water by rabbit ileum, in vitro and in vivo. Nature (London) 206: 1158
21. Lund P, Watford M (1976) Glutamine as a precursor of urea. In: Grisolia S, Báguena R, Mayor F (eds) The urea cycle. Wiley, New York, pp 479–488
22. Matsutaka H, Aikawa T, Yamamoto H, Ishikawa E (1973) Gluconeogenesis and amino acid metabolism. III. Uptake of glutamine and output of alanine and ammonia by non-hepatic splanchnic organs of fasted rats and their metabolic significance. J Biochem 74: 1019–1029
23. Neptune EM Jr (1965) Respiration and oxidation of various substrates by ileum in vitro. Am J Physiol 209: 329–332
24. Pinkus LM, Berkowitz JM (1980) Utilization of glutamine by canine pancreas in vivo and acinar cells in vitro. Fed Proc 39: 1902
25. Pinkus LM, Windmueller HG (1977) Phosphate-dependent glutaminase of small intestine: localization and role in intestinal glutamine metabolism. Arch Biochem Biophys 182: 506–517
26. Pitts RF, Pilkington LA, MacLeod MB, Leal-Pinto E (1972) Metabolism of glutamine by the intact functioning kidney of the dog: studies in metabolic acidosis and alkalosis. J Clin Invest 51: 557–565
27. Riklis E, Quastel JH (1958) Effects of cations on sugar absorption by isolated surviving guinea pig intestine. Can J Biochem Physiol 36: 347–362
28. Ross BD, Hems R, Krebs HA (1967) The rate of gluconeogenesis from various precursors in the perfused rat liver. Biochem J 102: 942–951
29. Schröck H, Goldstein L (1981) Interorgan relationships for glutamine metabolism in normal and acidotic rats. Am J Physiol 240: E519–E525
30. Sevdalian DA, Ozand PT, Zielke HR (1980) Increase in glutaminase activity during the growth cycle of cultured human diploid fibroblasts. Enzyme 25: 142–144
31. Squires EJ, Brosnan JT (1983) Measurements of the turnover rate of glutamine in normal and acidotic rats. Biochem J 210: 277–280

32. Uchiyama C, Mori M, Tatibana M (1981) Subcellular localization and properties of N-acetylglutamate synthase in rat small intestinal mucosa. J Biochem (Tokyo) 89: 1777–1786
33. Wakabayashi Y, Henslee JG, Jones ME (1983) Pyrroline-5-carboxylate synthesis from glutamate by rat intestinal mucosa: subcellular localization and temperature stability. J Biol Chem 258: 3873–3882
34. Wakabayashi Y, Jones ME (1983) Pyrroline-5-carboxylate synthesis from glutamate by rat intestinal mucosa. J Biol Chem 258: 3865–3872
35. Watford M, Lund P, Krebs HA (1979) Isolation and metabolic characteristics of rat and chicken enterocytes. Biochem J 178: 589–596
36. Weber FL Jr, Veach GL (1979) The importance of the small intestine in gut ammonium production in the fasting dog. Gastroenterology 77: 235–240
37. Windmueller HG (1980) Enterohepatic aspects of glutamine metabolism. In: Mora J, Palacios R (eds) Glutamine: metabolism, enzymology, and regulation. Academic Press, London New York, pp 235–257
38. Windmueller HG (1982) Glutamine utilization by the small intestine. In: Meister A (ed) Advances in enzymology, vol 53. Wiley, New York, pp 201–237
39. Windmueller HG, Spaeth AE (1972) Fat transport and lymph and plasma lipoprotein biosynthesis by isolated intestine. J Lipid Res 13: 92–105
40. Windmueller HG, Spaeth AE (1974) Uptake and metabolism of plasma glutamine by the small intestine. J Biol Chem 249: 5070–5079
41. Windmueller HG, Spaeth AE (1975) Intestinal metabolism of glutamine and glutamate from the lumen as compared to glutamine from blood. Arch Biochem Biophys 171: 662–672
42. Windmueller HG, Spaeth AE (1976) Metabolism of absorbed aspartate, asparagine and arginine by rat small intestine in vivo. Arch Biochem Biophys 175: 670–676
43. Windmueller HG, Spaeth AE (1978) Identification of ketone bodies and glutamine as the major respiratory fuels in vivo for postabsorptive rat small intestine. J Biol Chem 253: 69–76
44. Windmueller HG, Spaeth AE (1980) Respiratory fuels and nitrogen metabolism in vivo in small intestine of fed rats: quantitative importance of glutamine, glutamate, and aspartate. J Biol Chem 255: 107–112
45. Windmueller HG, Spaeth AE (1981) Source and fate of circulating citrulline. Am J Physiol 241: E473–E480
46. Windmueller HG, Spaeth AE (1981) Vascular autoperfusion of rat small intestine in situ. In: Jakoby, WB (ed) Detoxication and drug metabolism: conjugation and related systems. Academic Press, London New York, pp 120–129
47. Windmueller HG, Spaeth AE (1984) Vascular perfusion of rat small intestine for permeation and metabolism studies. In: Csáky TZ (ed) Pharmacology of intestinal permeation. Handbook of experimental pharmacology). Springer, Berlin Heidelberg New York (in press)
48. Windmueller HG, Spaeth AE, Ganote CE (1970) Vascular perfusion of isolated rat gut: norepinephrine and glucocorticoid requirement. Am J Physiol 218: 197–204
49. Wolff JE, Bergman EN, Williams HH (1972) Net metabolism of plasma amino acids by liver and portal-drained viscera of fed sheep. Am J Physiol 223: 438–446

Hepatic Glutamine and Ammonia Metabolism

Nitrogen and Redox Balance and the Intercellular Glutamine Cycle

H. SIES[1] and D. HÄUSSINGER[2]

Introduction

Our interest in glutamine metabolism arose from an investigation of mitochondrial NADPH/NADP$^+$ as affected during the metabolism of ammonia in the process of ureogenesis [1, 2]. Glutamine served as a compound yielding ammonia plus glutamate within the mitochondria via the activity of mitochondrial glutaminase, whereas added ammonia yielded glutamate only at the expense of 2-oxoglutarate via reductive amination [3]. Clearly, such model experiments on redox compartmentation have a physiological bearing, because the effects occur at portal concentrations of ammonia and glutamine within the physiological range of about 0.3 mM and 0.6 mM, respectively [4]. Thus, this article will present the redox transitions related to this important sector of nitrogen metabolism.

Liver was identified early on in the work by Krebs in 1935 [5] as having a particular role in glutamine metabolism, contrasting other organs like brain or kidney in properties of glutaminase and glutamine synthetase. Much effort has since been devoted to this problem [e.g., 6–12], identifying the role of the liver in glutamine metabolism with that of net glutamine synthesis or net glutamine degradation, depending on the metabolic conditions. This field has developed recently from a more descriptive stage into a conceptual stage, (a) by focusing on the role of the liver in the regulation of pH by coordinated action in urea synthesis and glutamine disposition review by Atkinson and Camien [13]), and (b) by identifying the cycling of glutamine [4] as an intercellular coordination between a more periportally located capacity of ureogenesis and glutaminase [15] and a perivenously located capacity of glutamine synthetase. This was detected by the histochemical localization of glutamine synthetase in a small layer of cells at the perivenous end by Gebhardt and Mecke [14] and by steady state metabolic studies in antegrade and retrograde perfusions of the liver by Häussinger [15]. Thus, the second part of this article will present the current knowledge on properties of hepatic glutamine metabolism at the level of the intact organ.

1 Institut für Physiologische Chemie I, Universität Düsseldorf, Moorenstr. 5, 4000 Düsseldorf, FRG
2 Medizinische Universitätsklinik, Hugstetterstr. 55, 7800 Freiburg, FRG

Glutamine Metabolism in Mammalian Tissues
Edited by D. Häussinger and H. Sies

Redox Relations in Hepatic Ammonia and Glutamine Metabolism

Ammonia. Whereas the redox transitions during the metabolism of ammonia have been studied in detail, there is little information so far on glutamine in this respect. Briefly, the effect of ammonia is essentially an oxidation of mitochondrial NADPH to $NADP^+$, leaving the mitochondrial NADH system only slightly affected. The underlying reaction is the reductive amination of 2-oxoglutarate by glutamate dehydrogenase [1, 16, 17]. In view of the cellular concentrations of NADPH and NADH, the enzyme preferentially utilizes NADPH in this direction of the glutamate dehydrogenase reaction [16], as resulted from intensive studies with isolated mitochondria and the intact organ. It is of interest to note that energy-linked transhydrogenase is involved in urea formation from ammonia, because the return of the 4-carbon skeleton from the aspartate that originally was exported from the mitochondrial space towards the cytosol to serve as substrate in the argininosuccinate synthase reaction comes

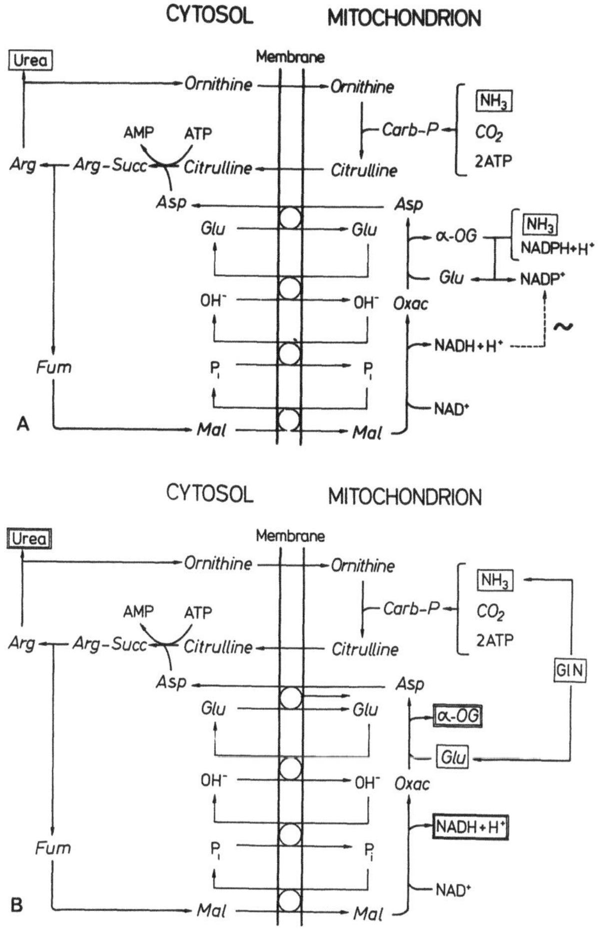

Scheme 1. A Metabolic scheme of urea synthesis from ammonium ions [17]; **B** metabolic scheme of urea synthesis from glutamine

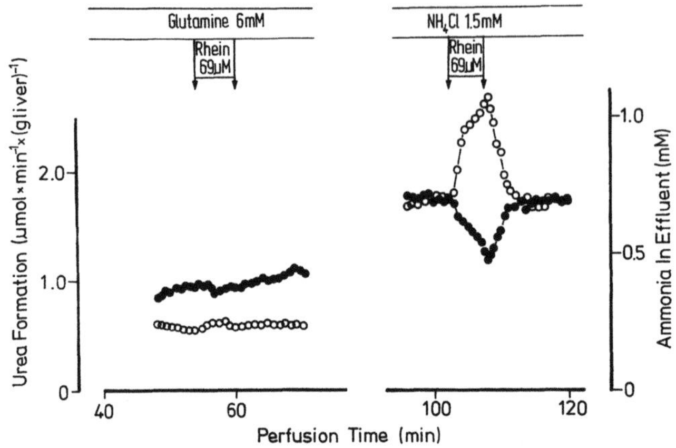

Fig. 1. Effect of the transhydrogenase inhibitor rhein on urea production from ammonium ions and glutamine [17]. Urea ●; ammonia ○

back in the form of malate into the mitochondrial space to regenerate oxaloacetate (Scheme 1A). This reaction yields NADH which, by way of the energy-linked transhydrogenase reaction, participates in providing NADPH for the maintenance of the glutamate dehydrogenase reaction. Evidence for this comes from the inhibitory effect of an inhibitor of energy-linked transhydrogenase, rhein, on ammonia-supported urea formation (Fig. 1).

The net effect of ammonia on the mitochondrial pyridine nucleotides is a slight oxidation of NADH and a pronounced oxidation of NADPH, as demonstrated by chemical analyses and organspectrophotometry as well as by the indicator metabolite couple for NADH/NAD^+, β-hydroxybutyrate/acetoacetate (Fig. 2A). Interestingly, the couple employed for indicating the cytosolic NADH/NAD^+ system, lactate/pyruvate, showed an increase which is not expected from reactions linked to the urea cycle and ancillary steps. Rather, this is due to an activation of pyruvate dehydrogenase from the inactive phospho form to the active dephospho form [1, 3] (see below).

Glutamine. With the glutaminase reaction providing ammonia and glutamate, there is no requirement for reductive amination, and the net result will be an increase in the NADH/NAD^+ ratio in the mitochondrial matrix, together with an increase in the production of 2-oxoglutarate, as schematically shown in Scheme 1B. The indicator metabolite couple, β-hydroxybutyrate/acetoacetate, reaches a higher value after the glutaminase reaction has been stimulated by a short-term pulse of ammonia. Thus, as is depicted by the reaction sequence in Scheme 1B, the formation of 2-oxoglutarate and NADH from glutamine is catalyzed by the steps related to the urea cycle and the related enzymes, such as fumarase, malate dehydrogenase and aspartate aminotransferase, and glutamate dehydrogenase is not required. The cytosolic NADH system, indicated in its redox state by the lactate/pyruvate ratio, is not affected by the increase in flux through glutaminase (Fig. 2B).

Fig. 2A, B. Profile of the indicator metabolite ratios lactate/pyruvate and 3-hydroxybutyrate/acetoacetate during the metabolism of ammonium ions (**A**) and glutamine (**B**) [1]

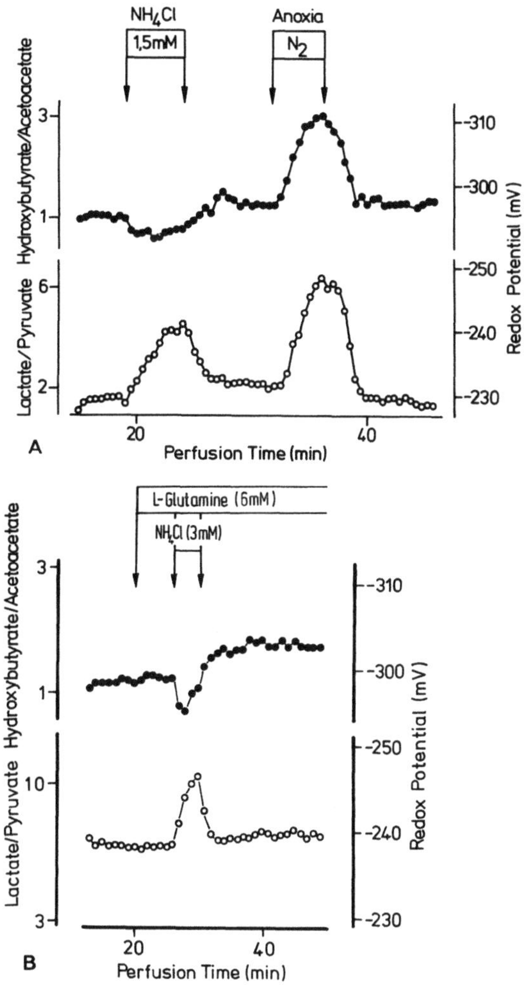

Pyruvate Dehydrogenase and 2-Oxoglutarate Dehydrogenase in Ammonia and Glutamine Metabolism

The increase in the lactate/pyruvate ratio observed upon the addition of ammonia (Fig. 2) is due to a loss of pyruvate (1,3), and was attributed to an activation of pyruvate dehydrogenase. This activation of the enzyme feeding carbon into the citric acid cycle occurs in a situation when the cycle is drained of 2-oxoglutarate because of the requirement of glutamate formation from 2-oxoglutarate (Scheme 1A). It is of interest to note that, conversely, the addition of glutamine leads to an inhibition of pyruvate dehydrogenase (Fig. 3) following the activation of glutaminase. This occurs in a situation when the citric acid cycle is supplied with the extra carbon skeleton coming from glutamine (Scheme 1B), and similar effects can also be observed when the carbon skeleton comes from proline instead of glutamine [18].

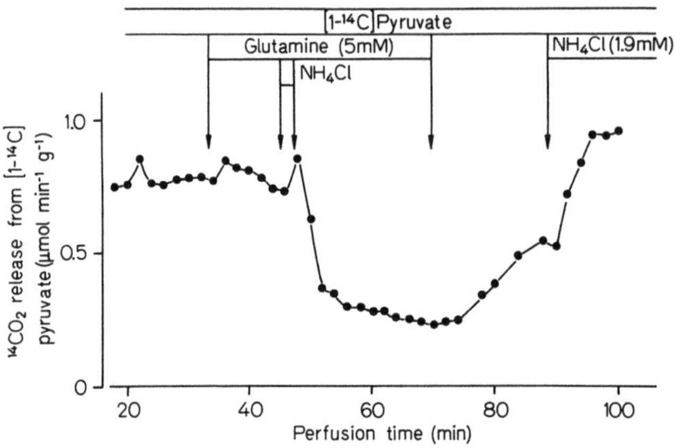

Fig. 3. Inhibition of $^{14}CO_2$ production from 1-^{14}C-pyruvate during the metabolism of glutamine [18]

Further there is a complementary pattern in the regulation of flux through pyruvate and 2-oxoglutarate dehydrogenases. More recently, the α-adrenergic stimulation by phenylephrine was also shown to diminish flux through pyruvate dehydrogenase [19] and increase flux through 2-oxoglutarate dehydrogenase [20].

Regulation of Glutaminase and Glutamine Synthetase in Liver

Until a few years ago it was accepted that glutamine in concentrations below 1 m*M* is not metabolized by perfused rat liver [21]. On the other hand, glutamine synthesis by perfused liver or isolated hepatocytes from different substrates was found to be very low and was in the range of about 0.1 μmol/g/min [10]. This is surprising in view of the high assayable activity contents of glutaminase and glutamine synthetase and obviously points to a suitable regulatory pattern of glutaminase and glutamine synthetase in the intact liver as is underlined by the occurrence of glutamine cycling, i.e., the simultaneous activity of glutaminase and glutamine synthetase [4, 22]. In view of these findings, studies on the regulation of glutaminase and glutamine synthetase in the intact liver or in isolated liver cells, especially in the presence of physiological glutamine concentrations, had to consider two problems: (a) the accumulation of ammonium ions during glutamine degradation, which are known to activate glutaminase, and (b) the separate determination of simultaneous fluxes through glutaminase and glutamine synthetase. The first problem was solved by use of the open, nonrecirculating system of liver perfusion, preventing an accumulation of ammonium ions and allowing maintainance of steady state metabolite concentrations and determination of metabolic flux rates during metabolic steady states [23]. The simultaneous fluxes through glutaminase and glutamine synthetase in the intact liver can separately be determined by use of 1-^{14}C-glutamine, if added at concentrations below 2 m*M* [22]. In experiments without and with methionine sulfoximine, an inhibitor of glutamine synthetase [24] but not of glutaminase, it was shown that $^{14}CO_2$ production from 1-^{14}C-glutamine represents glutaminase flux in the intact liver [22]. Consequently, glutamine synthetase flux can be measured as the rate of net glutamine production plus glutaminase flux [22].

The following review on the regulation of glutamine metabolism in liver will only briefly consider the transport of glutamine across biological membranes (see Kovačević and McGivan, this volume). Although there is no evidence at present for a rate limitation of glutamine degradation by the uptake step, a possible limitation of glutamine synthesis by export from the cell will be discussed in a following section.

Regulation by Ammonium Ions

Activation of glutaminase by ammonium ions was first shown in studies with isolated rat liver mitochondria [25], mitochondrial homogenate [26], and isolated perfused rat liver [3, 4, 26] and was later confirmed by studies on gluconeogenesis from glutamine in isolated hepatocytes [27]. Further, in vivo studies showed a considerable decrease of hepatic glutamine content upon an ammonia load [28]. Whereas ammonium ions activate glutaminase in the sense of a feed-forward [4], the other product of glutaminase reaction, glutamate, is without effect on hepatic glutaminase activity. This was shown in liver homogenates [5] and later in isolated perfused liver by the lack of effect of intracellular glutamate accumulation on nitrogen production from glutamine added in unphysiologically high concentrations [29]. Activation of glutaminase by endogenous ammonium ions is also taken to explain the so-called lag phase in glutamine metabolism [11, 12, 25, 26, 29] (about 40 min in presence of 5 m*M* glutamine). This was shown by the extreme shortening of this lag phase in perfused rat liver by a 3 min pulse of ammonium chloride (3 m*M*) resulting in a subsequent activation of glutaminase and maintainance of this activation by ammonium ions produced by glutaminase reaction [3, 26, 29], even after the washout of added ammonium ions. However, these studies were generally carried out at high concentrations of glutamine as well of ammonium ions. In the presence of a physiological portal glutamine concentration of 0.5–0.6 m*M* [7, 21, 31], but in the absence of added ammonium ions, there is no extra urea or ammonia production from glutamine in perfused rat liver [4]. This is in agreement with the reported strict requirement of glutaminase for ammonium ions [30]. However, nitrogen production from glutamine is observed when portal ammonium ions are present (Fig. 4). This activating effect of portal ammonium ions on glutamine degradation was half-maximal at concentrations of 0.2–0.3 m*M* and was maximal at 0.6 m*M*, pointing to a physiological significance of this effect in view of the physiological portal ammonia concentration of 0.2–0.3 m*M* [21, 32]. Although studies on nitrogen production from glutamine represent net turnover rates of hepatic glutamine metabolism, the separate determination of glutaminase flux in the liver also shows a similar activation kinetics of glutaminase by ammonium ions [22]. However, with a physiological portal

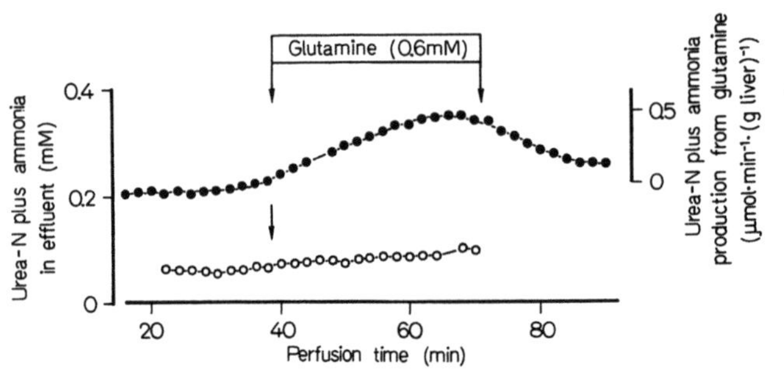

Fig. 4. Urea-nitrogen plus ammonia production from glutamine (0.6 m*M*) in absence (○) and presence (●) of a physiological portal ammonia concentration (0.24 m*M*) [4]

glutamine concentration, glutamine synthetase flux simultaneously exceeded glutaminase flux, resulting in a net glutamine production by the liver in the presence of ammonium ions (Table 1). Also, with subphysiological concentrations of glutamine, ammonium ions stimulate glutaminase flux, whereas simultaneous stimulation of glutamine synthetase flux by ammonium ions decreased with increasing portal glutamine concentrations as a consequence of glutamine synthetase inhibition by glutamine [33]. Therefore, with ammonium ions a net glutamine production by the liver is observed at portal glutamine of up to 0.7 m*M* turning to a net glutamine uptake at higher glutamine.

This shows that portal ammonium ions within the physiological concentration range effectively modify fluxes through glutaminase and glutamine synthetase with the consequence of either net glutamine uptake or release.

Regulation by Hormones

A stimulation of hepatic glutamine degradation was observed by glucagon [27, 34], cyclic AMP [35], phenylephrine [36], vasopressin [37], and adrenaline [37]. With physiological portal glutamine, glucagon stimulated glutaminase flux about three fold. Simultaneously, glutamine synthetase flux was stimulated by glucagon at portal glutamine below 0.5 m*M*, but was strongly inhibited at higher concentrations [22]. The activating effects of glucagon and ammonium ions on glutaminase flux are additive, suggesting different sites of activation (Table 1).

Phenylephrine increases glutaminase flux by about 30% and inhibits glutamine synthetase flux to a similar extent [20]. Especially in the presence of ammonium ions glutamine synthetase flux is largely inhibited by phenylephrine, whereas the activating effects on glutaminase flux by phenylephrine and ammonium ions are additive (Table 1). Inhibition of glutamine synthesis by phenylephrine might be due to decreased tissue levels of glutamate and oxoglutarate as a consequence of flux stimulation through glutamate dehydrogenase and oxoglutarate dehydrogenase by phenylephrine [20] in view of a comparatively high K_M(glutamate) of 5 m*M* for glutamine synthetase with 10

Table 1. Effect of ammonium ions, glucagon, phenylephrine on flux through glutaminase and glutamine synthetase in presence of a physiological portal glutamine concentration (0.6 m*M*)[a]

Condition	Glutaminase flux	Glutamine synthetase flux	Net glutamine release	Rate of glutamine cycling
	$nmol \cdot g^{-1} \cdot min^{-1}$			
Control (15)	67 ± 3	148 ± 10	81 ± 11	67 ± 3
NH_4Cl (0.6 m*M*) (7)	168 ± 9	248 ± 12	80 ± 8	168 ± 9
Glucagon (0.1 μ*M*) (7)	231 ± 32	84 ± 20	−147 ± 30	84 ± 20
plus NH_4Cl (5)	379 ± 36	122 ± 27	−254 ± 33	122 ± 27
Phenylephrine (5 μ*M*) (6)	88 ± 4	113 ± 9	25 ± 8	88 ± 4
plus glucagon (4)	152 ± 15	64 ± 13	− 88 ± 22	64 ± 13
plus NH_4Cl (4)	299 ± 7	83 ± 5	−216 ± 15	83 ± 5

[a] Data are given as means ± SEM from (*n*) perfusion experiments

m*M* Mg^{2+} [33]. This shows that stimulation of glutamine uptake by phenylephrine at physiological glutamine and ammonia concentrations is primarily due to an inhibition of glutamine synthetase flux.

The activation of glutaminase flux by glucagon is largely diminished in the presence of phenylephrine (Table 1). This may be due to depletion of hepatic Ca^{2+} by phenylephrine [38] and the requirement for Ca^{2+} of glucagon or cAMP induced stimulation of gluconeogenesis [36, 37, 39] and glutaminase activation.

Also the hormonal regulation of hepatic glutamine metabolism involves simultaneously flux changes of glutaminase and glutamine synthetase at similar extents, resulting in either glutamine synthesis or uptake. This explains the contradictory data on the role of the liver in glutamine metabolism in the literature [21, 31, 40–42], thus, identifying its role with either uptake [31, 40, 41] or release [21] of glutamine depending on experimental or nutritional conditions.

Regulation by pH

In isolated perfused rat liver, the rate of glutamine utilization was shown to be restricted when perfusate pH was lowered by infusion of sulfuric [43] or hydrochloric acid [29] or by decreasing the HCO_3^-/CO_2 ratio [11, 29]. This may serve as a model for metabolic acidosis, where hepatic glutamine utilization was relatively more lowered than glutamine synthesis and when renal glutamine utilization is enhanced (see Gerok and Häussinger, this volume).

Separate determination of the fluxes through glutaminase and glutamine synthetase in the presence of physiological portal glutamine concentrations in perfused rat liver showed not only an inhibition of glutaminase flux at lowering the perfusate pH, but also a stimulation of hepatic glutamine synthetase flux to a similar extent [22]. Conversely, alkalinization of the perfusate increased glutaminase flux, but decreased glutamine synthetase flux, resulting in an increased net glutamine uptake by the liver. However, the response of both enzymes to pH changes is modified by the presence of glucagon or ammonium ions [22].

Inhibition of hepatic glutaminase activity upon lowering the pH is due to an increased concentration of ammonium ions required for activation of glutaminase [30]. An activation of glutaminase by bicarbonate in isolated hepatocytes [35] or isolated liver mitochondria [44] did not occur in perfused liver [29]. The reason for this discrepancy is thought to reside in the more difficult control of pH in the cell incubations, in contrast to the maintenance of the pH steady state in the open perfused system. In support of this, recent studies on glutaminase activity is isolated mitochondria [45] could not demonstrate an activating effect of bicarbonate. However, bicarbonate was also reported to decrease the ammonium ion concentration required for half-maximal activation of glutaminase [30].

Hepatocyte Heterogeneity and Intercellular Metabolic Interactions

The functional units of the liver are the so-called acini [46], extending from the terminal portal venule along the sinusoids to the terminal hepatic venule. Hepatocyte heterogeneity in oxygen, substrate and hormone supply, and the different enzyme equipment of hepatocytes from the periportal and perivenous area led to the concept of metabolic zonation, i.e., the different distribution of metabolic pathways in the liver acinus (for review see [47]).

Periportal Location of Urea Synthesis and Perivenous Location of Glutamine Synthetase

Hepatocyte heterogeneity has mainly been studied with histochemical, immunohistochemical, microdissection, and microbiochemical techniques (for review see [47]). Because these techniques either imply the structural or metabolic desintegration of the liver, noninvasive methods, like microlightguides for registration of surface fluorescence, have been developed [48]. More recently, hepatocyte heterogeneity was studied in the metabolically and structurally intact perfused rat liver by comparing metabolic flux rates depending on the direction of perfusion [15]. If two metabolic pathways are differently distributed in the liver lobule and are competing for the same substrate, under conditions of a rate-limiting substrate supply, one of these pathways will be favored against the other, dependent on the direction of perfusion. Such competing pathways are urea and glutamine synthesis from ammonium ions, and it was shown that added ammonium ions are mainly converted to urea during physiological antegrade (portal to hepatic vein) perfusion, but are converted to glutamine during retrograde (hepatic vein to portal vein) [15] perfusion (Fig. 5). When competition of urea and glutamine synthesis for ammonium ions was abolished by inhibition of one of these pathways or by excess addition of ammonium ions,

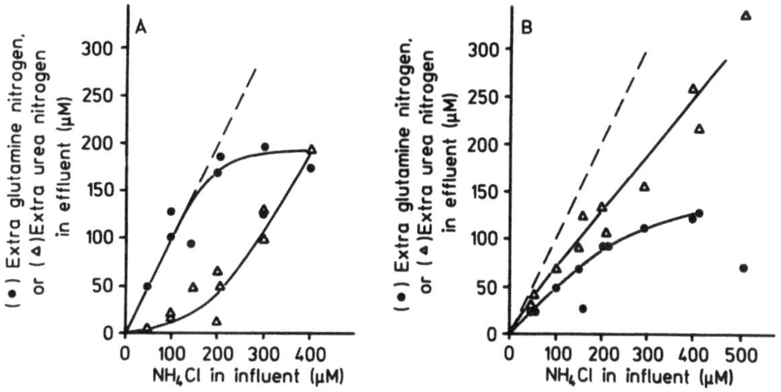

Fig. 5A, B. Effect of perfusion direction on and glutamine urea formation from added ammonium ions. **A** retrograde perfusion; **B** antegrade perfusion [15]

the perfusion direction was without effect on the rates of glutamine or urea synthesis [15]. Ammonium ions derived from the endogenous protein and amino acid breakdown in the liver were utilized for glutamine synthesis in antegrade perfusion, but were washed out during retrograde perfusion or in the presence of methionine sulfoximine, regardless of the direction of perfusion (Table 2). These studies demonstrate the periportal localization of urea synthesis and the perivenous localization of glutamine synthetase in the liver lobule by direct measurements of metabolic flux rates [15] and are in agreement with the immunohistochemical localization of carbamoyl phosphate synthetase [49] and glutamine synthetase [14], shown to be periportal and perivenous, respectively. The data further identified a predominantly periportal localization of endogenous ammonia production, in agreement with the histochemical localization of alanine aminotransferase [50] and tyrosine aminotransferase [51], predominantly in the periportal area. Ammonia production from endogenous sources during catabolic conditions implies an increased oxidative deamination of glutamate and an increased NADPH formation via glutamate dehydrogenase in the periportal area of the liver lobule. On the other hand, perivenous glutamine synthesis from the ammonia which is released from periportal hepatocytes requires increased glutamate formation at glutamate dehydrogenase and, consequently, NADPH utilization. These postulated different directions of flux point to converse roles of glutamate dehydrogenase at the sublobular periportal and perivenous compartments during urea and glutamine synthesis from endogenous substrates.

As shown in Table 2, urea synthesis, but not glutamine synthesis, in perfused liver is restricted to ammonia concentrations above 50 μM, as evidenced by effluent ammonia concentrations during inhibition of glutamine synthesis by methionine sulfoximine or during retrograde perfusions [15]. On the other hand, ammonia release by the liver is about 18 nmol/g/min during antegrade perfusion and with glutamine synthetase active, corresponding to an ammonia concentration in the effluent of about 5 μM. The tenfold difference between ammonia removal by glutamine synthesis and urea synthesis was explained by an approx.

Table 2. Effect of direction of perfusion flow on release of glutamine, glutamate, urea nitrogen and ammonia by the liver in the absence and presence of inhibitors of glutamine or urea synthesis[a]

Condition	Glutamine release	Glutamate release	Urea nitrogen release	Ammonia release	Total nitrogen release
	$nmol \cdot min^{-1} \cdot g^{-1}$				
Control					
Antegrade	152 ± 15 (8)	56 ± 8 (9)	188 ± 19 (8)	18 ± 2 (9)	569 ± 51 (8)
Retrograde	54 ± 8 (12)	62 ± 7 (12)	183 ± 22 (7)	124 ± 15 (7)	507 ± 40 (7)
Methionine sulfoximine					
Antegrade	15 ± 4 (5)	79 ± 16 (5)	343 ± 27 (5)	164 ± 16 (5)	615 ± 55 (5)
Retrograde	12 ± 3 (6)	90 ± 15 (6)	337 ± 21 (6)	190 ± 17 (6)	640 ± 44 (6)
Bicarbonate-free					
Antegrade	155 ± 26 (4)	32 ± 8 (4)	63 ± 14 (4)	48 ± 2 (4)	429 ± 74 (4)
Retrograde	61 ± 15 (5)	40 ± 8 (5)	64 ± 18 (5)	200 ± 26 (5)	427 ± 75 (5)

[a] Inhibition of glutamine synthetase and carbamoyl-phosphate synthetase was performed by addition of methionine sulfoximine and replacement of the bicarbonate buffer by Hepes, respectively. Data from different perfusion experiments are given as means ± SEM, numbers of experiments in parenthesis

ten fold higher $K_M(NH_4^+)$ for carbamoylphosphate synthetase (2 m*M*) [52] compared to glutamine synthetase (0.3 m*M*) [33]. Further, the more alkaline intramitochondrial pH [53] will favor accumulation of ammonium ions in the cytosol and, thus, cytosolic glutamine synthesis instead of mitochondrial carbamoylphosphate synthesis.

The different sublobular distribution of glutamine synthetase and urea synthesis was further illustrated functionally in studies on ammonia metabolism in perfused livers from CCl_4-pretreated rats. CCl_4 pretreatment of rats leads to a fatty degeneration and necrosis characteristic of the perivenous area of the liver lobule [54] due to increased activities of drug metabolism and the more hypoxic conditions favoring lipid peroxidation (for review [47]). When perivenous liver cell necrosis was induced by CCl_4, a decreased ammonia uptake by these livers was due to an almost complete inhibition of glutamine synthetase, whereas urea synthesis was not affected [55].

The perivenous and periportal localization of glutamine synthetase and urea synthesis, respectively, as well as the exclusive role of glutamine synthesis for removal of low concentrations of ammonium ions, implies that perivenous glutamine synthesis must be regarded as an important mechanism for elimination of sinusoidal ammonia, which escaped urea synthesis. Thus, the enzyme has a scavenger role for ammonia before reaching systemic circulation.

Periportal Location of Glutaminase

The periportal location of hepatic glutaminase was identified by the finding that endogenously synthesized glutamine is available for glutaminase in retrograde perfusions, but not in antegrade perfusions, whereas glutaminase flux was unaffected by the direction of perfusion when glutamine synthetase was inhibited by methionine sulfoximine [15]. A different localization of glutamine synthetase (perivenous) and glutaminase (periportal) has also been suggested from the finding that during glutamine cycling labeled glutamate derived from added 1-^{14}C-glutamine was not utilized for glutamine synthesis [22]. A joint periportal localization of glutaminase and urea synthesis also explains that added glutamine is a good substrate of urea synthesis. Further, the three enzymes glutaminase [56], carbamoylphosphate synthetase [52], and ornithine transcarbamylase have a joint intramitochondrial localization. From this and in view of glutaminase activation by ammonium ions at physiological concentrations, portal glutamine will increasingly be utilized for urea synthesis in the presence of increasing portal ammonia concentrations.

Glutamine Transport: Rate Limitation for Perivenous Glutamine Synthesis?

Subfractionation studies on the liver in vivo showed mitochondrial and cytosolic glutamine concentrations of about 15 and 7 m*M*, respectively [57]. Similar concentrations were found in the isolated perfused rat liver in the presence of a physiological portal glutamine concentration of 0.6 m*M*, whereas in the absence of added glutamine the subcellular concentrations were decreased by about 90%. Glutamine is transported across the plasma membrane of the liver by a specific transport system, the so-called system N [58]. The restriction of glutamine synthetase to a small population of perivenous hepatocytes and the periportal localization of glutaminase implies that this glutamine transport system will bring about glutamine input in the periportal area, but glutamine output in the perivenous area of the liver lobule. Studies on the glutamine content of the perfused rat liver in the absence and presence of methionine sulfoximine, the inhibitor of glutamine synthetase, have shown that the glutamine content of the perivenous glutamine synthesizing hepatocytes is about 17 μmol/g dry cells and indicated a similar high intra/extracellular glutamine concentration gradient for perivenous as well periportal hepatocytes. Therefore, the difference between periportal glutamine utilizing hepatocytes and perivenous glutamine synthesizing hepatocytes regarding glutamine movements across the plasma membrane can only be explained by the existence of two different glutamine transport systems in the plasma membrane. Whereas there is no evidence for a rate limitation of periportal glutamine degradation at the level of glutamine import (see Kovačević and McGivan, this volume), a limitation of glutamine export out of the glutamine synthetase containing hepatocytes may be expected in view of the high intra/extracellular glutamine concentration gradient also observed in these hepatocytes. Consequently, this will result in an inhibition of glutamine synthetase by its intracellularly accumulating product glutamine (Ki = 5 m*M*) [33]. Such a mechanism of regulating the rate of glutamine synthesis by glutamine export out of the hepatocytes will also explain the low rate of glutamine synthesis of 0.4 μmol/g/min in intact hepatocytes or liver compared to the rate of glutamine synthesis in liver homogenates [10].

Hepatocyte Heterogeneity in Glutamate Metabolism

The restriction of glutamine synthetase to a small perivenous population of hepatocytes involves a considerable loss of oxoglutarate and glutamate from these cells during glutamine synthesis from ammonia. This is partly overcome by utilization of portal glutamate for glutamine synthesis [59]. Glutamate is transported across the plasma membrane of the liver by a sodium-dependent, specific high affinity, but low capacity transport system [60]. Similar to the metabolism of alanine [61], metabolism of portal glutamate is under the control of its transport across the plasma membrane [59]. Studies with isolated perfused rat liver showed that more than 60% of total hepatic glutamate uptake occurs in the glutamine synthesizing perivenous hepatocytes [59], and from the size of this hepatocyte pool [14] an approx. 20-fold higher glutamate transport capacity was calculated for these hepatocytes compared to others. In agreement with the predominantly perivenous localization of the glutamate transport system in the plasma membrane, $^{14}CO_2$ production from portal 1-^{14}C-glutamate was decreased by about 70% in the presence of a perivenous CCl_4 induced liver cell necrosis, whereas $^{14}CO_2$ production from U-^{14}C-proline was not inhibited under these conditions; this suggests a predominantly periportal localization of proline metabolism [59].

Glutamate is simultaneously taken up and released by the liver [59]. Therefore, it is of interest, whether this represents a simple exchange of glutamate by the transport system or whether glutamate uptake and release are differently localized within the liver lobule.

The common perivenous localization of glutamine synthetase and the glutamate transport system in the plasma membrane and the predominantly perivenous localization of glutamate dehydrogenase [62, 63] is in line with the concept of a similar localization of functionally linked processes [47] (Table 3). This concept is also preserved in the periportal localization of urea synthesis, glutaminase, endogenous ammonia production, alanine aminotransferase, tyrosine aminotransferase, and proline degradation in the liver, indicating that glutamate, ammonia, and urea formation from amino acids occurs predominantly in the periportal area of the liver lobule. The data are further consistent with the existence of different sublobular glutamate pools in liver, as also suggested earlier [64].

Table 3. Metabolic zonation in ammonia and glutamine metabolism

Periportal zone	Perivenous zone
Carbamoylphosphate synthetase [49]	Glutamine synthetase [14, 15]
Urea synthesis from amino acid and ammonia nitrogen [15]	Glutamate transport system in plasma membrane [59]
Endogenous ammonia production [15]	
Glutaminase [15]	
Proline degradation [59]	

The Intercellular Glutamine Cycle

The simultaneous activity of periportal glutaminase and perivenous glutamine synthetase in liver results in an energy consuming intercellular cycling of glutamine with zero net glutamine turnover [4, 15, 22]. Energy requirements of this glutamine cycle have been shown in perfused rat liver by an increased extra oxygen uptake, which could not be accounted for energy requirements of urea synthesis [4]. With physiological portal glutamine concentrations, the ATP utilization of this cycle was calculated to be of the order of about 0.1 μmol/g/min [4], in close agreement with the cycling rate of about 0.1 μmol/g/min under these conditions. Such energy consuming cycles without net substrate turnover are so-called futile cycles and their role in the regulation of metabolism has been pointed out [65, 66]. However, the glutamine cycle can not be designated as futile, because glutaminase and glutamine synthetase are differently localized in the liver lobule, resulting in an intercellular, as opposed to intracellular, glutamine cycle [15]. The rate of this cycle is determined by the flux through the enzyme with the lower activity and is equal to glutaminase flux if glutamine synthetase flux exceeds glutaminase flux, und vive versa. Because the effectors on glutamine metabolism, ammonium ions, portal glutamine, hormones, and pH involve flux changes of both enzymes, the rate of the intercellular glutamine cycle is also under control of these effectors (Fig. 6). In the presence of a physiological portal glutamine concentration, highest rates of the cycle are observed at physiological pH and during stimulation of urea synthesis by ammonium ions or glucagon [22]. Regarding the role of the intercellular glutamine cycle, evidence has been presented for a role of transport [22], a role during ureogenesis from portal ammonia [15], and a role in hepatic pH regulation [67].

Role as a Transport System

Regarding the net reaction and metabolite transfers during intercellular glutamine cycling, there is an energy driven translocation of glutamate and

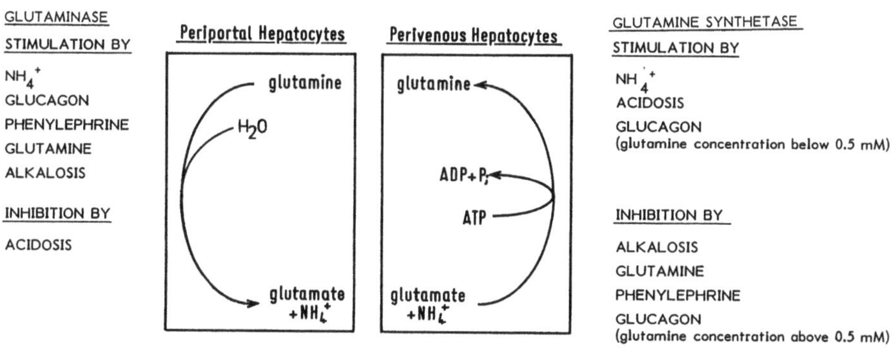

Fig. 6. Regulation of the intercellular glutamine cycle

ammonia from the cytosol of perivenous hepatocytes into the mitochondrial compartment of periportal hepatocytes. In view of oxoglutarate depletion during ureogenesis from ammonium ions [1] such a mechanism should favor citric acid cycle homeostasis by providing additional glutamate, especially in periportal mitochondria. Although two potent glutamate transport systems across the mitochondrial membrane are known (for review [68]), glutamate transport by intercellular glutamine cycling should operate regardless of the mitochondrial/cytosolic concentration gradient.

Role During Ureogenesis

During ureogenesis from portal ammonium ions, an increased glutamate and ammonia formation at mitochondrial glutaminase from portal glutamine favors urea synthesis by increased substrate supply of carbamoylphosphate synthetase and argininosuccinate synthase, whereas glutamine is resynthesized in the perivenous area of the liver lobule from ammonia not available for periportal urea synthesis (Scheme 2). This was shown in isolated perfused rat liver, where urea production from portal ammonium ions at physiological concentrations was increased by about 50% by further addition of glutamine (0.6 mM) at the expense of net glutamine production [15]. Therefore, the intercellular glutamine cycle provides an effective means for complete conversion of portal ammonia to urea by periportal urea synthesis from portal ammonia and glutamine and the perivenous resynthesis of glutamine from ammonia not available for periportal urea synthesis. In line with such a mechanism is also the finding that glutamine cycling increases with increasing portal ammonia concentrations [22].

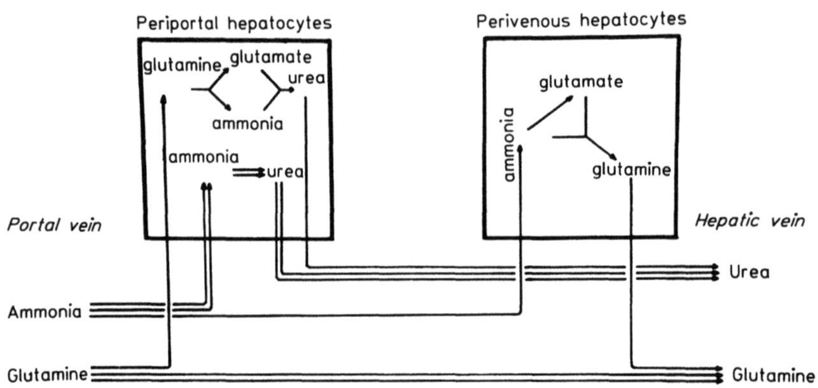

Scheme 2. Role of the intercellular glutamine cycle during ureogenesis from portal ammonia [15]. The *"intercellular cycle"* is defined as a sequence of metabolic reactions in an organ, leading to the regeneration of an initial substrate, whereby the metabolic steps are heterogeneously distributed among two or more parenchymal cell populations in this organ. Compared to intracellular cycles, the pool of products formed in a given reaction of an intercellular cycle must not necessarily be identical with the substrate pool for the following reaction

Role in Hepatic Regulation of pH Homeostasis

The significance of hepatic urea synthesis lies not only in the removal of potentially toxic ammonium ions, but also possibly even more important, in the removal of bicarbonate, which is generated during the oxidation of the carboxyl group of amino acids [13]. In contrast to urea synthesis, ammonia removal by hepatic glutamine synthesis and ammonium ion excretion by renal glutaminase will not remove bicarbonate. Therefore, a regulatory means is required for shifting hepatic ammonia detoxification from urea synthesis to glutamine synthesis during metabolic acidosis.

Urea Synthesis:

$$HCO_3^- + 2\,NH_4^+ \longrightarrow H_2NCONH_2 + 2\,H_2O + H^+$$

$$H^+ + HCO_3^- \longleftrightarrow CO_2 + H_2O$$

$$2\,HCO_3^- \quad 2\,NH_4^+ \longrightarrow H_2NCONH_2 + CO_2 + 3\,H_2O$$

Glutamine synthesis:

$$^-OOC(CH_2)_2CH(NH_3^+)COO^- + NH_4^+ \xrightarrow[ATP \; ADP + P_i]{} H_2NOC(CH_2)_2CH(NH_3^+)COO^- + H_2O$$

Lowering the perfusate pH is followed by a decreased urea production from portal ammonium ions and glutamine being accompanied by an increased net glutamine production by the liver (Fig. 7A) as also observed by Oliver et al. [69].

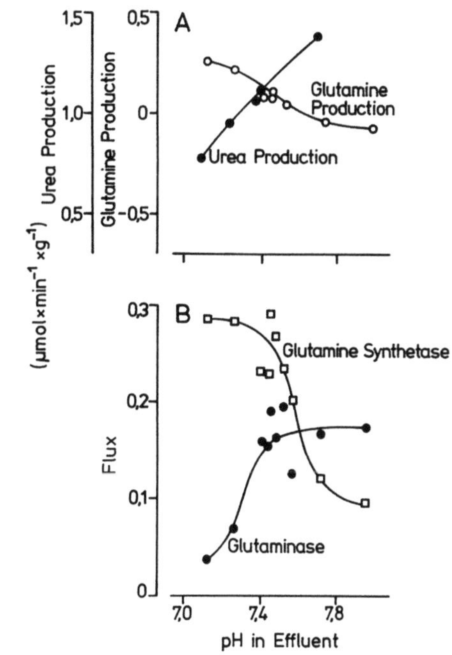

Fig. 7A, B. Effect of perfusate pH on the contribution of urea and glutamine synthesis during ammonia detoxification **(A)** and glutaminase and glutamine synthetase flux **(B)**. Influent perfusate contained NH_4Cl (0.6 m*M*) and glutamine (0.6 m*M*)

This increased net glutamine production by the liver at the expense of urea production is brought about by an inhibition of glutaminase flux and an activation of glutamine synthetase flux (Fig. 7B). Conversely, during metabolic alkalosis increased urea synthesis from portal ammonia and glutamine is due to an inhibition of glutamine synthetase flux and an activation of glutaminase flux with the consequence of an increased bicarbonate removal by urea synthesis. Thus, the two regulatory sites of the intercellular glutamine cycle, glutaminase and glutamine synthetase, provide an effective means for adaptive changes in the relative rates of urea and glutamine production during detoxification of portal ammonia and point to a role of the intercellular glutamine cycle for maintainance of pH homeostasis in the organism [67].

Acknowledgements. Fruitfull discussions with Theodor Bücher, Wolfgang Gerok, Joseph Tager, Theo Akerboom and Sybille Soboll during the course of our studies are gratefully acknowledged. Supported by Deutsche Forschungsgemeinschaft, Schwerpunktsprogramm "Regulationsmechanismen des Kohlehydrat- und Lipidstoffwechsels" and Sonderforschungsbereich 154 "Klinische und experimentelle Hepatologie".

References

1. Sies H, Häussinger D, Grosskopf M (1974) Mitochondrial nicotinamide nucleotide systems: ammonium chloride responses and associated metabolic transitions in hemoglobin-free perfused rat liver. Hoppe-Seyler's Z Physiol Chem 355: 305–318
2. Sies H, Summer KH, Bücher Th (1975) A process requiring mitochondrial NADPH: urea formation from ammonia. Febs Lett 54: 274–278
3. Häussinger D, Weiss L, Sies H (1975) Activation of pyruvate dehydrogenase during metabolism of ammonium ions in hemoglobin perfused rat liver. Eur J Biochem 52: 421–431
4. Häussinger D, Sies H (1979) Hepatic glutamine metabolism under the influence of the portal ammonia concentration in the perfused rat liver. Eur J Biochem 101: 179–184
5. Krebs HA (1935) Metabolism of amino acids. The synthesis of glutamine from glutamic acid and ammonia and the enzymic hydrolysis of glutamine in animal tissues. Biochem J 29: 1951–1959
6. Horowitz ML, Knox WE (1968) A phosphate activated glutaminase in rat liver different from that in kidney and other tissues. Enzymol Biol Clin 9: 241–255
7. DuRuisseau JP, Greenstein JP, Winitz M, Birnbaum SM (1957) Studies on the metabolism of amino acids and related compounds in vivo. VI. Free amino acid levels in the tissues of rats protected against ammonia toxicity. Arch Biochem Biophys 68: 161–171
8. Schimassek H, Gerok W (1965) Control of the levels of free amino acids in plasma by the liver. Biochem Z 343: 407–415
9. Addae SK, Lotspeich WD (1968) Glutamine balance in metabolic acidosis as studied with the artificial kidney. Am J Physiol 215: 278–281
10. Lund P (1971) Control of glutamine synthesis in rat liver. Biochem J 124: 653–660
11. Saheki T, Katunuma N (1975) Analysis of regulatory factors for urea synthesis by isolated perfused rat liver. J Biochem 77: 659–669
12. Chamalaun RAFM, Tager JM (1970) Nitrogen metabolism in the perfused rat liver. Biochim Biophys Acta 222: 119–134
13. Atkinson DE, Camien MN (1982) The role of urea synthesis in the removal of metabolic bicarbonate and the regulation of blood pH. Curr Top Cell Regul 21: 261–302
14. Gebhardt R, Mecke D (1983) Heterogeneous distribution of glutamine synthetase among rat liver parenchymal cells in situ and in primary culture. Embo J 2: 567–570
15. Häussinger D (1983) Hepatocyte heterogeneity in glutamine and ammonia metabolism and the role of an intercellular glutamine cycle during ureogenesis in perfused rat liver. Eur J Biochem 133: 269–275

16. Bücher Th, Sies, H (1976) Mitochondrial and cytosolic redox states in perfused rat liver: methods and problems in metabolic compartmentation. In: Tager JM, Söling HD, Williamson JR (eds) Use of isolated liver cells and kidney tubules in metabolic studies. Elsevier/North Holland, Amsterdam, pp 41–64
17. Sies H, Summer KH, Häussinger D, Bücher Th (1976) NADPH utilisation in mitochondria: urea synthesis from ammonia in rat liver cells. In: Tager JM, Söling HD, Williamson JR (eds) Use of isolated liver cells and kidney tubules in metabolic studies. Elsevier/North Holland, Amsterdam, pp 311–316
18. Häussinger D, Gerok W, Sies H (1982) Inhibition of pyruvate dehydrogenase during the metabolism of glutamine and proline in hemoglobin-free perfused rat liver. Eur J Biochem 126: 69–76
19. Sies H, Graf P, Crane D (1983) Decreased flux through pyruvate dehydrogenase during calcium ion movements induced by vasopressin, α-adrenergic agonists and the ionophore A 23187 in perfused rat liver. Biochem J 212: 271–278
20. Häussinger D, Sies H (1984) Effect of phenylephrine on hepatic glutamine and glutamate metabolism. Biochem J (in press)
21. Lund P, Watford M (1976) Glutamine as a precursor of urea. In: Grisolia S, Baguena R, Mayor F (eds) The urea cycle. Wiley, New York London, pp 479–488
22. Häussinger D, Gerok W, Sies H (1983) Regulation of flux through glutaminase and glutamine synthetase in isolated perfused rat liver. Biochim Biophys Acta 755: 272–278
23. Sies H (1978) The use of perfusion of liver and other organs for the study of microsomal electron transport and cytochrom P 450 systems. Methods Enzymol 52: 48–59
24. Meister A (1980) Catalytic mechanism of glutamine synthetase; overview of glutamine metabolism. In: Mora J, Palacios R (eds) Glutamine: metabolism, enzymology and regulation. Academic Press, London New York, pp 1–40
25. Charles R (1968) Mitochondrial citrulline synthese: een ammoniak fixerend en ATP verbruikend proces. PhD thesis, Univ Amsterdam, Rotatype, Amsterdam
26. Häussinger D, Sies H (1975) Activation of glutaminase by ammonium ions in perfused rat liver. Abstr Commun 10th FEBS Meet 1497
27. Joseph S, McGivan JD (1978) The effect of ammonium choride and glucagon on the metabolism of glutamine in isolated liver cells from starved rats. Biochim Biophys Acta 543: 16–28
28. Nordmann R, Petit MA, Nordmann J (1972) Recherches sur le mècanisme de l'accumulation intra-hépatique d'acides aminès dicarboxyliques au cours de l'intoxication ammoniacale. Biochimie 54: 1473–1478
29. Häussinger D, Akerboom TPM, Sies H (1980) The role of pH and the lack of a requirement for hydrogencarbonate in the regulation of hepatic glutamine metabolism. Hoppe-Seyler's Z Physiol Chem 361: 995–1001
30. Verhoeven AJ, Van Iwaarden JF, Joseph SK, Meijer AJ (1983) Control of rat liver glutaminase by ammonia and pH. Eur J Biochem 133: 241–244
31. Yamamoto H, Aikawa T, Matsutaka H, Okuda T, Ishikawa E (1974) Interorganal relationships of amino acid metabolism in fed rats. Am J Physiol 226: 1428–1433
32. Lund P, Brosnan JT, Eggleston LV (1970) The regulation of ammonia metabolism in mammalian tissues. In: Bartley W, Kornberg HA, Quayle JR (eds) Essays in cell metabolism. Wiley, New York London, pp 167–188
33. Deuel TF, Louie M, Lerner A (1978) Glutamine synthetase from rat liver. J Biol Chem 253: 6111–6118
34. Lacey JH, Bradford NM, Joseph SK, McGivan JD (1981) Increased activity of phosphate-dependent glutaminase in liver mitochondria as a result of glucagon treatment of rats. Biochem J 194: 29–33
35. Baverel G, Lund P (1979) A role for bicarbonate in the regulation of mammalian glutamine metabolism. Biochem J 184: 599–606
36. Joseph SK, Verhoeven AJ, Meijer AJ (1981) Effect of trifluoperazine on the stimulation by Ca^{2+}-dependent hormones of gluconeogenesis from glutamine in isolated hepatocytes. Biochim Biophys Acta 677: 506–511
37. Corvera S, Garcia-Sainz JA (1983) Hormonal stimulation of mitochondrial glutaminase. Biochem J 210: 957–960

38. Blackmore PF, Dehaye JP, Exton JH (1979) Studies on α-adrenergic activation of hepatic glucose output. J Biol Chem 254: 6945–6950
39. Siess EA, Wieland OH (1980) Early kinetics of glucagon action in isolated hepatocytes at the mitochondrial level. Eur J Biochem 110: 203–210
40. Schröck H, Goldstein L (1981) Interorgan relationships for glutamine metabolism in normal and acidotic rats. Am J Physiol 240: E519–E525
41. Aikawa T, Matsutaka H, Yamamoto H, Okuda T, Ishikawa E, Kawano T, Matsumara E (1973) Gluconeogenesis and amino acid metabolism. Inter-organal relations and roles of glutamine and alanine in the amino acid metabolism of fasted rats. J Biochem 74: 1003–1017
42. Rémésy C, Demigné C, Aufrére J (1978) Interorgan relationships between glucose, lactate and amino acids in rats fed on high-carbohydrate or high-protein diets. Biochem J 170: 321–329
43. Lueck JD, Miller LL (1970) The effect of perfusate pH on glutamine metabolism in the isolated perfused rat liver. J Biol Chem 245: 5491–5497
44. Joseph SK, McGivan JD (1978) The effects of ammonium chloride and bicarbonate on the activity of glutaminase in isolated liver mitochondria. Biochem J 176: 837–844
45. McGivan JD, Bradford NM (1983) Properties of rat liver glutaminase at the submitochondrial level. Hoppe-Seyler's Z Physiol Chem 364: 1240
46. Rappaport AM (1976) The microcirculatory acinar concept of normal and pathological hepatic structure. Beitr Pathol 157: 215–243
47. Jungermann K, Katz N (1982) Metabolic heterogeneity of liver parenchyma. In: Sies H (ed) Metabolic compartmentation. Academic Press, London New York, pp 411–435
48. Ji S, Lemasters JJ, Thurman RG (1980) A non-invasive method to study metabolic events within sublobular regions of hemoglobin-free perfused liver. Febs Lett 113: 37–41
49. Gaasbeek Janzen JW, Moorman AFM, Lamers WH, Los JA, Charles R (1981) The localization of carbamoyl-phosphate synthase in adult rat liver. Biochem Soc Trans 9: 279P
50. Welsh FA (1972) Changes in distribution of enzymes within the liver lobule during adaptive increases. J Histochem Cytochem 20: 107–111
51. Morrison GR, Brock FE, Karl I, Shank RE (1965) Quantitative analysis of regenerating and degenerating areas within the lobule of the carbon tetrachloride-injured liver. Arch Biochem Biophys 111: 448–464
52. Lusty CJ (1978) Carbamylphosphate synthetase I of rat liver mitochondria. Eur J Biochem 85: 373–383
53. Soboll S, Elbers R, Scholz R, Heldt HW (1980) Subcellular distribution of di- and tricarboxylates and pH gradients in perfused rat liver. Hoppe-Seyler's Z Physiol Chem 361: 69–76
54. Slater TF (1966) Necrogenic action of carbon tetrachloride in the rat: a speculative mechanism based on activation. Nature (London) 209: 36–40
55. Häussinger D, Gerok W (1984) Hepatocyte heterogeneity in ammonia metabolism: impairment of glutamine synthesis in carbon tetrachloride-induced liver cell necrosis with no effect on urea synthesis. Chem Biol Interact 48: 191–194
56. Kalra J, Brosnan JT (1973) Localization of glutaminase in rat liver. Febs Lett 37: 325–328
57. Soboll S, Keim M, Häussinger D (1983) Subcellular distribution of glutamine in liver. Hoppe-Seyler's Z Physiol Chem 364: 1242
58. Kilberg MS, Handlogten ME, Christensen HN (1980) Characteristics of an amino acid transport system in rat liver for glutamine, asparagine, histidine and closely related analogs. J Biol Chem 225: 4011–4019
59. Häussinger D, Gerok W (1983) Hepatocyte heterogeneity in glutamate uptake by isolated perfused rat liver. Eur J Biochem 136: 421–425
60. Sips HJ, De Graaf PA, Van Dam K (1982) Transport of L-aspartate and L-glutamate in plasma-membrane vesicles from rat liver. Eur J Biochem 122: 259–264
61. Sips HJ, Groen AK, Tager JM (1980) Plasma-membrane transport of alanine is rate-limiting for its metabolism in rat-liver parenchymal cells. Febs Lett 119: 271–274
62. Shank RE, Morrison G, Cheng CH, Karl I, Schwartz R (1959) Cell heterogeneity within the liver lobule (quantitative histochemistry). J Histochem Cytochem 7: 237–239

63. Wimmer M, Pette D (1979) Microphotometric studies on intraacinar enzyme distribution in rat liver. Histochemistry 64: 23–33
64. Ui M, Exton JH, Park CR (1973) Effects of glucagon on glutamate metabolism in the perfused rat liver. J Biol Chem 248: 5350–5359
65. Newsholme EA, Start C (1973) Regulation in metabolism. Wiley, New York London
66. Hue L (1982) Futile cycles and regulation of metabolism. In: Sies H (ed) Metabolic compartmentation. Academic Press, London New York
67. Häussinger D, Gerok W, Sies H (1984) Hepatic role in pH regulation. Role of the intercellular glutamine cycle. Trends Biochem Sci (in press)
68. LaNoue KF, Schoolwerth AC (1979) Metabolite transport in mitochondria. Annu Rev Biochem 48: 871–922
69. Oliver J, Koelz AM, Costello J, Bourke E (1977) Acid-base induced alterations in glutamine metabolism and ureogenesis in perfused muscle and liver of the rat. Eur J Clin Invest 7: 445–449

Cellular Distribution and Regulation of Glutamine Synthetase in Liver

R. GEBHARDT and D. MECKE[1]

Introduction

In *Escherichia coli* and in other prokaryotic organisms glutamine is the precursor for syntheses of nitrogenous compounds and is required especially for biosyntheses of nucleic acids and proteins. Consequently, glutamine synthetase is regulated by a complex control system, including repression/derepression of enzyme synthesis [1], cumulative feedback inhibition by metabolic effectors [2], and by metabolite controlled chemical interconversion of the enzyme [3, 4]. This highly sophisticated regulation seems to be necessary to adapt these organisms to substrate dependent alterations of energy and nitrogen metabolism [5].

In mammalian cells negative results in attempts to demonstrate a chemical modification of glutamine synthetase in vivo and in vitro [6], (Ch. v. Kries and D. Mecke, unpublished results), as well as the uniformity of the enzyme exclude a regulation similar to that of bacteria. On the other hand, the role of glutamine on enzyme formation has also been established in cultures of mammalian cells. Since the first report on the influence of glutamine on glutamine synthetase in Hela cells [7], the regulation of enzyme synthesis and degradation by metabolic conditions or by hormones have been reported for other cultured cells (e.g., [8–13]), including those of hepatomas.

Although the validity of these results, particularly of those obtained with hepatomas for the conditions in normal livers has to be discussed with caution, it seems to be well-established that hepatic glutamine synthetase is controlled effectively, thus, adapting the capacity for glutamine formation to variations in the physiological conditions.

The scope of this article is to review the different levels of regulation of hepatic glutamine synthesis and to evaluate their importance for adaptation of ammonia metabolism. Based on recent findings of an intercellular compartmentation of glutamine synthetase, of its hormonal control, and of controlled transport processes, a model will be discussed which allows an independent regulation of glutamine synthesis and urea formation.

Heterogeneous Cellular Distribution of Glutamine Synthetase in Liver Parenchyma

Glutamine synthetase is an ubiquitous enzyme present in various tissues [14–18]. In some of those which are composed of a variety of different cell types,

1 Physiologisch-Chemisches Institut der Universität, Hoppe-Seyler-Straße 1, 7400 Tübingen, FRG

Glutamine Metabolism in Mammalian Tissues
Edited by D. Häussinger and H. Sies

such as kidney, neural retina, or brain, the enzyme has been shown to be heterogeneously distributed among the different cells [19–22]. In contrast, the distribution of glutamine synthetase in the liver whose morphologic appearance seems to be rather uniform has been unknown until recently. It is known, however, that many enzymes and metabolic functions vary in activity in different areas of the liver acinus [23–25]. Since a possible heterogeneous distribution of glutamine synthetase in liver parenchyma would be of importance for the understanding of the role of this enzyme in ammonia detoxification and in glutamine homeostasis, we have recently investigated the localization of this enzyme by means of immunofluorescence [26, 27].

Surprisingly, we found that glutamine synthetase is exclusively localized within a small subpopulation of pericentral liver parenchymal cells (Fig. 1A, B). These results were obtained using an antiserum against pork brain glutamine synthetase, which crossreacted specifically with rat liver glutamine synthetase [27]. As demonstrated in Fig. 1A, B the cells containing glutamine synthetase form a continuous cell-layer (1–3 cells thick) surrounding the terminal hepatic venules and facing the endothelial cells of these vessels. Other parts of the parenchyma are virtually free of enzyme specific immunofluorescence. Furthermore, nonparenchymal cells do not stain with the antiserum used.

This peculiar distribution of glutamine synthetase does not seem to be a unique feature of rodent liver. Using the same antibody which also crossreacted with human liver glutamine synthetase, a similar distribution was demonstrated in cryotome sections of human liver (Fig. 1C, D).

During ontogeny and in chemically induced liver injury the localization of glutamine synthetase is remarkably invariant. Thus, the distribution of the enzyme in the liver of a 10-day-old rat closely resembled that in the adult liver with the exception of a much weaker immunofluorescence (Fig. 2A, B), corresponding to a low enzyme activity (not shown). There is also no change in the localization of the enzyme in the allylformate-treated animal, where the periportal areas are selectively damaged (Fig. 2C, D). Conversely, after treatment with CCL_4, resulting in pericentral cell injury and necrosis, the intensity of immunofluorescence is reduced and appears somewhat diffuse, but does not lose its close association with the terminal hepatic venules (Fig. 2E, F).

In agreement with these immunohistochemical findings, biochemical measurements revealed that the enzyme activity is strongly reduced in CCL_4-treated liver (Table 1), corroborating the results of O'Donovan [28, 29]. Furthermore, Häussinger and Gerok have demonstrated the impairment of glutamine synthesis in CCl_4-induced liver cell necrosis with almost no effect on urea synthesis [30]. On the other hand, glutamine synthetase is virtually maintained after short treatment with allylformate (Table 1). In addition, these measurements demonstrated an inverse behavior of urea cycle enzymes with respect to maintenance or decline of enzyme activity in response to both treatments, whereas other enzyme activities (cytosolic or membrane bound) were affected in a more uniform manner (Table 1).

The inverse response of glutamine synthetase and of the urea cycle enzymes suggests a complementary distribution of these enzymes, in accordance with

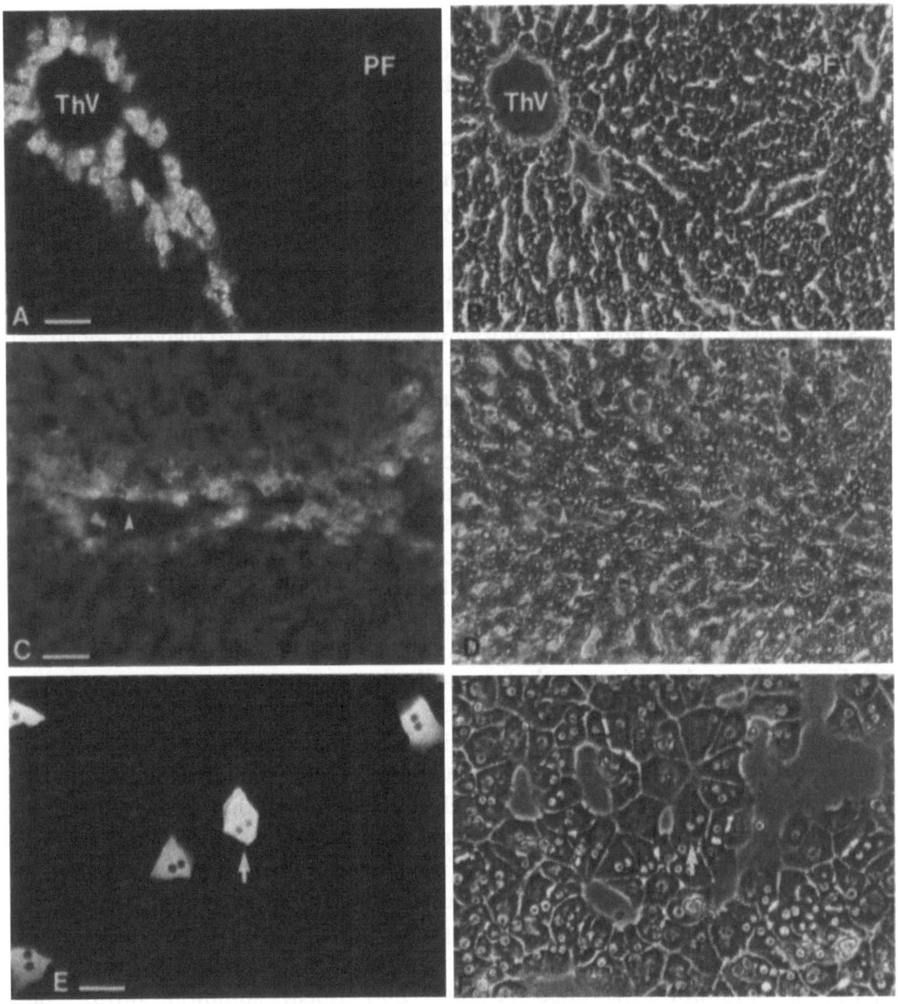

Fig. 1A–F. Localization of glutamine synthetase in liver in situ and in hepatocyte cultures revealed by indirect immunofluorescence according to Gebhardt and Mecke [27]. **A** Fluorescence micrograph of a cryotome section of normal rat liver. Fluorescence can be seen only within a small subpopulation of parenchymal cells surrounding a terminal hepatic venule *(ThV)*. **B** Phase contrast micrograph corresponding to **(A)**. A portal field *(PF)* is shown in the upper right corner. **C, D** Section of human liver (10 h postmortem) showing the fluorescent parenchymal cell layer around a terminal hepatic venule *(ThV)*. Note the absence of specific staining in the endothelial cells *(arrow-head)*. Unspecific background staining is much stronger with human than with rat liver. **E, F** Rat hepatocyte culture after 3 days. Single fluorescent parenchymal cells are spread over the monolayer (*arrow*). (*bars* represent 60 μm)

recent measurements by Häussinger [31] of glutamine and urea synthesis in the isolated perfused rat liver. This picture is stressed even more if one compares the cellular distribution of glutamine synthetase with those of mitochondrial carbamoylphosphate synthetase in the rat [32] and of argininosuccinate synthetase in human liver [33] which seem to be present throughout the

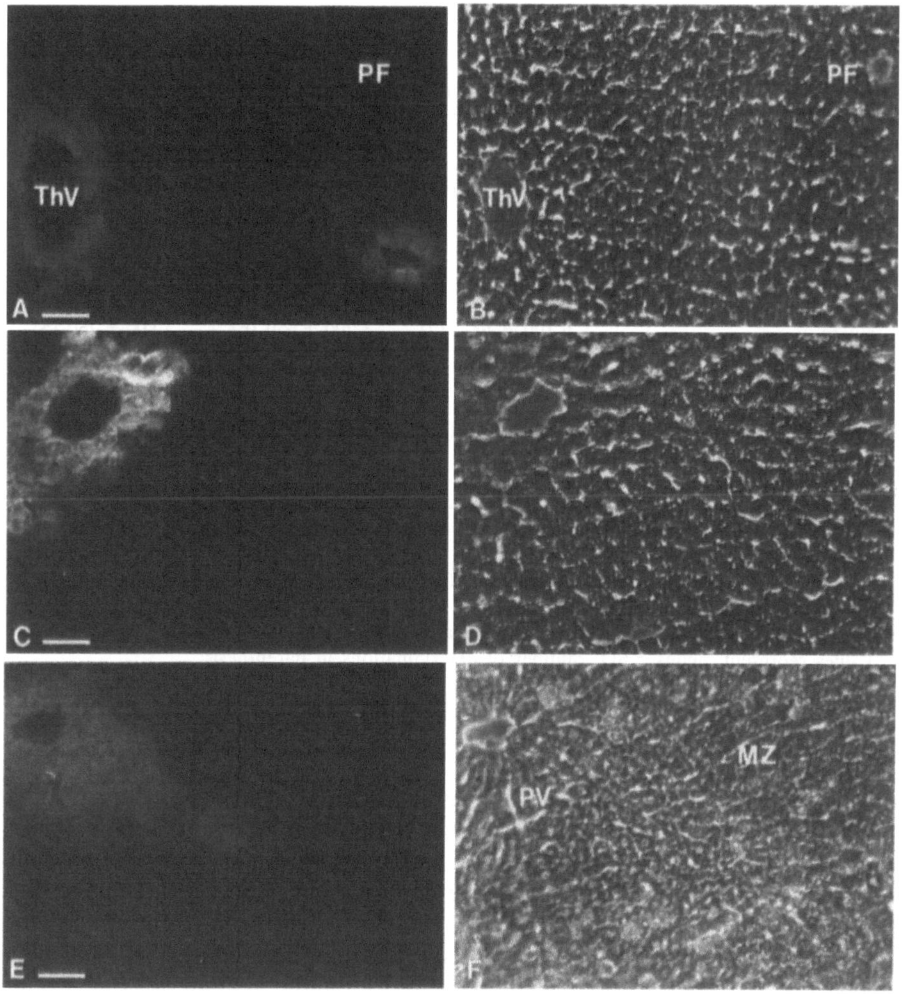

Fig. 2A–F. Localization of glutamine synthetase in developing rat liver and subsequent to liver injury by allylformate and CCl_4. **A, B** Fluorescence and phase contrast micrographs of a liver section from a 10-day-old rat. Note the weak fluorescence surrounding the terminal hepatic venule *(THV)* compared to Fig. 1A. *(PF)* portal field. **C, D** Liver of an adult rat after treatment with allylformate (cf. Table 1). Intensity and distribution of fluorescence are virtually not changed. **E, F** Liver of an adult rat after treatment with CCl_4 (cf. Table 1). The intensity of glutamine synthetase specific fluorescence is strongly reduced and appears diffuse. Note the periveneous *(PV)* necrotic zone and steatosis in the midzonal *(MZ)* area (*bars* represent 60 μm)

parenchyma except some few cells around the terminal hepatic venules, suggesting a strict complementary intercellular compartmentation of these enzymes. If this view is corroborated by direct experimental evidence and can be extended also to the other urea cycle enzymes, it would imply that the pathways of glutamine and urea synthesis are clearly separated in cellular terms, thus, preventing the competition of both pathways for ammonia fixation within the same parenchymal cell as has been hypothesized previously [27].

Table 1. Influence of liver injury caused by allylformate or CCl_4 on the levels of selected hepatic enzymes

Enzymes	Specific activity (mU/mg protein)[a]		
	Control	Allylformate[b]	CCl_4[c]
Glutamine synthetase	24 ± 3 (12)[d]	21 ± 3 (6)	3 ± 1 (5)
Carbamoylphosphate synthetase	30 ± 8 (5)	11 ± 2 (6)	26 ± 7 (5)
Ornithine transcarbamoylase	1,530 ± 210 (5)	890 ± 190 (6)	1,180 ± 210 (5)
Argininosuccinate synthetase	6 ± 2 (6)	3 ± 1 (4)	6 ± 1 (5)
Argininosuccinate lyase	20 ± 4 (7)	n.d.	17 ± 6 (4)
Arginase	6,100 ± 700 (7)	3,200 ± 650 (6)	5,800 ± 750 (5)
5′-Nucleotidase	83 ± 19 (8)	127 ± 46 (6)	112 ± 25 (5)
Alkaline phosphatase	27 ± 4 (5)	53 ± 12 (4)	n.d.
Lactate dehydrogenase	2,470 ± 440 (10)	2,180 ± 370 (8)	2,125 ± 310 (7)
Aspartate aminotransferase	595 ± 95 (8)	525 ± 83 (8)	455 ± 105 (7)
Alanine aminotransferase	77 ± 14 (8)	59 ± 22 (4)	65 ± 19 (7)

[a] Enzyme assays were carried out as described by Gebhardt and Mecke [47, 50, 96] or by Jung et al. [105]

[b] i.p. injection of 100 μl saline containing 10 μl allyformate/100 g body weight; treatment for 2 h

[c] i.p. injection of 100 μl/100 g body weight; treatment for 16 h

[d] Values represent means ± SD; number of determinations in parenthesis

The apparently exclusive localization of glutamine synthetase within a small subpopulation of parenchymal cells raises the question of the determinants of this distribution. Recent ideas to explain heterogeneity among parenchymal cells (e.g., for enzymes of carbohydrate metabolism) have focused mainly on the differential supply of oxygen, substrates, and hormones for the different regions of the liver acini [22, 23] as well as on their different innervation [34]. If such an explanation would also hold in the case of glutamine synthetase, one would expect the heterogeneous distribution to disappear in culture, where all hepatocytes are exposed to almost the same environment. However, the heterogeneity among liver parenchymal cells, which seems to reflect an all-or-none phenomenon, was preserved during enzymatic dissociation of liver tissue and persisted in culture for at least 3 days (Fig. 1E, F) without change in the percentage of hepatocytes staining for the enzyme [27]. Furthermore, hormonal induction of the enzyme in cultured hepatocytes (cf. Sect. on Hormonal Regulation of Hepatic Glutamine Synthetase) also caused no change in the number of fluorescent cells, indicating that induction of the enzyme is restricted to the same cell population [27].

These observations together with the invariance of the distribution of glutamine synthetase in liver injury mentioned above suggest that the heterogeneity of liver parenchyma with respect to glutamine synthetase is stable rather than dynamic and is not subject to adaptive changes as reported for carbohydrate metabolism [34]. (For further discussion see later sections.)

Developmental and Adaptive Changes in Hepatic Glutamine Synthetase Activity

In fetal rat liver glutamine synthetase activity is low [16, 35, 36] and shows a small decay in activity immediately after birth [35]. The adult level is subsequently reached with a gradual increase in activity over a period of several weeks [16, 35, 36]. Because of this developmental behavior glutamine synthetase can be appointed according to Greengard [37] to the cluster of enzymes appearing during the late suckling period.

In the adult animal the enzyme level is rather stable and compared to many other enzymes of amino acid metabolism, adaptive variations in enzyme activity in response to nutritional changes occur slowly. This is most probably due to the relatively long half-life of the enzyme of about 5 days which was measured by immunoprecipitation in adult rats [37]. This, however, may be much longer than that found in young rats, where the apparent half-life determined after administration of puromycin [35] was about 12 h.

When animals are shifted to a protein-free diet, the enzyme activity is decreased to about one-half of that found in rats maintained on a normal protein diet (25% casein) within 3 days [38, 39]. These changes seem to be associated with a shortening of the half-life of the enzyme [38]. On the other hand, increasing the protein content of the diet may not affect the liver enzyme [39] or may even decrease it [40]. Only during repletion from a fast, which transiently decreases glutamine synthetase activity in relation to liver protein [18, 39, 40], does the enzyme activity seem to be further increased by a high protein diet [39].

In pregnant and lactating rats which also exhibit an altered nitrogen metabolism [41, 42], a decrease in enzyme activity during late pregnancy and lactation has been described [43] which is not seen in early pregnancy or during the peak lactation period. These developmental and adaptational alterations in enzyme activity seem to be regulated by hormones and perhaps other factors.

Hormonal Regulation of Hepatic Glutamine Synthetase

The involvement of glucocorticoids in the regulation of glutamine synthetase has been suggested by many laboratories on the basis of results obtained with adrenalectomy [40] or by administration of cortisol [35, 44]. However, these results are not consistent, since several investigators have reported that the decay of glutamine synthetase activity after adrenalectomy could not be reversed by cortisol [36, 45]. A possible explanation of these differences may be an age-dependent responsiveness, allowing an induction of the enzyme only in young rats [44].

Alternatively, in one of these studies the participation of growth hormone was made probable because this hormone partially reversed the effect of hypophysectomy which diminishes enzyme activity [45]. Since hypophysectomy is associated with an impaired function of the adrenals and the thyroid gland, an

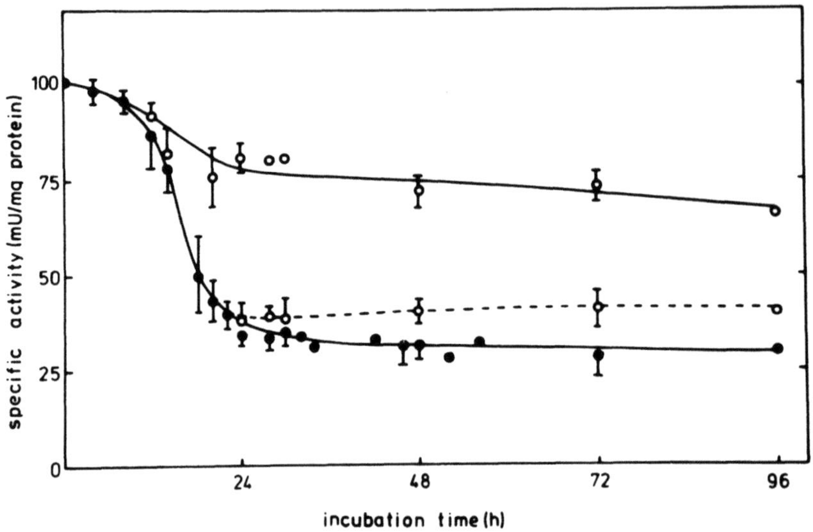

Fig. 3. Time course of glutamine synthetase activity in perifused (○) and discontinuously cultured (●) hepatocytes. Details for perifusion and discontinuous cultivation see [47, 50, 96]. Start of perifusion: 4 h, *continuous line;* 20 h, *dotted line*

involvement of glucocorticoids and thyroxine cannot be ruled out. Indeed, thyroxin was found to increase liver glutamine synthetase to a small extent in young rats [16, 35], but not in adult ones [44]. It has not been reported whether insulin or glucagon affect the liver enzyme activity in the intact animal. Only a short report exists demonstrating a decrease in glutamine synthetase activity during alloxan diabetes [46].

The involvement and interaction of the individual hormones suspected of participating in the regulation of glutamine synthetase was elucidated with greater efficiency using primary monolayer cultures of adult rat hepatocytes [47–49]. As depicted in Fig. 3, the basal activity of glutamine synthetase in cultured hepatocytes shows either an early rapid decay (in the discontinuous culture system [47]) or only a gradual decline (in the perifusion system [50]). The reason for this difference is not known, but it is likely that the continuous supply of medium in the perifusion affects the initial adaptation to the new environment. Despite these differences in the time course of basal enzyme activity, the cells in both systems responded to added hormones in a similar way [48]. According to these findings the activity of glutamine synthetase is controlled mainly by a coordinated action of dexamethasone and growth hormone. The response to these hormones is further modified by trijodothyronine, insulin, and perhaps glucagon, whereas some other hormones, such as estradiol, testosterone, prolactin, and epinephrine have so far been found to exert no effect, neither when added separately nor in combination with the main stimulators ([48], R. Gebhardt, unpublished observations).

The main function of the glucocorticoids seems to comprise a permissive effect allowing the onset of stimulation by growth hormone (Fig. 4) subsequent

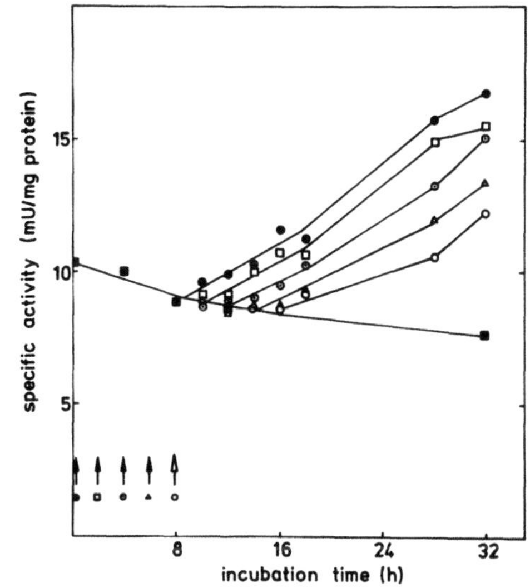

Fig. 4. Effect of preincubation with dexamethasone on the stimulation of glutamine synthetase in response to a mixture of growth hormone, dexamethasone, and trijiodothyronine. At zero time, (30 h after inoculation) (●——●), or after 2 h (□——□), 4 h (⊙——⊙), and 6 h (△——△) dexamethasone (2 μ*M*) was added to different cultures *(closed arrows)*. At 8 h (○——○) all cultures received a mixture of growth hormone (4 μg/ml), dexamethasone (2 μ*M*), and trijodothyronine (10 μg/ml) *(open arrow)*. Cells were harvested at the times indicated and enzyme activity was assayed in homogenates. (■——■) Control, no additions [47]

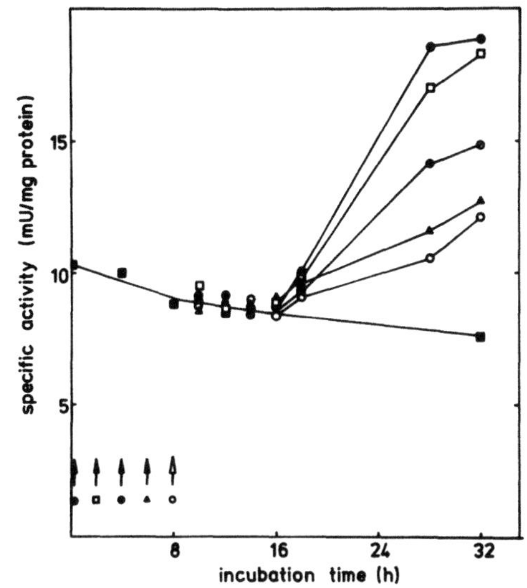

Fig. 5. Effect of preincubation with growth hormone on the stimulation of glutamine synthetase. Experimental details are the same as in Fig. 4 except that dexamethasone was replaced by growth hormone (4 μg/ml) for the different preincubations [47]

to a long lag-phase involving protein synthesis [47]. In accordance with this function cultured hepatocytes must be continuously exposed to dexamethasone which is effective in very low concentrations (down to 10^{-9} *M*) [48]. Growth hormone shows a refractory period of about 20 h [47] and triggers not only the onset of stimulation ("onset" function), but also its slope ("promoting" function) (Fig. 5). Although the molecular nature of the increase in enzyme activity has not been elucidated, it seems most likely that the synthesis of new enzyme molecules is involved in this process.

The coordinated action of dexamethasone and growth hormone is strongly dependent upon the sequence and time at which both hormones are added to the

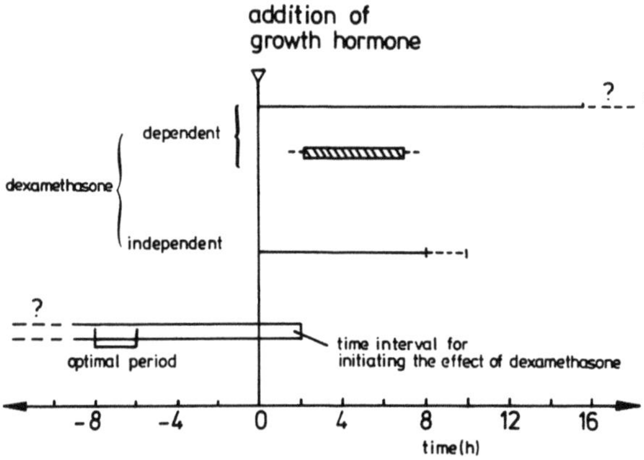

Fig. 6. Temporal relationship and coordination of the different effects of growth hormone and dexamethasone on glutamine synthetase in cultured hepatocytes. Time scale is chosen relative to the addition of growth hormone

culture medium [47, 48]. The main features of this time-dependent interaction are illustrated schematically in Fig. 6. According to this scheme, dexamethasone is effective in an interval of at least 10 h preceding the addition of growth hormone with an optimal period between 6 and 8 h. Furthermore, growth hormone must not be given more than 2 h before dexamethasone, since the onset function lasts no longer than about 10 h and the hormone exerts a refractory period lasting from 2 to 8 h after its addition [48]. On the other hand, the ability of growth hormone to increase the rate of stimulation (promoting function) which is not dependent upon the presence of dexamethasone seems not to be restricted to a limited period.

It is interesting to note that the optimal coordination of the exposure to dexamethasone and growth hormone coincides closely with the sequence of the peak values of these hormones in the circulation [51, 52] and fits also to the short-term rhythmicity of growth hormone [53]. The findings obtained with cultured hepatocytes, therefore, suggest a regulatory scheme of physiological significance rather than an artificial requirement of cells in culture. They are also in accordance with the view that the establishment of the pituitary-adrenal axis [37] may be the major stimulus for the enzymes appearing during the late suckling period.

The action of the modulator hormones trijodothyronine and insulin is much less clear. Trijodothyronine seems to enhance the rate of the stimulatory process [47] in agreement with the stimulatory influence of thyroxine in vivo [16, 35]. The response to insulin seems to be rather complex. When cultures are exposed to insulin during the early hours of cultivation, glutamine synthetase declines somewhat faster than in untreated controls (Fig. 7). On the other hand, insulin given subsequently to the stimulation of glutamine synthetase by dexamethasone and growth hormone reinforces the stimulatory process in a similar manner to trijodothyronine (Table 2). Although this action is exerted within a small interval (6 h), limiting the possibility that insulin is totally degraded, it is seen

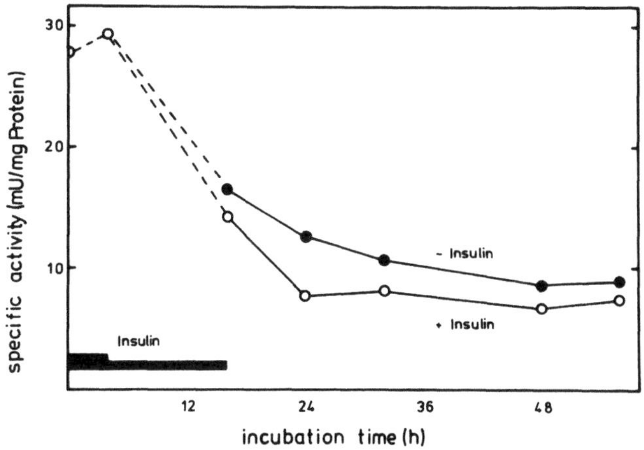

Fig. 7. Effect of insulin on the basal activity of glutamine synthetase in cultured hepatocytes. Insulin (10^{-6} *M*) was present either from 0–4 h (●) or from 0–16 h (○)

Table 2. Influence of short-term exposure to insulin during different cultivation periods on the stimulation of glutamine synthetase by dexamethasone and growth hormone[a]

Insulin[b]				Dexamethasone growth hormone[c]	Specific activity (% control)	
Cultivation period (h)						
0–4	4–16	16–24	28–34	24–48	48 h	56 h
–	–	–	–	–	100	100
–	–	–	–	+	223	212
+	–	–	–	+	217	207
+	+	–	–	+	150	141
+	+	+	–	+	110	147
–	–	–	+	+	264	208

[a] Hepatocytes were cultured as described [47]; dexamethasone was present only between 24 and 48 h also in control cultures
[b] (10^{-6} *M*)
[c] Dexamethasone (10^{-6} *M*); growth hormone (4 μg/ml)

only in the presence of high concentrations of this hormone, pointing to an artificial rather than a physiological effect of insulin itself. Whether this is linked to a somatomedin-like action known to occur at high insulin concentrations [54] cannot be decided at present. However, it is possible that some functions of growth hormone are also mediated by somatomedins which have been found to be produced by cultured hepatocytes in response to this hormone [55].

Regulation of Glutamine Synthetase in Hepatoma Cells

The regulation of glutamine synthetase in hepatomas and hepatoma cell lines seems to be rather different from that observed in rat liver in vivo or in hepatocytes in primary culture and shall be briefly reviewed here.

Most hepatomas grown in vivo have been shown to contain glutamine synthetase although the levels may sometimes be very low [38, 39, 56, 57]. Almost no correlation has been found between the degree of differentiation or the growth rate and the activity of glutamine synthetase [36, 44]. There is also no uniform response of glutamine synthetase in the host liver to different tumor types and sizes [44, 56–60]. It has also been observed that the enzymes of hepatomas and of host liver show striking differences in their response to different diets [38], hormones, or glutamine [44].

In cultured hepatoma cell lines (HTC cells, Chinese hamster liver cells) glucocorticoids seem to be the major stimulators of glutamine synthetase [60–63] inducing the synthesis of new enzyme protein [62, 63]. The second potent regulator of glutamine synthetase activity in these cells, glutamine [60–62], has been shown in many studies by means of immunochemical techniques to enhance selectively the degradation of this enzyme [64–66]. The modulation of glutamine synthetase by glutamine seems to involve a labile polypeptide factor stabilized by the amino acid [67] and enucleation-sensitive steps not shared with the degradative pathways of most other cellular proteins [68]. It has recently been suggested that lysosomal proteases participate in the glutamine-provoked degradation of the enzyme [69].

Metabolic and Environmental Control of Glutamine Synthetase Activity

Liver glutamine synthetase is considered to be a soluble rather than a membrane bound enzyme [6, 70] in contrast for instance to the brain enzyme [71]. This view has emerged from the fact that physiological salt concentrations release the enzyme from its association to microsomal membranes [6, 70]. However, a weak association with membranous components has not only been found with fractionation in isotonic sucrose [6, 70], but also under other conditions of fractionation (E. Schmincke-Ott, R. Gebhardt, D. Mecke, unpublished observations). Therefore, it appears too early to exclude the possibility of a weak association of the enzyme with subcellular organelles which could also be of regulatory significance. However, it seems settled that glutamine synthetase is present in a compartment separate from glutaminase.

The activity of purified liver glutamine synthetase has been shown to be affected by a variety of metabolites ([6, 72, 73], for review see [74]) only the most important of which should be mentioned here. The enzyme is markedly activated by α-ketoglutarate [72, 75] and is inhibited by glycine, alanine, and histidine [6, 72, 75]. Both the apparent K_m values for the normal substrates as well as the K_i values for alanine and histidine are strongly dependent upon the concentrations of divalent cations, particularly of Mn^{2+} [6, 72]. In addition, Mn^{2+} ions by themselves markedly inhibit the enzyme [72, 76] suggesting a central role in the regulation of glutamine synthetase activity.

As suggested by the experiments with hepatoma cells (see preceding Sect.) and by the experiments with other cultured mammalian cells [7, 8], the idea that the level of hepatic glutamine synthetase could be influenced by blood glutamine

Table 3. Influence of culture media and different amino acids on basal and stimulated glutamine synthetase activity

Medium[a]	Amino acid[b]			Specific activity (mU/mg protein)[c]	
				20 h	46 h
W/AB 77	Gln	+		12.4 ± 0.8[d]	7.3 ± 0.4
		+	+ Growth hormone[e]	–	15.1 ± 0.3
		–		11.1 ± 0.8	7.4 ± 0.5
		–	+ Glutaminase[f]	13.2 ± 0.6	7.2 ± 0.4
L-15	Gln	+		13.1 ± 0.5	7.8 ± 0.7
		–		13.5 ± 0.9	7.2 ± 0.4
		–	+ Glutaminase	12.3 ± 0.3	7.4 ± 0.3
W/AB 77	Glu			12.7 ± 0.4	7.4 ± 0.5
	Leu			11.8 ± 0.9	7.9 ± 0.6
	Met			12.3 ± 0.5	7.3 ± 0.6
	Trp			13.1 ± 0.6	8.5 ± 0.7
	Tyr			12.5 ± 0.7	7.3 ± 0.3

[a] W/AB 77 and L-15 were free of glutamine; dexamethasone (10^{-7} *M*) was present throughout except for initial 4 h
[b] Final concentration 5 m*M*
[c] Determined according to Gebhardt and Mecke [47]
[d] Values are means ± SD of three determinations
[e] 4 µg/ml; added after 22 h
[f] 0.1 U/ml; present throughout cultivation

concentration has been attractive for many years. All studies carried out so far in the intact animal argue against such a possibility [35, 36, 39, 44]. However, these studies in vivo suffer from the fact that plasma glutamine concentrations have not been determined. Therefore, it remains doubtful, particularly if the excellent homeostasis of glutamine concentration is taken into account, whether significant and, thus, effective changes in blood glutamine could be brought about by either oral administration [39] or intraperitoneal injection [35]. Indirect evidence against a possible inverted correlation between glutamine concentration and the level of glutamine synthetase is provided by the observation made in developing or some tumor-bearing rats of parallel alterations of both parameters [35, 44, 58].

The experimental difficulties inherent in studies in vivo in demonstrating or excluding a direct effect of glutamine can be evaded by using cultured hepatocytes. As shown in Table 3 glutamine synthetase was found not to be affected by the extracellular concentrations of glutamine, some other amino acids, or by different culture media. Furthermore, a possible effect of different oxygen tensions on the basal or induced level of glutamine synthetase in cultured hepatocytes could also be ruled out (Table 4). In addition, no alterations in the relative number of glutamine synthetase-containing cells determined according to Gebhardt and Mecke [27] were observed in response to these culture conditions.

Table 4. Hormonal stimulation of glutamine synthetase at different oxygen tensions

Oxygen[a]	Growth hormone[b] (22–46 h)	Specific activity (mU/mg protein) (46 h)
5%	+	13.6[c]
	–	6.9
13%	+	14.3
	–	7.1
20%	+	13.8
	–	6.6

[a] Oxygen content of the atmosphere for cell cultivation
[b] 4 μg/ml; dexamethasone (10^{-7} *M*) was present in all incubations except during the initial 4 h
[c] Values are means of 2–4 determinations

Hormonal Integration of Glutamine Synthesis

It seems obvious from the localization of glutamine synthetase in liver tissue and from the extensive regulation of this enzyme that the liver must play an important role in glutamine synthesis. In accordance, different conditions have been described resulting in a small but significant net-synthesis of glutamine in isolated liver preparations [cf. 31, 77–81]. However, various aspects of hepatic glutamine synthesis, particularly the interplay of the different regulatory events in this process, are as yet poorly understood [cf. 72].

In order to investigate whether the hormonal stimulation of glutamine synthetase seen in the culture system gives rise to an enhanced synthesis of glutamine, the capacity of cultured hepatocytes for glutamine synthesis was measured under different hormonal conditions (R. Gebhardt, E. Kleemann, D. Mecke, manuscript in preparation). As shown in Table 5, net consumption of glutamine could be measured in controls and in glucagon-treated cultures during a 24 h incubation period in the presence of 2 m*M* glutamine. However, despite this high initial concentration of glutamine a net production of this amino acid occurred in cultures exposed to dexamethasone. During the same incubation interval a decrease in glutamate concentration in the culture medium could be detected which was most prominent after pretreatment with dexamethasone (Table 5). This finding fits well with the observation that in contrast to the former results of Hems et al. [83] obtained with liver slices, glutamate transport across the plasma membrane operates in cultured hepatocytes [84, 85] as well as in hepatocyte plasma membrane vesicles [86]. Extraction from the culture medium of amino acids other than glutamate which could possibly be utilized for the synthesis of glutamine was only little affected by the different hormones. In particular, their utilization did not respond in a comparable manner to the glucocorticoid treatment (not shown), suggesting that glucocorticoids favor the conversion of glutamate into glutamine presumably by stimulating glutamate uptake [85].

To test this hypothesis similar experiments were carried out in the presence of radiolabeled glutamate using glutamine-free W/AB 77 medium for cultiva-

Table 5. Influence of hormonal treatment of hepatocyte cultures on the extracellular concentrations of glutamine, glutamate, and urea[a]

Hormone[b]	Concentration (mM)[c]		
	Glutamine	Glutamate	Urea
	Initial concentrations		
	2.00	1.00	0
None	1.90[d]	0.83	1.9
Dexamethasone	2.46	0.12	2.4
Glucagon	1.54	0.48	3.9

[a] Hepatocytes were cultured according to [47] in W/AB 77 (+ 10% calf serum) for 2 h and then without serum for 40 h. (At 42 h cultures contained approx. 1.7 mg of cellular protein per 2 ml of medium)
[b] Hormones (10^{-6} M each) were added 2 h after inoculation
[c] Determined by HPLC; incubation period from 18 to 42 h
[d] Values are means of duplicate determinations with a variation less than 5%

tion. Within a period of 5 h glutamine synthesis which proceeded almost linear with time increased nearly two fold in cultures exposed to dexamethasone and about fourfold in those exposed to dexamethasone and growth hormone (Table 6). A substantial amount of glutamine was derived from labeled glutamate in agreement with recent results of Häussinger [87] obtained with perfused rat liver. As determined from the specific radioactivity of glutamine given in Table 6, glutamine formed within 5 h from extracellular glutamate amounted to 14, 80, and 114 nmol per mg protein in controls, dexamethasone-treated cultures, and cultures exposed to dexamethasone and growth hormone, respectively. The increase seen under the different conditions roughly correlates with the respective transport capacities for glutamate measured at the end of the incubation period (Table 6), suggesting that the utilization of extracellular glutamate for the synthesis of glutamine is limited by glutamate uptake. This assumption seems also to be true for the conversion of glutamate into aspartate, the only amino acid in addition to glutamine which was labeled to a significant extent, although the concentration of aspartate in the medium declined during cultivation (Table 6). Conversion of glutamate into aspartate was blocked by 0.5 mM aminooxyacetate, while glutamine synthesis was further enhanced (not shown).

As obvious from Table 6 cultivation in the presence of both dexamethasone and growth hormone strongly reinforced the synthesis of glutamine compared to dexamethasone alone. This observation is in accordance with the hormonal stimulation of glutamine synthetase in the culture system, indicating that an enhanced enzyme activity measured in vitro results in an enhanced synthesis of glutamine in the intact cell. The observed stimulation of glutamine synthesis under these conditions, however, resulted only to a minor extent from an increased utilization of labeled extracellular glutamate, but was mainly due to an enhanced conversion of unlabeled substrates not yet identified. Thus, in addition

Table 6. Hormonal influence on glutamate metabolism and transport in hepatocyte cultures[a]

	Control	Dexa-methasone[b]	Growth hormone[c]
	(μmol × 5 h^{-1} × mg protein^{-1})		
Glutamate[d]	−0.57[e]	−0.70	−0.83
Glutamine[f]	+0.14	+0.25	+0.50
Aspartate[g]	−0.06	−0.06	−0.06
	Specific radioactivity (% Glu)[h]		
Glutamine	10.0	32.5	22.8
Aspartate	3.2	22.6	23.2
	(nmol × min^{-1} × mg protein^{-1})		
Glutamate transport[i]	0.51	1.31	1.42

[a] Hepatocytes were cultured as in Table 5, footnote a, for 66 h. During the incubation in the presence of L-[U-^{14}C]-glutamate (0.2 μCl/μmol) for 5 h glutamine-free W/AB 77 medium was used

[b] Present at 10^{-6} *M* throughout cultivation except initial 4 h

[c] Growth hormone (4 μg/ml) was added 20 h after inoculation

[d] Initial conc. 1.00 m*M*

[e] Values are means of duplicate determinations by HPLC; (−) net consumption; (+) net production

[f] Initial conc. 0 m*M*

[g] Initial conc. 0.45 m*M*

[h] Determined at the end of the incubation period

[i] Actual rate of transport at the end of the incubation period determined directly in the culture medium as described similarly [85]

to the flux through glutamine synthetase the flux of some reaction(s) leading to glutamate seems also to be favored by dexamethasone and growth hormone.

Hepatocyte cultures established from a complete suspension of parenchymal cells may not be an ideal system to determine the optimal conditions for glutamine synthesis, because of the random intermixing of cells with different ability for glutamine synthesis. However, if the extent of the simultaneous glutamine utilization (determined to be in the range of 50 to 75 nmol × 5 h^{-1} × mg protein^{-1}) is taken into account, an approximate total amount of 0.6 μmol glutamine per mg of protein may have formed in growth hormone treated cultures within a 5 h period. Thus, glutamine is formed in the cultured cells with approximately one-sixth of the maximal rate of glutamine synthesis measured in vitro (cf. Fig. 4).

In conclusion, it appears from this study that glucocorticoids and growth hormone function as integrating regulators of glutamine synthesis operating on several crucial steps such as transport across the plasma membrane and enzyme activity.

Discussion

In view of the low net production or even net consumption of glutamine by the liver seen under most physiological conditions which contrasts with the relatively high specific activity of hepatic glutamine synthetase, it has been assumed that the activity of this enzyme may be largely inhibited in vivo [77]. Potent factors responsible for this inhibition have not been identified with the possible exception of Mn^{2+} ions [76]. However, this explanation does not provide a satisfying answer for the question of why the liver contains a higher enzyme activity than many other tissues. Furthermore, it is hard to understand why the level of an enzyme which is inhibited to a large extent should be adapted to various physiological conditions.

A full explanation for these problems does not seem available at present. However, as obvious from the facts reviewed herein the view has changed dramatically during the last few years and the puzzling picture is somewhat clarified. Above all, the unexpected localization of the enzyme within liver parenchyma described previously [26, 27] has introduced a new point of view into the discussion of hepatic glutamine synthesis. The main implications of this finding are:

a) the requirements of the predominant part of liver parenchyma for glutamine in biosynthetic pathways must be met by extraction of this amino acid from the circulation;

b) glutamine synthesized by the liver is exclusively destined for export via the circulation.

As a further consequence of the preceding points degradation and synthesis of glutamine occur simultaneously, but in separate cells, providing the basis for the glutamine cycle reported previously [80] which is now recognized to function unidirectionally and intercellularly [27, 31].

This view, dividing the liver parenchyma into a cellular "compartment for urea synthesis" comprising the periportal, midzonal, and the outer part of the periveneous area of the liver acinus and a "compartment of glutamine synthesis" restricted to the most inner part of the periveneous area is illustrated schematically in Fig. 8. It is obvious that this peculiar distribution cannot be described by the ordinary zonal terminology: since more than 90% of the liver parenchyma are involved in urea synthesis it seems inadequate to refer to urea synthesis as a "periportal" function. Therefore, we would like to propose the terms "afferent urea compartment" and "efferent glutamine compartment" in order to describe this type of functional partition of liver parenchyma.

The indication in this scheme of a high uptake of glutamine in the compartment for urea synthesis is corroborated by the previous identification of a distinct glutamine transport system known so far as unique to the liver [88–90]. Interestingly, this transport system is under adaptive [89, 90] and hormonal (R. Gebhardt, unpublished observation) control. The importance of amino acid transport is further demonstrated by the existence of a hormonally inducible transport system for glutamate [85]. Both transport systems may serve to regulate the availability of glutamine and glutamate to the respective compartments of the parenchyma. It is an interesting question whether these

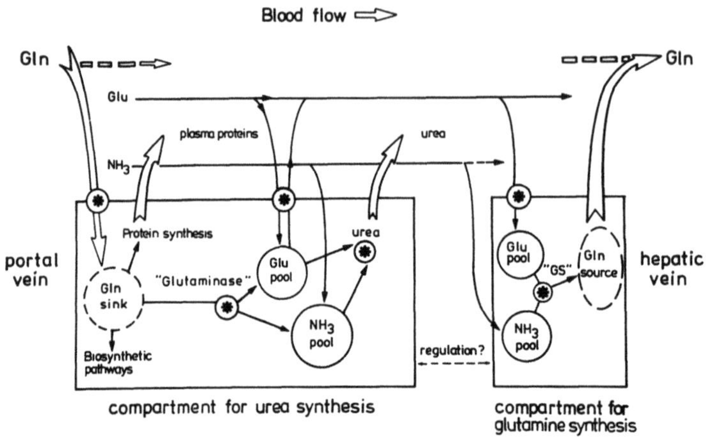

Fig. 8. Cellular compartmentation of glutamine metabolism and urea formation across the liver acinus (After [31]). Main points of adaptive and hormonal control are marked by ⊛ · *"GS"* glutamine synthetase

systems exhibit different transport capacities in the different parts of the liver acinus. Evidence for such a heterogeneity obtained with isolated perfused rat liver has recently been presented [87].

In view of the sequential events of uptake and release, the rates at which glutamine degradation and synthesis proceed in the liver may far exceed the rates derived from net consumption or net synthesis seen under different physiological conditions. In this respect it should be mentioned that the liver is the major source of plasma proteins [91] and is also the site of an intensive synthesis of nucleotides, amino sugars, etc. As depicted in Fig. 8, the compartment lacking glutamine synthetase accounts for the quantitatively dominant part of these synthetic functions which, thus, have to utilize glutamine derived from extrahepatic sources. In order to maintain the plasma glutamine concentration, the consumption of glutamine must be compensated by a high glutamine synthesis in the periveneous cells competent for this process. Therefore, the rate of the glutamine cycle in the liver in vivo may exceed the rate seen in isolated perfused rat liver even under stimulated conditions [cf. 80, 81], because the biosynthetic functions in that system may be limited for a variety of reasons. The presumably high rate of the glutamine cycle may provide an explanation for the high specific activity of hepatic glutamine synthetase. In addition, a high capacity of this enzyme may provide a safety device to prevent this reaction from being rate limiting for glutamine synthesis [27], particularly if one assumes that the protein synthesizing capacity of the few enzyme containing cells may not be sufficient to allow a rapid adaptation of the amount of glutamine synthetase to rapidly changing requirements for glutamine synthesis. Furthermore, as reported here (preceding Sect.), glutamine synthesis in cultured hepatocytes proceeds at a reasonable rate compared to the maximal activity of glutamine synthetase present. Therefore, it does not seem very likely that the enzyme is inhibited to a great extent inside the cell, but that glutamine synthesis may be limited by substrate availability in most cases.

Under normal conditions the liver seems to be near to equilibrium between consumption and synthesis of glutamine (cf. [92, 93] for review [82]). Mechanisms for short-term control have to be postulated to explain such an effective balancing. Effector control of glutamine synthetase activity (cf. Sect. on Metabolic and Environmental Control of Glutamine Synthetase Activity) may be involved, as may be control of substrate availability. However, it remains doubtful whether such a "homeostasis" could be brought about merely by altering the intracellular concentrations of metabolites, the more, since not only glutamine synthetase flux has to be regulated, but also the fluxes through pathways providing glutamate as a precursor. Amongst others, possible candidates which could carry information from the compartment for urea synthesis to the compartment for glutamine synthesis are ammonium ions and blood glutamate. In either case the possible link to the control of glutamine synthesis is speculative, but it should be borne in mind that ammonium ions are well-known as potent regulators in the afferent urea compartment acting on pyruvate dehydrogenase [94], glutaminase [95], and affecting glutamine cycling in perfused rat liver [80]. The full elucidation of the regulatory interrelationship between both compartments of the parenchyma, however, must await the separation of the respective cell populations.

In contrast to the unresolved problems concerning short-term balancing, the understanding of the hormonal regulation of glutamine synthetase and of other steps possibly involved in glutamine synthesis seems to be on the threshold of a unifying concept of the long-term control of these processes. As pointed out in the preceding sections, glucocorticoids are of the utmost importance in this regulatory scheme either by directly influencing the activity of the respective components (e.g., glutamate transport [85]) or by exerting a permissive action (e.g., glutamine synthetase [47], urea cycle [96, 97]). A second important regulator with respect to glutamine synthesis may be growth hormone, the involvement of which in the regulation of related metabolic functions has been known for several years [98, 99]. In combination with insulin and glucagon this hormone couple may provide a general means superimposed to short-term control mechanisms determining the extent to which ammonia is excreted in the form of urea or served in the form of glutamine. This may be of particular importance in situations in which high amounts of fixed nitrogen are needed, such as in development, lactation, or acidosis. In this context it should be briefly noted that other enzymes of ammonia metabolism are affected in different ways by different sets of hormones. For instance, glutaminase is controlled by glucagon [100], vasopressin, and adrenalin [101], but not by growth hormone [102]; whereas the urea cycle enzymes are regulated mainly by glucagon under the permissive influence of glucocorticoids [96, 97]. (Effects of insulin [103] and growth hormone [102, 104] on urea cycle enzymes seen in vivo could not be found in primary cultured hepatocytes [97]). Thus, the different patterns of metabolite flow which characterize each physiological state could be caused by the different influence of the various hormones on transport systems, enzyme activities, etc.

A third problem comprising both the aspect of localization and that of regulation concerns the question of how the expression of glutamine synthetase is controlled and whether the heterogeneous distribution reflects the existence of parenchymal cells with different phenotypes. This problem has been addressed at several points in this article and only some additional aspects will be discussed here.

The major arguments for the existence of different phenotypes are: (a) the apparent all-or-none phenomenon of the expression of glutamine synthetase and (b) the stability of the distribution among parenchymal cells under various in vivo and in vitro conditions and despite large alterations in enzyme level.

In order to explain the heterogeneous distribution of glutamine synthetase on the basis of a single parenchymal cell type one has to assume the existence of a threshold concentration of some factor(s) which must be reached only one to three cells apart from the terminal hepatic venules. In view of the lack of effect of various environmental conditions on the activity and on the distribution of the enzyme among parenchymal cells (Sect. on Metabolic and Environmental Control of Glutamine Synthetase Activity, Tables 3, 4), such a possibility seems to be rather unlikely.

The striking stability of the cellular distribution of glutamine synthetase could alternatively be explained by the long half-life of the enzyme in adult rats [38] which would prevent that significant changes occur within a period of 3 days or less. Although the half-life of the enzyme in cultured hepatocytes is not known, there is evidence that it is shortened under these conditions, thus, approximating that found in young rats [35]. Therefore, the reported stability of distribution during hormonal stimulation of the enzyme activity in the cultured cells argues in favor of cell populations exhibiting different phenotypes.

Many questions, however, cannot be answered at present. Firstly, there is the problem at what time during ontogenesis the different cell types appear and what stimuli trigger their development. Secondly, it is not known why the glutamine synthetase containing cells are located in the ultimate vicinity of the terminal hepatic venules. Possible explanations offered so far [27], such as an interaction with endothelial cells of these vessels or with particular elements of the nervous system, are still hypothetical.

Acknowledgements. The authors are indebted to Dr. H. Schmid (Pathologisches Institut, Universität Tübingen) for providing the cryotome sections of human liver and to Dr. Ch. Klessen (Anatomisches Institut, Universität Tübingen) for his cooperation in the histochemical experiments. The skillful technical assistance of Mrs. M. Behringer, Mrs. M. Landesvatter, and Mrs. A Schneck is gratefully acknowledged. This work was supported by the Deutsche Forschungsgemeinschaft and the Fonds der Chemischen Industrie.

References

1. Mecke D, Holzer H (1966) Repression und Inaktivierung von Glutamin Synthetase in Escherichia coli durch NH_4^+. Biochim Biophys Acta 122: 341–351
2. Woolfolk CA, Stadtman ER (1964) Cumulative feedback inhibition in the multiple end product regulation of glutamine synthetase activity in Escherichia coli. Biochem Biophys Res Commun 17: 313–319

3. Mecke D, Wulff K, Holzer H (1966) Metabolit-induzierte Inaktivierung von Glutamin Synthetase aus Escherichia coli im zellfreien System. Biochim Biophys Acta 128: 559–567
4. Stadtman ER, Ginsburg A (1974) The glutamine synthetase of Escherichia coli: structure and control. In: Boyer PD (ed) The enzymes, vol 8. 3rd edn. Academic Press, London New York, pp 755–807
5. Schutt H, Holzer H (1972) Biological function of the ammonia induced inactivation of glutamine synthetase in Escherichia coli. Eur J Biochem 26: 68–72
6. Deuel TF, Louie M, Lerner A (1978) Glutamine synthetase from rat liver. Purification, properties and preparation of specific antisera. J Biol Chem 253: 6111–6118
7. De Mars R (1958) The inhibition by glutamine of glutamyl transferase formation in cultures of human cells. Biochim Biophys Acta 27: 435–436
8. Kirk DL, Moscona AA (1963) Synthesis of experimentally induced glutamine synthetase glutamotransferase activity in embryonic chick retina in vitro. Dev Biol 8: 341–357
9. Barnes PR, Youngberg D, Kitos PA (1971) Factors affecting the production of glutamine in cultured mouse cells. J Cell Physiol 77: 135–144
10. Reif-Lehrer (1968) Development of glutamine synthetase activity in chick embryo retina cultures in the absence of added hydrocortisone. Biochem Biophys Res Commun 33: 984–989
11. Viceps D, Christofalo VJ (1975) Glutamine synthetase activity in WI-38 cells. J Cell Physiol 86: 9–14
12. Miller RE, Hackenberg R, Gershman H (1978) Regulation of glutamine synthetase in cultured 3T3-L1 cells by insulin, hydrocortisone, and dibutyryl cyclic AMP. Proc Natl Acad Sci USA 75: 1418–1422
13. Miller RE, Pope SR, DeWille JW, Burns DM (1983) Insulin decreases and hydrocortisone increases the synthesis of glutamine synthetase in cultured 3T3-L1 adipocytes. J Biol Chem 258: 5405–5413
14. Wu C (1963) Glutamine synthetase. I. A comparative study of its distribution in animals and its inhibition by DL-allo-α-hydroxylysine. Comp Biochem Physiol 8: 335–351
15. Lund P (1970) A radiochemical assay for glutamine synthetase and activity of the enzyme in rat tissues. Biochem J 118: 35–39
16. Chader GJ (1971) Hormonal effects of the neural retina. I. Glutamine synthetase development in the retina and liver of the normal and trijodothyronine-treated rat. Arch Biochem Biophys 144: 657–662
17. Herzfeld A, Estes NA (1973) The distinction between γ-glutamylhydroxamate synthetase and L-glutamine-hydroxylamine glutamyltransferase activities in rat tissues. Studies in vivo. Biochem J 133: 59–66
18. Arola LI, Palou A, Remesar X, Alemany M (1981) Glutamine synthetase activity in the organs of fed and 24 hours fasted rats. Horm Metab Res 13: 199–202
19. Burch HB, Choi S, McCarthy WZ, Wong PY, Lowry OH (1978) The location of glutamine synthetase within the rat and rabbit nephron. Biochem Biophys Res Commun 82: 498–505
20. Riepe RE, Norenburg MD (1977) Müller cell localization of glutamine synthetase in rat retina. Nature (London) 268: 654–655
21. Linser P, Moscona AA (1979) Induction of glutamine synthetase in embryonic neural retina: localization in Müller fibers and dependence on cell interactions. Proc Natl Acad Sci USA 76: 6476–6480
22. Hallermayer K, Harmening C, Hamprecht B (1981) Cellular localization and regulation of glutamine synthetase in primary cultures of brain cells from newborn mice. J Neurochem 37: 43–52
23. Rappapaort AM (1976) The microcirculatory acinar concept of normal and pathological hepatic structure. Beitr Pathol 157: 215–243
24. Jungermann K, Sasse D (1978) Heterogeneity of liver parenchymal cells. Trends Biochem Sci 3: 198–202
25. Gumucio JJ, Miller DL (1981) Functional implications of liver cell heterogeneity. Gastroenterology 80: 393–403

26. Gebhardt R, Mecke D (1982) Liver cell heterogeneity: the exclusive localization of glutamine synthetase within a small population of rat liver parenchymal cells in situ and in primary culture. Abstr Commun Basel Liver Week, Liver Metab Diseases, No 62
27. Gebhardt R, Mecke D (1983) Heterogeneous distribution of glutamine synthetase among rat liver parenchymal cells in situ and in primary culture. EMBO J 2: 567–570
28. Dunne TM, O'Donovan DJ (1976) The metabolic effects of liver destruction in the rat. J Physiol 263: 255–256P
29. O'Donovan DJ (1980) Glutamine metabolism in carbon tetrachloride-treated acidotic rats. Int J Biochem 12: 85–87
30. Häussinger D, Gerok W (1984) Hepatocyte heterogeneity in ammonia metabolism: impairment of glutamine synthesis in CCL_4-induced liver cell necrosis with no effect on urea synthesis. Chem Biol Interact 48: 191–194
31. Häussinger D (1983) Hepatocyte heterogeneity in glutamine and ammonia metabolism and the role of an intercellular glutamine cycle during ureogenesis in perfused rat liver. Eur J Biochem 133: 269–275
32. Gaasbeek-Janzen JW, Lamers WH, Moorman AFM, deGraaf A, Los JA, Charles R (1984) Immunohistochemical localization of carbamoyl-phosphate synthetase (ammonia) in adult rat liver; evidence for a heterogeneous distribution. J Histochem Cytochem (in press)
33. Saheki T, Yagi Y, Sase M, Nakano K, Sato E (1983) Immunohistochemical localization of argininosuccinate synthetase in the liver of control and citrullinemic patients. Biomed Res 4: 235–238
34. Jungermann K, Katz N (1982) Functional heterogeneity of liver parenchyma. Hepatology 2: 385–395
35. Wu C (1964) Glutamine synthetase III. Factors controlling its activity in the developing rat. Arch Biochem Biophys 106: 394–401
36. Knox WE, Kupchik HZ, Liu LP (1971) Glutamine and glutamine synthetase in fetal, adult, and neoplastic rat tissues. Enzyme 12: 88–98
37. Greengard O (1970) The developmental formation of enzymes in rat liver. In: Litwack G (ed) Biochem Actions Horm, vol I. Academic Press, London New York, pp 53–87
38. Louie M, Deuel TF, Morries HP (1978) Abnormal dietary regulation of glutamine synthetase in rat hepatomas. J Biol Chem 253: 6119–6124
39. Wu C (1964) Glutamine synthetase IV. Its formation in rat liver following partial hepatectomy and during repletion. Arch Biochem Biophys 106: 402–409
40. Raina PN, Rosen F (1968) Induction of glutamine synthetase by cortisol. Biochim Biophys Acta 165: 470–475
41. Freinkel N, Metzger BE, Nitzam N, Hare JW, Shambough GE, Marshall RT, Surmaczynska RZ, Hagel TC (1972) Accelerated starvation, a mechanism for the conversation of maternal nitrogen during pregnancy. Isr J Med Sci 2: 426–439
42. Palou A, Arola L, Alemany M (1977) Plasma amino acid concentrations in pregnant rats and in 21 day old fetuses. Biochemical J 166: 49–55
43. Remesar X, Arola LI, Palou A, Alemany M (1982) Glutamine synthetase activity in rat tissues during pregnancy and lactation. Horm Metab Res 14: 419–421
44. Wu C, Morris HP (1970) Responsiveness of glutamine-metabolizing enzymes in Morris hepatomas to metabolic modulations. Cancer Res 30: 2675–2684
45. Wong BS, Dunn A (1977) Possible growth hormone regulation of rat liver glutamine synthetase activity. Biochem Biophys Res Commun 79: 876–884
46. Kennan AL (1964) Effect of Insulin on glutamine synthetase activity in rat liver. Endocrinology 74: 518–520
47. Gebhardt R, Mecke D (1979) The role of growth hormone, dexamethasone and trijodothyronine in the regulation of glutamine synthetase in primary cultures of rat hepatocytes. Eur J Biochem 100: 519–525
48. Gebhardt R (1980) Hormonale Wechselwirkungen bei der Kontrolle hepatischer Enzymspiegel. Primärkulturen von Hepatocyten als Modellsysteme für Untersuchungen zur Stoffwechselregulation. Thesis, Univ Tübingen
49. Mecke D, Gebhardt R (1980) Interactions of hormones in regulation of nitrogen metabolism in cultured hepatocytes. Hoppe-Seyler's Z Physiol Chem 361: 616

50. Gebhardt R, Mecke D (1979) Perifused monolayer cultures of rat hepatocytes as an improved in vitro system for studies on ureogenesis. Exp Cell Res 124: 349–359
51. Wurtman RJ, Axelrod J (1967) Daily rhythmic changes in tyrosine aminotransferase activity of the rat liver. Proc Natl Acad Sci USA 57: 1594–1598
52. Simon ML, George R (1975) J Neuroendocrinol 17: 125–138
53. Cocchi D, Gil-Ad I, Panerai AE, Locatelli V, Müller EE (1976) Circadian variations in plasma growth hormone and prolactin in the infant rat: comparison with the adult pattern. Life Sci 19: 825–836
54. Zapf J, Froesch ER, Humbel RE (1981) The insulin-like growth factors (IGF) of human serum: chemical and biological characterization and aspects of their possible physiological role. Curr Top Cell Regul 19: 257–309
55. Spencer EM (1979) Synthesis by cultured hepatocytes of somatomedin and its binding protein. FEBS Lett 99: 157–161
56. Wu C, Roberts EH, Bauer JM (1965) Enzymes related to glutamine metabolism in tumor-bearing rats. Cancer Res 25: 677–684
57. Wu C (1967) "Minimal deviation" hepatomas: a critical review of the terminology, including a commentary on the correlation of enzyme activity with growth rate of hepatomas. J Nat Cancer Inst 39: 1149–1154
58. Knox WE (1976) Enzyme patterns in fetal, adult and neoplastic rat tissues. Karger, Basel
59. Wu C, Bauer JM (1960) A study of free amino acids and of glutamine synthesis in tumor-bearing rats. Cancer Res 20: 848–857
60. Tiemeier DC, Milman G (1972) Regulation of glutamine synthetase in cultured chinese hamster cells. Induction and repression by glutamine. J Biol Chem 247: 5722–5727
61. Kulka RG, Cohen H (1973) Regulation of glutamine synthetase activity of hepatoma tissue culture cells by glutamine and dexamethasone. J Biol Chem 248: 6738–6743
62. Arad G, Kulka RG (1978) Effects of glutamine, methionine sulfon and dexamethasone on rates of synthesis of glutamine synthetase in cultured hepatoma cells. Biochim Biophys Acta 544: 153–162
63. Crook RB, Louie M, Deuel TF, Tomkins GM (1978) Regulation of glutamine synthetase by dexamethasone in hepatoma tissue culture cells. J Biol Chem 253: 6125–6131
64. Milman G, Portnoff LS, Tiemeier DC (1975) Immunochemical evidence for glutamine-mediated degradation of glutamine synthetase in cultured chinese hamster cells. J Biol Chem 250: 1393–1399
65. Arad G, Freikopf A, Kulka RG (1976) Glutamine-stimulated modification and degradation of glutamine synthetase in hepatoma tissue culture cells. Cell 8: 95–101
66. Crook RB, Tomkins GM (1978) Effect of glutamine on the degradation of glutamine synthetase in hepatoma tissue culture cells. Biochem J 176: 47–52
67. Freikopf A, Kulka RG (1976) Involvement of protein synthesis in the post translational control of glutamine synthetase activity of cultured hepatoma cells. Biochem Biophys Res Commun 72: 1195–1200
68. Freikopf-Cassel A, Kulka RG (1981) Role of the nucleus in the glutamine dependent degradation of glutamine synthetase in cultured hepatoma cells. FEBS Lett 124: 27–30
69. Freikopf-Cassel A, Kulka RG (1981) Regulation of the degradation of ^{125}I-labeled glutamine synthetase introduced into cultured hepatoma cells by erythrocyte ghost-mediated injection. FEBS Lett 128: 63–66
70. Wu C (1963) Glutamine synthetase. II. The intercellular localization in the rat liver. Biochim Biophys Acta 77: 428–493
71. Dierks-Ventling C, Cone AL, Bessman SP (1975) Attachment of glutamine synthetase to brain membranes. Biochem Med 13: 213–223
72. Tate SS, Meister M (1971) Regulation of rat liver glutamine synthetase: activation by α-ketoglutarate and inhibition by glycine, alanine, and carbamyl phosphate. Proc Natl Acad Sci USA 68: 781–785
73. Iqbal K, Wu C (1971) Glutamine synthetase. VIII. Purification and some properties of the enzyme from rat liver. Enzyme 12: 553–560
74. Meister A (1974) Glutamine synthetase of mammals. In: Boyer PD (ed) The enzymes, vol X. Academic Press, London New York, pp 699–754

75. Tiemeier DC, Milman G (1972) Chinese hamster liver glutamine synthetase. Purification, physical and biochemical properties. J Biol Chem 247: 2272–2277
76. Joseph SK, Bradford NM, McGivan JD (1979) Inhibition of glutamine synthetase activity by manganous ions in a cytosol extract of rat liver. Biochem J 184: 477–480
77. Lund P (1971) Control of glutamine synthesis in rat liver. Biochem J 124: 653–660
78. Lund P, Watford M (1976) Glutamine as a precursor of urea. In: Grisolia S, Baguena R, Mayor F (eds) The urea cycle. Wiley, London New York, pp 479–488
79. Lueck JD, Miller LL (1970) The effect of perfusate pH on glutamine metabolism in the isolated perfused rat liver. J Biol Chem 245: 5491–5497
80. Häussinger D, Sies H (1979) Hepatic glutamine metabolism under the influence of the portal ammonia concentration in the perfused rat liver. Eur J Biochem 101: 179–184
81. Häussinger D, Gerok W, Sies H (1983) Regulation of flux through glutamines and glutamine synthetase in isolated perfused rat liver. Biochim Biophys Acta 755: 272–278
82. Lund P (1980) Glutamine metabolism in the rat. FEBS Lett 117: K86–K92
83. Hems R, Stubbs M, Krebs HA (1968) Restricted permeability of rat liver for glutamate and succinate. Biochem J 107: 807–815
84. Gazzola GC, Dall Asta V, Bussolati O, Makowske M, Christensen HN (1981) A stereoselective anomaly in dicarboxylic amino acid transport. J Biol Chem 256: 6054–6059
85. Gebhardt R, Mecke D (1983) Glutamate uptake by cultured rat hepatocytes is mediated by hormonally inducible, sodium-dependent transport systems. FEBS Lett 161: 275–278
86. Sips HJ, DeGraaf PA, Van Dam K (1982) Transport of L-aspartate and L-glutamate in plasma-membrane vesicles from rat liver. Eur J Biochem 122: 259–264
87. Häussinger D, Gerok W (1983) Hepatocyte heterogeneity in glutamate uptake by isolated perfused rat liver. Eur J Biochem 136: 421–425
88. Joseph SK, Bradford NM, McGivan JD (1978) Characteristics of the transport of alanine, serine and glutamine across the plasma membrane of isolated rat liver cells. Biochem J 176: 827–836
89. Kilberg MS, Handlogten ME, Christensen HN (1980) Characteristics of an amino acid transport system in rat liver for glutamine, asparagine, histidine, and closely related analogs. J Biol Chem 255: 4011–4019
90. Handlogten ME, Kilberg MS, Christensen HN (1982) Incomplete correspondence between repressive and substrate action by amino acids on transport systems A and N in monolayered rat hepatocytes. J Biol Chem 257: 345–348
91. Arias I, Popper H, Schachter D, Shafritz DA (eds) (1982) The Liver, biology and pathology. Raven Press, New York
92. Ishikawa E (1975) The regulation of uptake and output of amino acids by rat tissues. Adv Enzyme Regul 14: 117–136
93. Rémésy C, Démigne C, Aufrère J (1978) Inter-organ relationship between glucose, lactate and amino acids in rats fed on high-carbohydrate or high-protein diets. Biochem J 170: 321–329
94. Häussinger D, Weiss L, Sies H (1975) Activation of pyruvate dehydrogenase during metabolism of ammonium ions in hemoglobin-free perfused rat liver. Eur J Biochem 52: 421–431
95. Verhoeven AJ, Van Iwaarden JF, Joseph SK, Meijer AJ (1983) Control of rat-liver glutaminase by ammonia and pH. Eur J Biochem 133: 241–244
96. Gebhardt R, Mecke D (1979) Permissive effect of dexamethasone on glucagon induction of urea-cycle enzymes in perifused primary monolayer cultures of rat hepatocytes. Eur J Biochem 97: 29–35
97. Lin RC, Snodgrass PJ, Rabier D (1982) Induction of urea cycle enzymes by glucagon and dexamethasone in monolayer cultures of adult rat hepatocytes. J Biol Chem 257: 5061–5067
98. Jefferson LS, Korner A (1967) A direct effect of growth hormone on the incorporation of precursors into proteins and nucleic acids of perfused rat liver. Biochem J 104: 826–832
99. Jefferson LS, Schworer CM, Tolman EL (1975) Growth hormone stimulation of amino acid transport and utilization by the perfused rat liver. J Biol Chem 250: 197–204

100. Lacey JH, Bradford NM, Joseph SK, McGivan JD (1981) Increased activity of phosphate-dependent glutaminase in liver mitochondria as a result of glucagon treatment of rats. Biochem J 194: 29–33
101. Corvera S, Garcia-Sainz JA (1983) Hormonal stimulation of mitochondrial glutaminase. Effects of vasopressin, angiotensin II, adrenalin and glucagon. Biochem J 210: 957–960
102. Palekar AG, Collip PJ, Maddaiah VI (1981) Growth hormone and rat liver mitochondria. Effects of urea cycle enzymes. Biochem Biophys Res Commun 100: 1604–1610
103. McLean P, Novello F (1965) Influence of pancreatic hormones on enzymes concerned with urea synthesis in rat liver. Biochem J 94: 410–422
104. McLean P, Gurney MW (1963) Effect of adrenalectomy and of growth hormone on enzymes concerned with urea synthesis in rat liver. Biochem J 87: 96–104
105. Jung W, Gebhardt R, Mecke D (1982) Alterations in activity and ultrastructural localization of several phosphatases on the surface of adult rat hepatocytes in primary monolayer culture. Eur J Cell Biol 27: 230–241

Liver Glutaminase

J. D. McGivan[1], N. M. Bradford[1], A. J. Verhoeven[2]
and A. J. Meijer[2]

Introduction

The original observation that extracts of liver contain considerable glutamine-hydrolysing activity was reported by Krebs in 1935 [1]. Errera [2] demonstrated that the major part of the glutaminase activity was dependent on added phosphate and was located in the particulate fraction of the liver. Definitive evidence that glutaminase is a mitochondrial enzyme was first obtained by Guha [3]. Liver extracts also contain a phosphate-independent glutamine-hydrolysing activity located in the supernatant fraction [2, 4–6]. This activity is stimulated by maleate and represents a partial reaction of γ-glutamyltransferase (EC 2.3.2.2) [7]. In a careful reinvestigation of the characteristics of glutamine hydrolysis in rat liver extracts, Horowitz and Knox [8] showed that 90% of the glutamine hydrolysing activity was phosphate-dependent and located in the mitochondria and was thus catalysed by the enzyme now classified as glutaminase (EC 3.5.1.2)

The properties of liver glutaminase differ markedly from those of the corresponding enzyme found in kidney. The enzymes from kidney and several other tissues are inhibited by glutamate, whereas the liver enzyme is not [1]. Liver glutaminase has a much higher K_m for glutamine than does the kidney enzyme, and is fully activated at lower concentrations of inorganic phosphate [8]. The enzymes have different molecular weights, and the kidney enzyme, but not the liver enzyme, polymerises in phosphate-borate buffer [9]. Curthoys et al. [10] found that antisera prepared against purified kidney glutaminase did not precipitate solubilised glutaminase from liver mitochondria, showing clearly that the liver enzyme is a different protein from that in kidney.

The submitochondrial location of glutaminase in rat liver mitochondria was studied by McGivan et al. [11]. It was found that when the mitochondria were disrupted by repeated freezing and thawing, glutaminase was recovered only in the pellet fraction, while matrix marker enzymes such as malate dehydrogenase were largely released. Sonication of the mitochondria led to solubilisation of the enzyme, in agreement with the earlier work of Kalra and Brosnan [12]. Solubilisation of the enzyme increased the apparent K_m for glutamine from 6 m*M* to 21 m*M*. These results were interpreted to indicate that glutaminase was loosely associated with the mitochondrial inner membrane.

1 Department of Biochemistry, University of Bristol Medical School, Bristol, England
2 Laboratory of Biochemistry, B.C.P. Jansen Institute, University of Amsterdam, P.O. Box 20151, 1000 HD Amsterdam, The Netherlands

Glutamine Metabolism in Mammalian Tissues
Edited by D. Häussinger and H. Sies

In this review on liver glutaminase we have chosen to discuss its properties in systems of increasing complexity, namely the enzyme in (partially) purified form, in various preparations of disrupted mitochondria, in intact liver mitochondria and finally the enzyme as it functions in the intact hepatocyte. However, the reader should bear in mind that historically most research was done in the reverse order.

Properties of Glutaminase at the Submitochondrial Level

The only report on the properties of purified rat liver glutaminase in the literature appears to be that of Huang and Knox [9]. In this study, glutaminase was partially purified by extracting lyophilised mitochondria with phosphate-borate buffer, followed by ammonium sulphate fractionation and chromatography on Sepharose 4B, giving a four-fold purification over the mitochondria extract. Glutaminase was found to have a mol. wt. of 170,000 as judged by sucrose density centrifugation. The K_a for phosphate was found to be 5 m*M*, and the K_m for glutamine was 28 m*M*.

We have recently partially purified glutaminase from rat liver by DEAE cellulose chromatography (J. D. McGivan and N. M. Bradford, unpublished work). This procedure leads to an eight- to ten-fold purification over intact mitochondria. The enzyme has a mol. wt. of approx. 320,000 as judged by chromatography on Sephacryl S-300, and readily loses activity on dilution, probably due to dissociation of subunits. The properties of this preparation will be compared with those of the enzyme in various preparations of disrupted mitochondria.

Dependence of Glutaminase Activity on the Concentration of Glutamine and Phosphate

In the presence of saturating concentrations of phosphate and NH_4Cl (see next section), the dependence of partially purified glutaminase on glutamine is highly sigmoidal, with little activity below 10 m*M* and a half-maximum rate at 22 m*M* glutamine. There was no activity in the absence of phosphate. Half-maximum activation was obtained at 5 m*M* phosphate.

The properties of glutaminase in mitochondria disrupted by freezing and thawing differed markedly from those of the purified enyzme. The dependence on glutamine was hyperbolic with a K_m of 6 m*M*. In this preparation, 70% of the maximal activity was obtained in the absence of added phosphate (cf. Fig. 1 and [11]).

Figure 1 also shows an experiment in which a suspension of frozen and thawed mitochondria was treated with phospholipase A_2, which had the effect of solubilising the enzyme. As a result of phospholipase treatment, the glutamine dependence changed from hyperbolic to sigmoidal, and the activity became completely dependent on added phosphate. Re-addition to the phospholipase extract of phosphatidyl choline or phosphatidyl ethanolamine (but not other

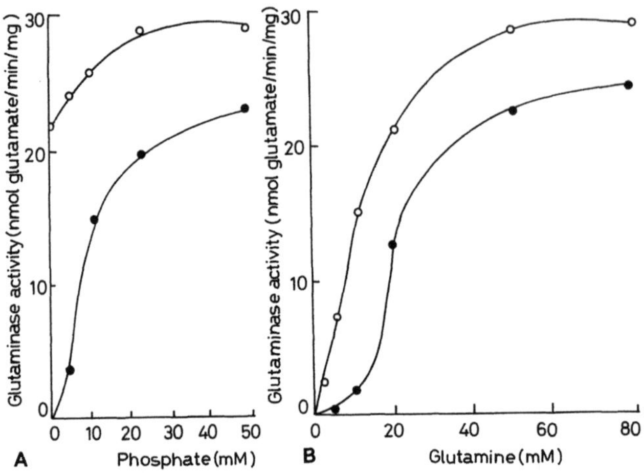

Fig. 1A, B. Effect of phospholipase A_2 on the activity of glutaminase in frozen and thawed liver mitochondria. Mitochondria were disrupted by freezing and thawing. An aliquot of the suspension was treated with 5 U/ml of phospholipase A_2 in the presence of 1 m*M* $CaCl_2$ for 10 min at 30° C. 10 m*M* EGTA was added, the suspension was centrifuged at 40,000 *g* for 20 min and glutaminase activity was assayed in the extract. The assay medium was 50 m*M* Tris-HCl, 2 m*M* NH_4Cl and **A** 50 m*M* glutamine plus various concentrations of phosphate; **B** 50 m*M* phosphate plus various concentrations of glutamine at pH 8.0 and 30° C. ○——○ frozen and thawed mitochondria; ●——● phospholipase supernatant

phospholipids) partially but not completely restored the kinetics of the enzyme to those seen in the frozen and thawed system [13]. Phospholipids had little effect on the activity of purified glutaminase. It appears that the kinetics of glutaminase are markedly different when this enzyme is associated with the mitochondrial membrane. Some membrane component in addition to phospholipids must be involved in the protein-membrane interaction.

Dependence on Ammonia

It was shown by Charles [14] in 1968 that added NH_4Cl increased the rate of glutamine hydrolysis in intact mitochondria. It was later shown that NH_4Cl activated glutaminase also in mitochondria disrupted by freezing and thawing, indicating that ammonia exerted a direct effect on the enzyme [15]. With the same preparation and using short incubation periods in order to avoid autoactivation of the enzyme due to accumulating ammonia, Verhoeven et al. [16] recently demonstrated that glutaminase is not only stimulated by, but actually completely dependent on the presence of ammonia for its activity.

The activation of glutaminase by NH_4Cl in sonicated mitochondria has also been studied in some detail [16a]. Investigation of the ammonia dependence is complicated by the fact that commercially available glutamine contains approximately 0.3% ammonia as an impurity. Figure 2 shows an experiment in

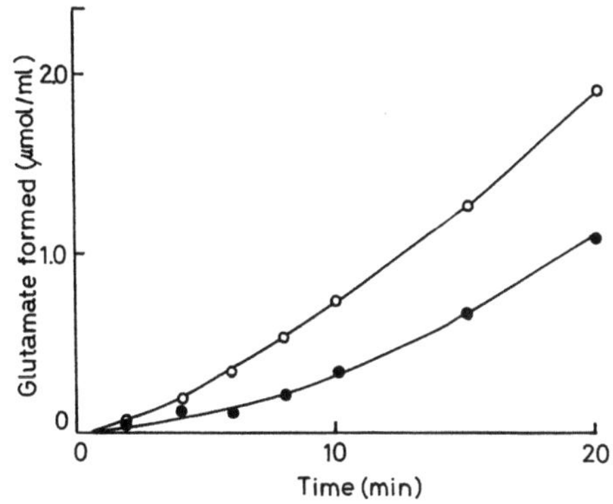

Fig. 2. Time course of the glutaminase reaction in sonicated mitochondria and effect of added NH_4Cl. Sonicated mitochondria (3 mg protein/ml) were incubated at pH 7.3 and 30° C with 50 m*M* glutamine plus 50 m*M* phosphate. The initial concentration of endogenous ammonia was 0.18 m*M*. Samples of the incubation were deproteinised at various times and glutamate was measured enzymatically in the neutralised extracts. ●——● control; ○——○ + 1 m*M* NH_4Cl

which sonicated mitochondria were incubated with saturating concentrations of glutamine and phosphate at pH 7.3 in the absence of added ammonia. The time course of glutamate production was not linear, and there was a prolonged lag phase, during which the rate gradually increased to a final steady value (cf. [16]). Addition of NH_4Cl decreased the lag phase without affecting the final steady rate. Under these conditions, the activity of glutaminase was dependent solely on the small amount of endogenous ammonia present in the glutamine and the mitochondria. The enzyme was subsequently autoactivated by the ammonia produced during the reaction.

The activating effect of a given concentration of NH_4Cl increased with increasing pH [16, 16a]. At the pH optimum of glutaminase (pH 8), the enzyme was already fully activated by the endogenous ammonia present in the glutamine and no effect of added NH_4Cl was observed. This finding probably accounts for the failure of earlier investigators to detect the requirement of the enzyme for ammonia. Analysis of the time courses of glutamate production at different pH values and different initial glutamine concentrations showed that, in the pH range 7.0–7.7, a half-maximum activation was obtained at a constant concentration of free NH_3 of 8 μ*M* which was independent of pH; but deviations below pH 7.0 are not excluded [16]. The concentrations of added NH_4Cl required to give half-maximum activation decreased markedly with increasing pH.

All these results are consistent with a mechanism in which ammonia acts as an obligatory activator of glutaminase, and the activating species is NH_3 rather than NH_4^+.

Preliminary work has indicated that partially purified glutaminase also requires ammonia as an obligatory activator (J. D. McGivan and N. M. Bradford, unpublished observations). However, much higher concentrations of NH_4Cl were necessary to activate the purified enzyme. Again the requirement for added NH_4Cl decreased with increasing pH. At pH 7.2, the endogenous ammonia present in the glutamine was insufficient to produce a significant activation, and the enzyme was completely inactive in the absence of added NH_4Cl.

Effect of Other Activators

It has been shown that addition of bicarbonate increases glutaminase activity in frozen and thawed mitochondria [11, 15]. However, recent work on partially purified glutaminase (J. D. McGivan and N. M. Bradford, unpublished observations) has shown that bicarbonate does not activate the enzyme when ammonia is excluded from the medium and also has no effect in the presence of a saturating ammonia concentration. It is probable that bicarbonate does not exert a direct activating effect on glutaminase. The activation seen in frozen and thawed mitochondria can be attributed to the effect of bicarbonate in increasing the pH of the incubation (due to loss of CO_2) and hence potentiating the activating effect of the endogenous ammonia present. Leucine, which stimulates flux through glutaminase in hepatocytes [17], was found to have no effect on the purified enzyme.

Glutamine Metabolism in Intact Liver Mitochondria

Glutamine Transport

Transport of glutamine across the liver mitochondrial membrane is electroneutral and driven by the concentration gradient [18]. It is presumably carrier-mediated as indicated by the stereospecificity of the transport and its sensitivity to certain sulphhydryl binding reagents [18, 19].

Transport of glutamine presumably plays only a minor role in the control of glutamine degradation by intact mitochondria since glutamine equilibrates very rapidly across the mitochondrial inner membrane, and rates of transport are too fast to be resolved by conventional silicone layer centrifugation techniques [15, 19, 20]. However, quantitative information about the control exerted by glutamine transport on mitochondrial glutamine degradation can only be obtained by careful titration with an inhibitor of glutamine transport, as has been carried out by Groen et al. [21] in order to quantitate the control exerted by the adenine nucleotide translocator on mitochondrial respiration. Unfortunately, the well-known inhibitor of glutamine transport in mitochondria, mersalyl [18], lacks specificity in liver mitochondria because it also inhibits glutaminase [19], whereas in kidney mitochondria mersalyl appears to inhibit glutamine transport only [22].

Dependence on the Concentration of Glutamine and Phosphate

Hird and Marginson [23] and Blackburn and Hird [24] observed that glutamine was a relatively poor respiratory substrate for rat liver mitochondria and that high concentrations of glutamine were necessary to obtain maximum rates of respiration. The rate of glutamine metabolism showed a sigmoidal dependence on glutamine concentration.

In later investigations, mitochondrial glutamine hydrolysis was measured by following the production of glutamate by isolated mitochondria in the presence of rotenone which inhibits further metabolism of this compound. In the presence of an energy source and NH_4Cl or $KHCO_3$ (see next section), the kinetics of glutamine hydrolysis were dependent on the osmolarity of the medium. At an osmolarity of 295 mOsmol the dependence on glutamine was highly sigmoidal with a $K_{0.5}$ of approx. 15 m*M* [15, 25]. As the osmolarity of the medium was decreased, the concentration dependence became less sigmoidal, until at 95 mOsmol the dependence was hyperbolic with a K_m of 6 m*M* [25]. At this osmolarity the mitochondria were massively swollen and the concentration dependence corresponded to that observed in frozen and thawed mitochondria.

The phosphate dependence was also a function of osmolarity. In the range 225–325 mOsmol, glutamine hydrolysis was entirely dependent on the addition of phosphate. The concentration of phosphate required for half-maximum velocity decreased from 6 m*M* at 325 mOsmol to 1 m*M* at 225 mOsmol. At lower osmolarities considerable activity was observed in the absence of added phosphate, and below 100 mOsmol 70% of the total activity was independent of added phosphate (J. D. McGivan and N. M. Bradford, unpublished work), again corresponding to the situation in mitochondria disrupted by freezing and thawing. These results could not be attributed to greatly altered distribution of glutamine or phosphate across the mitochondrial membrane.

Activation by Ammonia and Bicarbonate

As shown by Verhoeven et al. [16] glutaminase activity in intact rat liver mitochondria, like that in broken mitochondria (see Sect. Dependence on Ammonia), is completely dependent on the presence of ammonia for its activity. Activation by ammonia was strongly depressed by decreasing the pH of the medium from 7.0 to 6.8 and stimulated by bicarbonate. At a pH value of 7.0, the cytosolic pH, and in the presence of bicarbonate, half-maximum activation of glutaminase occurred at an extramitochondrial concentration of ammonia of about 0.4 m*M* [16] which is in the physiological range (see Sect. Ammonia). Deaciuc and Petrescu [26] failed to observe activation of glutaminase by added ammonia in isolated guinea pig liver mitochondria, incubated at pH 7.4 for 10 min. Although species differences cannot be ruled out, it may also be that the glutaminase was very rapidly autoactivated by accumulating ammonia, especially since at pH 7.4 glutaminase, at least in the case of the rat liver enzyme, has a very high affinity for ammonia (see Sect. Dependence on Ammonia).

Other Factors Influencing Glutaminase Activity in Intact Liver Mitochondria

Lacey et al. [25] and Joseph et al. [27] showed that glutaminase activity is greatly stimulated when liver mitochondria are incubated under hypoosmotic conditions

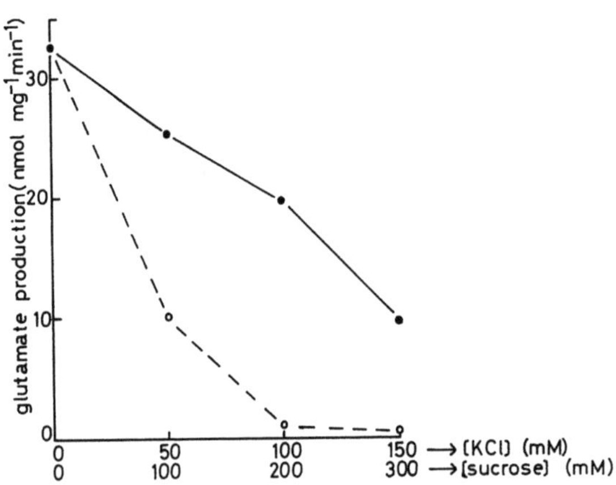

Fig. 3. Effect of increase in osmolarity of the incubation medium on glutaminase activity in intact rat liver mitochondria. Intact rat liver mitochondria were incubated as described in Ref. 16 in the presence of 5 mM NH_4Cl, except that in one set of incubations the KCl concentration was varied *(solid line)* and in another set KCl was replaced by different concentrations of sucrose *(dotted line)*

(cf. Sect. Dependence on the Concentration of Glutamine and Phosphate), which suggests that the activity of glutaminase is sensitive to the volume of the mitochondrial matrix space. Conversely, glutaminase is strongly inhibited when the total osmolarity of the medium is increased (see Fig. 3). Under these conditions a difference is observed between a sucrose-based and a KCl-based medium. Probably this difference reflects the greater volume of the mitochondrial matrix space in a KCl medium, due to energy-dependent influx of K^+ ions [28, 29]. When the permeability for K^+ ions is artificially increased by the addition of the ionophore valinomycin, glutaminase activity is also stimulated, due to swelling of the mitochondria [16]. The great sensitivity of glutaminase to internal volume changes of the mitochondria suggests that the known stimulation of glutaminase activity by energization of the mitochondrial membrane [30] is mediated by a change in the volume of the mitochondrial matrix. From a titration with FCCP in the presence of saturating concentrations of ammonia (see Fig. 4), it can be seen that under these conditions glutaminase is inhibited by 30% upon complete uncoupling of the respiratory chain. The possibility that the sensitivity to uncoupler reflects a small K^+-dependent swelling in the presence of a membrane potential, resulting in a higher glutaminase activity, can as yet not be excluded.

The underlying mechanism of the sensitivity of glutaminase to changes of the internal volume of the mitochondria remains unclear. However, the association of the enzyme with the mitochondrial inner membrane [11] and its interaction with membrane phospholipids (see Sect. Dependence of Glutaminase Activity on the Concentration of Glutamine and Phosphate) suggest that the activity of glutaminase reflects an interaction with the mitochondrial inner membrane. When volume changes are imposed on the mitochondria (for instance by changing the osmolarity of the medium, a change in properties of the membrane can be expected, due to a different arrangement of the membrane phospholipids. In this respect, glutaminase has regulatory features similar to those of the enzymes of the respiratory chain, which also have been shown to be influenced by the volume of the mitochondria (see e.g., [31]).

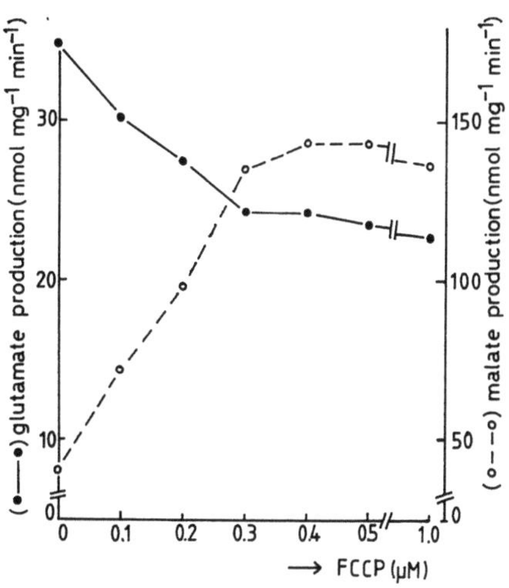

Fig. 4. Effect of uncoupler on glutaminase activity in intact rat liver mitochondria. Intact rat liver mitochondria were incubated for 10 min at 30° C in a medium containing 100 m*M* KCl, 20 m*M* Tris-HCl, 10 m*M* Tris-phosphate, 10 m*M* Tris-succinate, 2 μg/ml rotenone, 15 m*M* glutamine, 5 m*M* NH_4Cl, 1 m*M* EGTA and different concentrations of carbonyl cyanide p-trifluoromethoxyphenylhydrazone (FCCP). The pH was 7.4

Effects of Hormones on Mitochondrial Glutamine Hydrolysis

Glucagon has been shown to increase the rate of gluconeogenesis from glutamine in hepatocytes, and this effect was attributed to an activation of mitochondrial glutaminase [30]. The hormones vasopressin, adrenaline, phenylephrine and angiotensin II have also been found to stimulate glutamine metabolism in hepatocytes [32–34]. The same is true for thyroxine (A. J. Verhoeven, unpublished observations).

Lacey et al. [25] observed that glutaminase activity in mitochondria isolated from the livers of glucagon-injected rats was increased over that in control mitochondria. The increased activity was characterised by a decreased requirement for phosphate at a given osmolarity. The glutamine dependence also became less sigmoidal. In this respect, the effect of glucagon could be mimicked by suspending control mitochondria in a medium of slightly lower osmolarity. At very low osmolarities and in disrupted mitochondria, the effect of glucagon was lost.

Recently, Corvera and Garcia-Sanchez [35] have added various hormones to isolated hepatocytes and measured the activity of glutaminase in mitochondria subsequently isolated from these cells. Under these conditions, adrenaline, glucagon, vasopressin and angiotensin II were all found to lead to increased mitochondrial glutaminase activity.

The molecular basis of the activation of liver glutaminase by hormones has not yet been elucidated. It is probable that modification of the interaction of the enzyme with the mitochondrial membrane is an important factor in this mechanism. Such a modification could be the result of a hormone-induced change in membrane phospholipids, as described for thyroid hormone [36], or a hormone-induced increase in matrix volume, as described for glucagon [29].

Glutamine Degradation in Intact Liver Cells

Transport of Glutamine Across the Plasma Membrane and Intrahepatic Degradation of Glutamine

Before glutamine can be degraded by the hepatocyte, it must cross the plasma membrane. Glutamine transport across the plasma membrane of the hepatocyte is predominantly Na^+-dependent and is mediated by a system that is distinct from the amino acid transport system A, the Na^+-dependent system responsible for transport of amino acids like alanine, and system ASC, the Na^+-dependent transport system for transport of amino acids like cysteine [37–39].

From the published kinetic parameters of the glutamine transport system (K_m = 1.1 m*M*; V_{max} = 8 μmol per min per g intracellular water; [38]), it can be calculated that at physiological glutamine concentrations in the portal vein (0.4–0.6 m*M* [40, 41], transport of glutamine across the plasma membrane is sufficiently rapid to accommodate observed rates of glutamine degradation, which range from 0.07 to 0.38 μmol per g wet weight of liver per min [42]. In support of this, both in vivo [43] and in vitro with isolated hepatocytes [44, 30] the intracellular glutamine concentration greatly exceeds that in the extracellular fluid.

The glutamine transport system is not subject to regulation by either insulin [38] or glucagon [30, 38]. The lack of effect by glucagon is of particular interest since glucagon does stimulate glutamine degradation in hepatocytes [30]. Thus, the rate of glutamine breakdown appears not to be controlled by the rate of transport but by the rate of intracellular metabolism [30].

Apart from its degradation by mitochondrial phosphate-dependent glutaminase (known as "glutaminase I"), glutamine car also be degraded in the cytosol by the phosphate-independent glutaminase (see Introduction) or by the combined action of glutamine transaminase (EC 2.6.1.15) and α-keto acid ω-amidase (EC 3.5.1.3) (this combination is known as "glutaminase II"). The function of the latter system is presumably the salvage of essential amino acids by reamination of the 2-keto acids derived from them [44–46]. It is unlikely that these cytosolic glutaminases are involved in hepatic glutamine degradation, for two reasons. Firstly, the transaminase inhibitor aminooxyacetate, which inhibits glutamine transaminase [44], has no effect on the rate of conversion of glutamine to glutamate and ammonia, as observed in guinea pig liver [26] and in isolated rat hepatocytes (A. J. Meijer, unpublished results). Secondly, as will be discussed below, intrahepatic glutamine degradation is greatly accelerated by factors which are known to affect the activity of the mitochondrial glutaminase.

Factors Affecting the Activity of Mitochondrial Glutaminase in Perfused Liver and in Isolated Hepatocytes

At physiological glutamine concentrations (< 1 m*M*) hepatic degradation of glutamine is very slow [44, 47, 48], unless mitochondrial glutaminase is properly activated. As discussed in previous sections, several factors may stimulate the activity of this enzyme. Another factor that may play a role is the mitochondrial redox state which determines the fate of the products of glutaminase [26].

Because of the complexity of the regulation of glutaminase in the intact hepatocyte, a discussion of some of these various factors is given below.

Phosphate

Although mitochondrial glutaminase is absolutely dependent on phosphate for activity, it is unlikely that glutaminase activity in the intact hepatocyte is controlled by variations in intramitochondrial phosphate which appears to be rather constant [49]. Moreover, an increase in the intramitochondrial phosphate concentration brought about by increasing the extracellular phosphate concentration (cf. [50]) does not affect consumption of glutamine in the perfused liver [48]. However, it is possible that the activity of glutaminase is decreased under pathological conditions, such as in hypophosphatemia, or in experimental situations where intrahepatic phosphate is depleted by large doses of glycerol or fructose.

Ammonia

In contrast to phosphate, ammonia – the other essential activator of mitochondrial glutaminase (Sects. Dependence on Ammonia and Effect of Other Activators) – does play a very important role in controlling the activity of the enzyme in the intact hepatocyte. The concentration of ammonia required for half-maximum activation of glutaminase in the intact hepatocyte is about 0.2 m*M* [30, 42], similar to the concentration of ammonia required to give half-maximum activation of the enzyme in isolated liver mitochondria in the presence of HCO_3^- ions [16]; this value is about equal to the ammonia concentration in the portal vein [51]. The well-known lag in glutamine degradation in rat liver [48, 52] is likely to be due to autoactivation of glutaminase by accumulating ammonia, for the following reasons. Firstly, the lag is abolished upon addition of ammonia [30, 53, 54]. Secondly, exactly the same lag is observed in isolated liver mitochondria unless excess ammonia is added; at any point during the lag period the amount of ammonia accumulating in the mitochondrial suspension completely explains the observed glutaminase activity [16] (see also Sect. Dependence on Ammonia). In vivo activation of glutaminase by ammonia is indicated by the decrease in the intrahepatic glutamine concentration upon injection of NH_4Cl in the rat [43, 55].

Whereas the value of 0.2 m*M* reflects the K_a of glutaminase for added ammonia, in the intact hepatocyte it is ultimately the intramitochondrial concentration of ammonia that determines the activity of the enzyme. In this respect we recently made an unexpected observation (A. J. Meijer and G. M. van Woerkom, unpublished results). When rat hepatocytes were incubated with 5 m*M* glutamine and 3 m*M* ornithine, in the absence of added ammonia, glutamine was degraded at maximal speed after the initial lag period (about

20 min) was over; however, the concentration of ammonia in the extracellular fluid under these conditions was below 0.10 m*M*. The rate of urea production from glutamine was 250 µmol/g dry weight/h, which is about 25% of the V_{max} of carbamoylphosphate synthetase in our preparations [56]. In other words, one must conclude that the intramitochondrial concentration of ammonia must have been 1 m*M*, when one considers the fact that the K_m value for ammonia of carbamoylphosphate synthetase is about 2 m*M* for the rat liver enzyme [57]; this concentration of ammonia is sufficient for complete activation of mitochondrial glutaminase. In hepatocytes, the intramitochondrial pH is equal to, or slightly higher than, the extracellular pH [58, 59]. Apparently, under the condition of this experiment with ammonia generated inside the mitochondria, the intramitochondrial concentration of free ammonia was higher than the value expected on the basis of existing pH gradients across the mitochondrial and plasma membranes (cf. [60]). Unfortunately, direct measurement of the intramitochondrial concentration of free ammonia is technically very difficult due to the presence of a substantial pool of unreactive, inert, ammonia [60, 61].

Bicarbonate

Baverel and Lund [17] have shown that in isolated hepatocytes bicarbonate, when added at constant pH in the absence of added ammonia, accelerates glutamine degradation. On the other hand, Häussinger et al. [54] did not observe a stimulation by bicarbonate in the perfused liver; in their experiments, however, glutaminase was pre-activated with a high concentration of added ammonia.

These results are essentially in agreement with data obtained in isolated liver mitochondria, showing that bicarbonate stimulates flux through glutaminase only in the presence of low, but not of high, concentrations of ammonia [16].

Leucine

Leucine, and other branched-chain amino acids, have been reported to stimulate glutamine degradation in hepatocytes [17, 62]. The effect of leucine on glutaminase in isolated mitochondria has not yet been investigated. However, leucine has no effect on the activity of the partially purified enzyme (See Sect. Effect of Other Activators). Since leucine inhibits urea cycle activity by affecting one or more of the urea cycle enzymes [63–65], it is possible that stimulation of flux through glutaminase is due to accumulating ammonia. This idea is supported by the fact that leucine activation of glutamine degradation in hepatocytes is partly abolished by ornithine addition (A. J. Meijer and G. M. van Woerkom, unpublished results).

pH

Glutamine degradation in liver is strongly inhibited at low pH values [42, 48, 54, 66, 67]; this phenomenon is independent of a change in the bicarbonate concentration [54, 67]. It is likely that activation of hepatic glutamine degradation by OH^- is due to an increased affinity of glutaminase for ammonia (see Sect. Dependence of Ammonia and Ref. 16]. In support of this proposal is the recent demonstration by Häussinger et al. [42] that in livers perfused with glutamine plus ammonia at physiological concentrations (both at 0.6 m*M*) flux through glutaminase was increased five-fold when the pH of the perfusate was increased from 7.1 to 7.4. On the other hand, when ammonia was replaced by glucagon, flux through glutaminase remained very low unless the pH was raised above 7.5 [42].

Glucagon

Among the various hormones mentioned before, only glucagon has been studied extensively in relation to glutamine metabolism in intact liver cells. Evidence for activation of mitochondrial glutaminase by glucagon stems from the observation that addition of the hormone to isolated rat hepatocytes incubated with glutamine causes an increase in the rate of gluconeogenesis, accompanied by a decrease in intracellular glutamine and an increase in intracellular glutamate [30]. Also, in the perfused rat liver glucagon decreases intrahepatic glutamine [68]. Liver mitochondria isolated from rats injected with the hormone have an enhanced activity of glutaminase (see Sect. Effects of Hormones on Mitochondrial Glutamine Hydrolysis).

The mechanism of the effect of glucagon is unknown (see Sect. Effects of Hormones on Mitochondrial Glutamine Hydrolysis). From the fact that the effects of glucagon and ammonia on glutamine breakdown in the intact hepatocyte are additive, one may conclude that their modes of action are different [30, 42].

It has been suggested that the stimulation by glucagon of urea synthesis from glutamine plus ornithine in isolated rat hepatocytes is due to enhanced mitochondrial ATP production [69]. We consider this highly unlikely because in that case one would expect a large accumulation of ammonia in the absence of the hormone and this is not observed under these conditions [70].

Mitochondrial Redox State

The main carbon products of glutamine degradation in liver are glucose, glutamate, alanine and CO_2. Which of these products predominates depends on the experimental conditions [17, 26, 52]. It is important to note that during complete conversion of glutamine to urea and glucose more ATP is produced than is utilised so that, particularly at high glutamine concentrations when glutaminase flux is high, glutamate may accumulate [52]. Indeed, very low concentrations of the uncoupler 2,4-dinitrophenol increase production of urea and glucose from glutamine [26].

Concluding Remarks

From the results discussed it is clear that liver glutaminase is not only stimulated by but actually completely dependent on the presence of ammonia for its activity.

Recent findings concerning the exclusive localisation of liver glutamine synthetase in perivenous hepatocytes [71, 72] and the localisation of glutaminase primarily in periportal hepatocytes [72], provide a rationale for the physiological function of the specific ammonia dependence of liver mitochondrial glutaminase. Apparently, periportal hepatocytes cannot synthesize their own glutamine for processes such as purine and pyrimidine synthesis and for salvage of essential amino acids (cf. Sect. Transport of Glutamine Across the Plasma Membrane . . .), and are therefore completely dependent on the glutamine supply via the portal vein. Thus, when the amino acid supply by the portal vein is low, glutamine, instead of being converted to glucose, must be saved for the functions mentioned above. The intramitochondrial ammonia concentration, which can be expected to be relatively low under these conditions, could then act as a modulator of the activity of the degrading enzyme, glutaminase.

In addition to glutaminase, carbamoyl-phosphate synthetase (ammonia) (EC 6.3.4.16) also appears to be localized in periportal hepatocytes only [73]. Since both enzymes are mitochondrial, it is tempting to speculate that in case of a high nitrogen supply in the portal vein, glutaminase provides glutamate for the (mitochondrial) synthesis of N-acetylglutamate, the essential activator of carbamoyl-phosphate synthetase (see [56] for a review).

References

1. Krebs HA (1935) Metabolism of amino acids. The synthesis of glutamine from glutamic acid and ammonia, and the enzymic hydrolysis of glutamine in animal tissues. Biochem J 29: 1951–1959
2. Errera M (1949) Liver glutaminases. J Biol Chem 178: 483–493
3. Guha SR (1962) Glutaminases of gluinea-pig liver mitochondria: on the possible nature of phosphate and non-phosphate activated glutaminases of guinea-pig liver mitochondria. Enzymologia 24: 310–326
4. Katunuma N, Huzino A, Tomino I (1966) Organ specific control of glutamine metabolism. Adv Enzyme Regul 5: 55–69
5. Pestana A, Marco R, Sols A (1968) Allosteric glutaminase in rat liver of likely regulatory significance. FEBS Lett 1: 317–319
6. Katsunuma T, Temma M, Katunuma N (1968) Allosteric nature of a glutaminase isozyme in rat liver. Biochem Biophys Res Commun 32: 433–437
7. Tate SS, Meister A (1975) Identity of maleate-stimulated glutaminase with γ-glutamyl transpeptidase in rat kidney. J Biol Chem 250: 4619–4627
8. Horowitz ML, Know WE (1968) A phosphate activated glutaminase in rat liver different from that in kidney and other tissues. Enzymol Biol Clin 9: 241–255

9. Huang YZ, Knox WE (1976) A comparative study of glutaminase isozymes and rat tissues. Enzyme 21: 408–426
10. Curthoys NP, Kuhlenschmidt T, Godfrey SS, Weiss RF (1976) Phosphate-dependent glutaminase from rat kidney. Cause of increased activity in response to acidosis and identity with glutaminase from other tissues. Arch Biochem Biophys 172: 162–167
11. McGivan JD, Lacey JH, Joseph SK (1980) Localization and some properties of phosphate-dependent glutaminase in disrupted liver mitochondria. Biochem J 192: 537–542
12. Kalra J, Brosnan JT (1973) Localization of glutaminase in rat liver. FEBS Lett 37: 325–328
13. McGivan JD, Bradford NM (1983) Influence of phospholipids on the activity of phosphate-dependent glutaminase in extracts of rat-liver mitochondria. Biochem J 214: 649–652
14. Charles R (1968) Mitochondriale citrulline synthese: een ammoniak fixerend en ATP verbruikend proces. PhD thesis, Univ Amsterdam, Rototype, Amsterdam
15. Joseph SK, McGivan JD (1978) The effects of ammonium chloride and bicarbonate on the activity of glutaminase in isolated liver mitochondria. Biochem J 176: 837–844
16. Verhoeven AJ, Van Iwaarden JF, Joseph SK, Meijer AJ (1983) Control of rat-liver glutaminase by ammonia and pH. Eur J Biochem 133: 241–244
16a. McGivan JD, Bradford NM (1983) Characteristics of the activation of glutaminase by ammonia in sonicated rat-liver mitochondria. Biochim Biophys Acta 759: 296–302
17. Baverel G, Lund P (1979) A role for bicarbonate in the regulation of mammalian glutamine metabolism. Biochem J 184: 599–606
18. Kovacevic Z (1978) Mechanism of carrier-mediated glutamine transport across the inner mitochondrial membrane. In: Guder WG, Schmidt U (eds) Biochemical nephrology. Huber, Bern, pp 254–264
19. Joseph SK, Meijer AJ (1981) The inhibitory effects of sulphydryl reagents on the transport and hydrolysis of glutamine in rat-liver mitochondria. Eur J Biochem 119: 523–529
20. Kovacevic Z, Bajin K (1982) Kinetics of glutamine efflux from liver mitochondria loaded with the ^{14}C-labeled substrate. Biochim Biophys Acta 687: 291–295
21. Groen AK, Wanders RJA, Westerhoff HV, Van der Meer R, Tager JM (1982) Quantification of the contribution of various steps to the control of mitochondrial respiration. J Biol Chem 257: 2754–2757
22. Kovacevic Z (1976) Importance of the flux of phosphate across the inner membrane of kidney mitochondria for the activation of glutaminase and the transport of glutamine. Biochim Biophys Acta 430: 399–412
23. Hird FJR, Marginson MA (1968) The formation of ammonia from glutamine and glutamate by mitochondria from rat liver and kidney. Arch Biochem Biophys 127: 718–724
24. Blackburn EH, Hird FJR (1972) Metabolism of glutamine and glutamate by rat liver mitochondria. Arch Biochem Biophys 152: 258–264
25. Lacey JH, Bradford NM, Joseph SK, McGivan JD (1981) Increased activity of phosphate-dependent glutaminase in liver mitochondria as a result of glucagon treatment of rats. Biochem J 194: 29–33
26. Deaciuc IV, Petrescu I (1980) Regulation of glutamine catabolism in the perfused guinea-pig liver in relation to ureogenesis and gluconeogenesis. Int J Biochem 12: 605–618
27. Joseph SK, McGivan JD, Meijer AJ (1981) The stimulation of glutamine hydrolysis in isolated rat liver mitochondria by Mg^{2+} depletion and hypo-osmotic incubation conditions. Biochem J 194: 35–41
28. Azzi A, Azzone GF (1966) Swelling and shrinkage phenomena in liver mitochondria. Reversible swelling changes linked to transport of monovalent cations stimulated by valinomycin. Biochim Biophys Acta 113: 445–456
29. Halestrap AP (1978) Stimulation of pyruvate transport in metabolizing mitochondria through changes in the transmembrane pH gradient induced by glucagon treatment of rats. Biochem J 172: 389–398
30. Joseph SK, McGivan JD (1978) The effect of ammonium chloride and glucagon on the metabolism of glutamine in isolated liver cells from starved rats. Biochim Biophys Acta 543: 16–28
31. Armston AE, Halestrap AP, Scott RD (1982) The nature of the changes in liver mitochondrial function induced by glucagon treatment of rats. The effects of intramitochondrial volume, aging and benzyl alcohol. Biochim Biophys Acta 681: 429–439

32. Joseph SK, Verhoeven AJ, Meijer AJ (1981) Effect of trifluoperazine on the stimulation by Ca^{2+}-dependent hormones of gluconeogenesis from glutamine in isolated hepatocytes. Biochim Biophys Acta 677: 506–511
33. Corvera S, García-Sáinz JA (1981) α_1-Adrenoreceptor activation stimulates ureogenesis in rat hepatocytes. Eur J Pharmacol 72: 387–390
34. Corvera S, García-Sáinz JA (1982) Vasopressin and angiotensin II stimulate ureogenesis through increased mitochondrial citrulline production. Life Sci 31: 2493–2498
35. Corvera S, García-Sáinz JA (1983) Hormonal stimulation of mitochondrial glutaminase. Effects of vasopressin, angiotensin II, adrenaline and glucagon. Biochem J 210: 957–960
36. Hulbert AJ, Augee ML, Raison JK (1976) The influence of thyroid hormones on the structure and function of mitochondrial membranes. Biochim Biophys Acta 455: 597–601
37. Joseph SK, Bradford NM, McGivan JD (1978) Characteristics of the transport of alanine, serine and glutamine across the plasma membrane of isolated rat liver cells. Biochem J 176: 827–836
38. Kilberg MS, Handlogten ME, Christensen HN (1980) Characteristics of an amino acid transport system in rat liver for glutamine, asparagine, histidine, and closely related analogs. J Biol Chem 255: 4011–4019
39. Hayes MR, McGivan JD (1982) Differential effects of starvation on alanine and glutamine transport in isolated rat hepatocytes. Biochem J 204: 365–368
40. Aikawa T, Matsutaka H, Yamamoto H, Okuda T, Ishikawa E, Kawano T, Matsumara E (1973) Gluconeogenesis and amino acid metabolism. Inter-organal relations and roles of glutamine and alanine in the amino acid metabolism of fasted rats. J Biochem 74: 1003–1017
41. Rémésy C, Demigné C, Aufrère J (1978) Inter-organ relationships between glucose, lactate and amino acids in rats fed on high-carbohydrate or high-protein diets. Biochem J 170: 321–329
42. Häussinger D, Gerok W, Sies H (1983) Regulation of flux through glutaminase and glutamine synthetase in isolated perfused rat liver. Biochim Biophys Acta 755: 272–278
43. Brosnan JT, Williamson DH (1974) Mechanisms for the formation of alanine and aspartate on rat liver in vivo after administration of ammonium chloride. Biochem J 138: 453–462
44. Lund P, Watford M (1976) Glutamine as a precursor of urea. In: Grisolia S, Báguena R, Mayor F (eds) The urea cycle. Wiley, New York, pp 479–488
45. Cooper AJL, Meister A (1974) Isolation and properties of a new glutamine transaminase from rat kidney. J Biol Chem 249: 2554–2561
46. Meister A (1980) Catalytic mechanism of glutamine synthetase; overview of glutamine metabolism. In: Mora J, Palacios R (eds) Glutamine: metabolism, enzymology, and regulation. Academic Press, London New York, pp 1–39
47. Lund P (1971) Control of glutamine synthesis in rat liver. Biochem J 124: 653–660
48. Saheki T, Katunuma N (1975) Analysis of regulatory factors for urea synthesis by isolated perfused rat liver. Urea synthesis with ammonia and glutamine as nitrogen sources. J Biochem 77: 659–669
49. Akerboom TPM, Bookelman H, Zuurendonk PF, Van der Meer R, Tager JM (1978) Inramitochondrial and extramitochondrial concentrations of adenine nucleotides and inorganic phosphate in isolated hepatocytes from fasted rats. Eur J Biochem 84: 413–420
50. Sterman R, Decker K (1978) Differential response of ATP and orthophosphate in cytosol and mitochondria of rat hepatocytes to treatment with P_i and D-galactosamine. FEBS Lett 95: 214–216
51. Lund P, Brosnan JT, Eggleston LV (1970) The regulation of ammonia metabolism in mammalian tissues. In: Bartley W, Kornberg HA, Quayle JR (eds) Essays in cell metabolism. Wiley, London, pp 167–188
52. Chamalaun RAFM, Tager JM (1970) Nitrogen metabolism in the perfused rat liver. Biochim Biophys Acta 222: 119–134
53. Häussinger D, Weiss L, Sies H (1975) Activation of pyruvate dehydrogenase during metabolism of ammonium ions in hemoglobin-free perfused rat liver. Eur J Biochem 52: 421–431
54. Häussinger D, Akerboom TPM, Sies H (1980) The role of pH and the lack of a requirement for hydrogencarbonate in the regulation of hepatic glutamine metabolism. Hoppe-Seyler's Z Physiol Chem 361: 955–1001

55. Nordmann R, Petit MA, Nordmann J (1972) Recherches sur le mécanisme de l'accumulation intra-hépatique d'acides aminés dicarboxyliques au cours de l'intoxication ammoniacale. Biochimie 54: 1473–1478
56. Meijer AJ, Hensgens HESJ (1982) Ureogenesis. In: Sies H (ed) Metabolic compartmentation. Academic Press, London New York, pp 259–286
57. Lusty CJ (1978) Carbamoylphosphate synthetase I of rat-liver mitochondria. Purification, properties, and polypeptide molecular weight. Eur J Biochem 85: 373–383
58. Soboll S, Elbers R, Scholz R, Heldt HW (1980) Subcellular distribution of di- and tricarboxylates and pH gradients in perfused rat liver. Hoppe Seyler's Z Physiol Chem 361: 69–76
59. Hoek JB, Nicholls DG, Williamson JR (1980) Determination of the mitochondrial protonmotive force in isolated hepatocytes. J Biol Chem 255: 1458–1464
60. Sainsbury GM (1980) The distribution of ammonia between hepatocytes and extracellular fluid. Biochim Biophys Acta 631: 305–316
61. Wanders RJA, Hoek JB, Tager JM (1980) Origin of the ammonia found in protein-free extracts of rat-liver mitochondria and rat hepatocytes. Eur J Biochem 110: 197–202
62. Lund P (1980) Glutamine metabolism in the rat. FEBS Lett 117: K86–K92
63. Rognstad R (1977) Sources of ammonia for urea synthesis in isolated rat liver cells. Biochim Biophys Acta 496: 249–254
64. Krebs HA, Hems R, Lund P, Halliday D, Read WWC (1978) Sources of ammonia for mammalian urea synthesis. Biochem J 176: 733–737
65. Hensgens HESJ, Meijer AJ (1980) Inhibition of urea-cycle activity by high concentrations of alanine. Biochem J 186: 1–4
66. Lueck JD, Miller LL (1970) The effect of perfusate pH on glutamine metabolism in the isolated perfused rat liver. J Biol Chem 245: 5491–5497
67. Oliver J, Koelz AM, Costello J, Bourke E (1977) Acid-base induced alterations in glutamine metabolism and ureogenesis in perfused muscle and liver of the rat. Eur J Clin Invest 7: 445–449
68. Schworer CM, Mortimore GE (1979) Glucagon-induced autophagy and proteolysis in rat liver: mediation by selective deprivation of intracellular amino acids. Proc Natl Acad Sci USA 76: 3169–3173
69. Titheradge MA, Haynes RC (1980) The hormonal stimulation of ureogenesis in isolated hepatocytes through increases in mitochondrial ATP production. Arch Biochem Biophys 201: 44–55
70. Williamson JR, Meijer AJ, Ohkawa K (1974) Interrelation between anion transport, ureogenesis and gluconeogenesis in isolated rat liver cells. In: Lundquist F, Tygstrup N (eds) Regulation of hepatic metabolism. Munksgaard, Copenhagen, pp 457–479
71. Gebhardt R, Mecke D (1983) Heterogeneous distribution of glutamine synthetase among rat liver parenchymal cells in situ and in primary culture. EMBO J 2: 567–570
72. Häussinger D (1983) Hepatocyte heterogeneity in glutamine and ammonia metabolism and the role of an intercellular glutamine cycle during ureogenesis in perfused rat liver. Eur J Biochem 133: 269–275
73. Gaasbeek Janzen JW, Moorman AFM, Lamers WH, Los JA, Charles R (1981) The localization of carbamoyl-phosphate synthase in adult rat liver. Biochem Soc Trans 9: 279P

Mechanism and Control of Deprivation-Induced Protein Degradation in Liver: Role of Glucogenic Amino Acids

G. E. MORTIMORE[1] and A. R. PÖSÖ[2]

Introduction

In addition to the amino acid requirement for protein synthesis, all animals need an extra supply to replace irreversible amino acid losses by gluconeogenesis and other ongoing, oxidative and biosynthetic reactions. During intestinal absorption, this need is met by the breakdown of ingested proteins, but when absorption ceases, internal sources must be called into play. Because free amino acids are not stored in mammalian cells or extracellular spaces to any significant extent, this supply (or its nitrogen equivalent) can only come from the degradation of endogenous proteins.

Early investigators of protein metabolism recognized that this internal amino acid source was not a specific "storage" molecule, analogous to triglycerides and glycogen, but instead comprised the normal protein constituents of the cell and to a small extent, plasma proteins [1]. They also observed that the fractional decreases in total cellular protein which occurred after short periods of starvation varied widely among different tissues. Losses in liver, kidney, and intestine, for example, were among the highest whereas the average decrease in skeletal muscle was relatively small [2]. Later studies have in fact confirmed these differential effects in liver and muscle. After 48 h of starvation, rats and mice have been shown to lose 25–40% of their total liver protein [2–5] without significant net changes in skeletal muscle protein [6, 7]. Muscle protein degradation does accelerate, though, in prolonged starvation [7].

As interest developed in the mechanism and regulation of general protein turnover, it became apparent that the isolated liver or hepatocyte might prove to be a useful model because of the large magnitude of the regulatory effects and the relative ease of subcellular fractionation. Early studies with the perfused rat liver showed that the spontaneous net breakdown of cellular protein, which occurred in the absence of any additions to the medium, could be suppressed by insulin and amino acids with little or no stimulation of protein synthesis [8, 9]. Analogous findings were obtained in vivo during starvation and refeeding [10–12], and the conclusion was established that degradation rather than synthesis is the dominant site of regulation in liver for responses involving

1 Department of Physiology, The Milton S. Hershey Medical Center, The Pennsylvania State University, Hershey, PA 17033, USA

2 Department of Biochemistry, College of Veterinary Medicine, Hämeentie 57, SF-0050 Helsinki 55, Finland

Glutamine Metabolism in Mammalian Tissues
Edited by D. Häussinger and H. Sies

physiological alterations in total cell protein. The results of more recent investigations, utilizing a variety of experimental approaches, leave little doubt that cellular autophagy is the process responsible for the net protein breakdown observed in isolated liver cell preparations [13–17]. And because autophagy is primarily regulated by amino acids at physiological concentrations [16], this regulation undoubtedly has biological relevance.

The general aims of this chapter are to review the major lines of evidence that have led to these conclusions and to outline current problems concerning the amino acid control of proteolysis in the isolated liver. However, because the main focus of this paper is on the emerging role of glucogenic amino acids as a regulatory link between energy demands and intracellular protein degradation [18–20], somewhat more emphasis was placed on the authors' work than would have been done in a review of normal coverage.

Proteolytic and Lysosomal Responses of the Isolated Liver/Hepatocyte to Amino Acids

Protein Degradation

Our initial objective in approaching the mechanism of amino acid control of protein degradation was to establish an isolated liver cell preparation that would closely match the corresponding function in vivo. Prior experience with the cyclically perfused rat liver indicated that inhibitory responses to amino acid mixtures were highly reproducible and elicitable within the physiological range of plasma amino acids [9, 21, 22]. For these reasons we continued to use the perfused liver but employed the nonrecirculating or single-pass mode to insure constancy of perfusate amino acid concentrations during the course of the experiments [16]. At the conclusion of the single-pass phase, perfusion was switched to a second-stage, cyclic perfusion containing cycloheximide for the measurement of protein degradation. The measurement was based on the accumulation of free valine during the succeeding 5–15 min, a determination that has been shown to reflect closely the rate of protein breakdown that existed in the cell at the moment cycloheximide was introduced [4, 16, 23).

The foregoing determinations of total valine release include valine derived from the degradation of a short-lived protein component(s) as well as plasma proteins acquired by endocytosis. Because these protein fractions do not contribute to the pool of resident intracellular protein [4, 8, 16, 24] and are not subject to acute regulation [14, 25, 26], their contribution to valine release was subtracted from the total [16]. The resulting corrected value, shown by the broken line in Fig. 1, is an estimate of resident or long-lived protein breakdown and is in agreement with rates based on the linear release of label from previously labeled livers [16, 23, 24].

As depicted in Fig. 1, resident protein degradation in perfused fed rat livers can be regulated over approximately two-thirds of its full range by varying the level of amino acids in multiples or fractions of normal plasma concentrations. The rate of valine release during stringent amino acid deprivation in Fig. 1 was

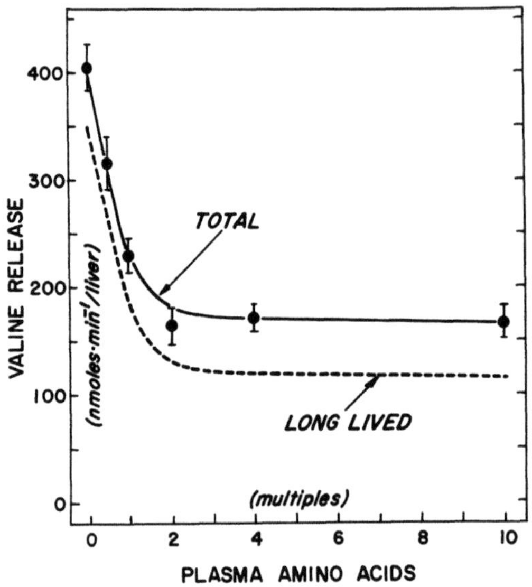

Fig. 1. Effects of graded plasma amino acid concentrations on rates of total and long-lived or resident protein degradation. Livers from fed rats were perfused in the single-pass mode for 40 min; protein degradation was determined from the release of valine in a second-stage cyclic perfusion containing cycloheximide. The *lower curve* (long-lived or resident protein degradation) has been corrected for short-lived valine release (see text) [16]

equivalent to 4.5% of total protein per h; corresponding basal rates at 4 and 10 times plasma amino acid levels were 1.5% h^{-1} [16]. Because the zone of highest responsiveness is within or slightly below the normal concentration of plasma amino acids, this function could play an important role in free amino acid homeostasis in vivo.

In considering our initial objective, it is instructive to compare the performance of the perfused liver with what we known of similar functions in the intact animal. Rates of resident protein synthesis in the perfused mouse liver have been found to equal those in vivo [4, 12], and from recent calculations by Pösö et al. [19], the same conclusion appears to hold for the rat liver as well. The fact that rates of resident protein synthesis are nearly identical to basal rates of degradation at full amino acid supplementation indicates that intracellular protein is capable of being brought into balance by levels of amino acids similar to those found in the portal vein during feeding [21, 22]. In the postabsorptive period, however, portal vein amino acids approach systemic levels, and the liver is known to lose protein at a relatively steady rate [4, 5]. In this connection, it is of interest that the increase in the rate of degradation at normal (1×) plasma amino acid levels over the estimated rate of synthesis that we have observed [9, 19] predicts a decrease in liver protein of 16–20% per day, a loss that is not greatly different from values observed in 24-h starved rats.

Inhibitory responses to amino acids have also been reported in experiments with isolated suspensions of hepatocytes and hepatocyte monolayers [27–29]. While there seems to be general agreement among investigators in the list of amino acids that have inhibitory activity (to be discussed later), significant quantitative differences appear to exist between hepatocyte responses [27–29] and effects in perfusion experiments [19]. Data of Seglen et al. [28] are especially noteworthy because the hepatocyte studies were carried out in a way that facilitates comparison with our perfusion experiments. Their freshly prepared hepatocyte suspensions exhibited almost the same maximal and basal rates of protein degradation as those in Fig. 1. In our view, the only major difference seems to lie in the sensitivity of the two systems. From our calculations, half-maximal inhibition by amino acids was elicited at about 0.5× normal plasma levels in the perfused liver whereas 3× levels were required for the hepatocyte suspensions, a difference of sixfold. Why hepatocytes are generally less responsive than the intact liver is not known. The question is important because even a partial answer could help explain the nature of amino acid recognition and the equally obscure connection between sites of recognition and the proteolytic machinery.

Lysosomal Alterations

As knowledge of the lysosomal-vacuolar system grew [30], with it came the realization that autophagy could play an important role in the breakdown of cellular constitutents. The discovery by Ashford and Porter [31] that glucagon is a potent inducer of autophagy in liver provided impetus for a series of investigations that eventually linked autophagy with the physiological turnover of intracellular proteins. Because glucagon was implicated as a protein catabolic agent in liver [32, 33], Ashford and Porter quite correctly suggested that the enhanced protein breakdown induced by glucagon could be mediated by autophagy.

This possibility raised an even more fundamental question concerning the mechanism of deprivation-induced proteolysis. Utilizing the osmotic shock technique of Deter and de Duve to monitor autophagy [34], Neely et al. [35] observed that autophagy increased spontaneously when livers from normal fed rats were cyclically perfused in the absence of additions to the medium. The autophagic response was completely reversible, regressing with a half-life of about 8 min after the addition of amino acids and insulin. Somewhat later, the morphologic features of this deprivation response were described in cultured cells [36] and in the perfused rat liver [37] and heart [38]; its autophagic nature was firmly established.

Dynamics of Deprivation-Induced Autophagy

The deletion of all amino acids from the medium induces an autophagic response in perfused livers of fed rats that is indistinguishable from that observed after glucagon [18, 37, 39, 40]. In a detailed study of the kinetics of autophagy, rat livers were perfused for varying periods of time in the single-pass mode and autophagy measured by stereologic methods [16]. New autophagic vacuoles (AVi) were induced rapidly on amino acid withdrawal, and after a lag of 7–8 min were transformed into digestive vacuoles (AVd) [16]. The sequestration and digestion of cytoplasm by this means appeared to be a rapid, ongoing process in which a steady state between the formation of AVi and the disappearance of AVd could be established within a comparatively short period of time (20 min). Because AVi formation is capable of being switched off or diminshed rapidly by the addition of amino acids or insulin [16, 41], the turnover of autophagic vacuoles can be indirectly determined from the ensuing exponential regression of the aggregate vacuolar volume. Estimates of the half-life, including the earlier measurement based on osmotic shock [35], have ranged from 8–9 min [16, 35, 41]. We have taken 0.087 min^{-1} as an approximation of the first-order rate constant for AV turnover in livers of rats and mice [5, 16].

From data in Figure 2A, rates of AVi formation and, hence, the volume quantities of all autophagic components appear to be strongly influenced by physiological concentrations of plasma amino acids [16]. Fractional cytoplasmic volumes of all three autophagic fractions (AVi, AVd, and total AV) fell sharply and proportionally as amino acid levels were increased. Nearly full suppression of autophagy, with values approximating only 5% of that observed at Ox, was achieved at amino acid concentrations within physiological limits (4×). No changes in the total aggregate volume of dense bodies were noted over the range of amino acid levels although there was a small but significant decrease in one dense body component, type A, at 1× and higher [16, 37]. This effect is important in the interpretation of microautophagy and has been discussed in detail elsewhere [5, 16].

Cellular and Subcellular Distribution of Autophagy

The cellular components of liver parenchyma are conventionally divided into two classes: (a) a large population of hepatocytes and (b) a group of much smaller cells (Kupffer, endothelial and lipid storage) that lie within or on the border of sinusoids [42]. Recent estimates of our own, based on data of Blouin et al. [42], indicate that hepatocytes comprise about 94% of the total cellular volume in livers of fed rats and mice [5]. Because the average protein content of sinusoidal cells appears to be much lower of that of the hepatocyte [5, 43, 44], hepatocytes would contain an even greater fraction of the total cellular protein of liver. From the foregoing data, we have calculated that the percentage could be as high as 99% [5]. But regardless of the exactness of the estimate, it seems clear that sinusoidal cells would contribute little to general protein turnover in the intact liver. Thus the additional observation that deprivation-induced autophagy is limited to the hepatocyte population in liver has little practical importance in interpreting the origin of the general regulatory effects since the bulk of demonstrable turnover is a function of the same cell group. In this respect, it is of interest that both maximal and basal rates of degradation reported by Seglen et al. in hepatocyte suspensions [28] are virtually identical with measurements in the perfused liver [16].

Two other points concerning the distribution of autophagic vacuoles in the hepatocyte should be made. First, all hepatocytes irrespective of their zonal position in the hepatic lobule appear to participate equally in the deprivation response [16]. In one control study, no difference was observed in the average fractional cytoplasmic volume of autophagic vacuoles between sections obtained from pericentral and periportal zones [16]. Second, cytoplasmic sequestration following deprivation is a relatively unselective process although Pfeifer has noted that the frequency of internalization of some subcellular organelles is at variance with their respective distributions in cytoplasm [41, 45]. At low rates of autophagy , where the average vacuole size is smaller than that noted at maximal rates [16], we have obtained evidence that fewer organelles of larger size, e.g., mitochondria, are sequestered [37].

Quantitative Relationship Between Overt Autophagy and Deprivation-Accelerated Protein Degradation

Although internalization of cytoplasm by the lysosomal-vacuolar system is generally believed to play a dominant role in the breakdown of liver proteins during deprivation, the response has been difficult to quantitate. Because protein internalization is an intracellular function, isotope labeling cannot be employed. The use of lysosomal inhibitors has yielded valuable information, but results have generally been variable and uncertain as to the extent of lysosomal inhibition [14, 15, 46–48]. However, an alternative approach, suggested by results in Fig. 2 and the previous estimates of autophagic vacuole turnover, has given consistent estimates [16]. The rationale behind it is quite straight-forward. We assumed that any degree of deprivation between basal and maximal rates would induce a steady state pool of autophagic vacuoles (AVi) whose aggregate size would be directly proportional to the quantity of protein internalized. This should be so because (a) AVi formation appears to be a discrete process that internalizes cytoplasm without disturbing the underlying cytoarchitecture [37, 41] and (b) local differences in cytoplasmic protein would be averaged out by the large number and variety of vacuoles formed. From these considerations it follows that one should be able to determine the fractional turnover of cytoplasmic volume during deprivation by multiplying the increase in AVi (above the basal level) by the first-order rate constant of autophagic vacuole turnover, 0.087 min^{-1} or 5.22 h^{-1} (see earlier discussion). When data from the experiments in Fig. 1A were used in this calculation and results compared with corresponding values of accelerated intracellular protein turnover, close quantitative agreement was obtained over the full range of deprivation [16].

This quantitative relationship was substantiated in an independent set of experiments in which pools of protein internalized within lysosomes were correlated with rates of intracellular protein degradation. Pool sizes were determined from the total quantity of free valine released during 120- to 180-min incubations of isotonic liver homogenates [17]. Virtually all of the released amino acid was shown to be derived from protein internalized within the lysosome prior to homogenization; less than 4% was generated in the particle-free supernatant and no detectable amount could be attributed to interactions between cytosolic proteins and lysosomes during the incubation [17]. The quantity of internalized protein was determined from the total increase of free valine after its release had leveled off [17].

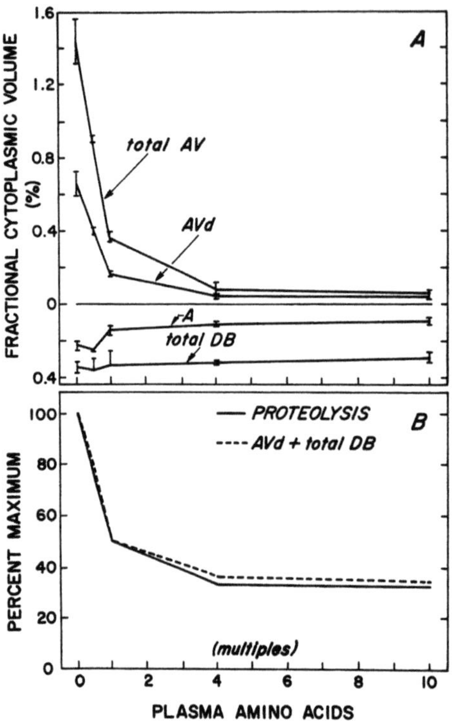

Fig. 2A, B. Effects of varying plasma amino acid concentrations on the degradation of resident proteins and fractional-cytoplasmic volumes of lysosomal-vacuolar elements. Single-pass liver perfusions were carried out for 40 min; at the end, livers were either fixed for electron microscopy or switched to a second-stage cyclic perfusion flask for the determination of proteolysis. **A** Values for AVi are represented *above* the zero base line by the differences between total autophagic vacuoles *(AV)* and *AVd;* similarly, values for residual bodies (type R dense bodies) are indicated *below* the zero base line by differences between total dense bodies *(DB)* and type A elements. The fractional volumes are expressed as a percentage of parenchymal cell cytoplasm and exclude nuclei, sinusoidal cells, and noncellular spaces. **B** Correlation between rates of resident protein breakdown and the fractional cytoplasmic volumes of degradative lysosomal components; maximal degradation (100%) equals 351 nmol valine min^{-1} per liver (100-g rat) [16]

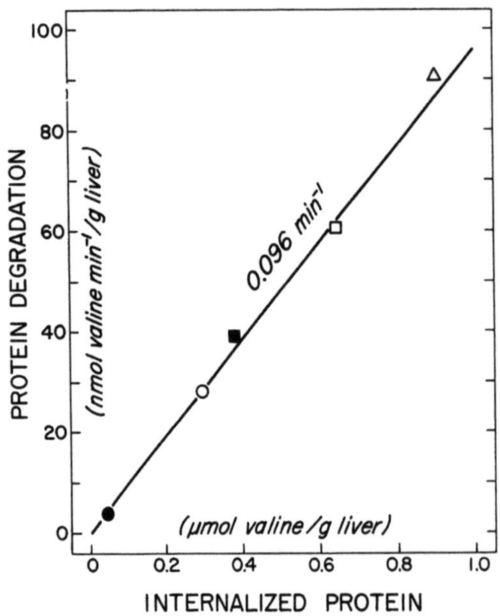

Fig. 3. Relationship between lysosomal pools of degradable protein and rates of resident protein breakdown. Pools of internalized protein determined as in [17], were plotted against corresponding average rates of hepatic protein degradation, ranging from maximal to near zero. The latter was obtained from 48-h starved/12-h refed mice [4]. Values of protein degradation between 30 nmoles ov valine min^{-1} and zero would represent basal turnover, a state associated with maximally suppressed overt autophagy [4]

As illustrated in Fig. 3, internalized degradable protein correlated directly with the rate of resident protein breakdown over the full range of cellular protein degradation. The upper two-thirds of the plot covers the deprivation-accelerated phase, a segment analogous to the range in Fig. 1; the value nearest zero was obtained from 48-h starved/12-h refed mice, a condition known to be associated with a dramatic adaptive suppression of basal protein degradation [4, 12]. While the overall relationship in Fig. 1 has implications concerning the mechanism of microautophagy that are beyond the scope of this article (the reader is referred to ref. 5,16 for additional interpretation), it is apparent that the acute deprivation-induced response can be satisfactorily explained by overt autophagy. It may be noted that the slope of the regression, 0.096 min^{-1}, is very close to our approximation of the first-order rate constant for autophagic vacuole turnover, 0.087 min^{-1} [16]. Such agreement would, of course, be expected if autophagy were the sole mechanism involved. While this does not constitute proof, autophagy is the only reasonable mechanism for general proteolysis that has currently been proposed to explain the deprivation response.

Amino Acid Specificity in the Control of Protein Degradation

Inhibitory and Noninhibitory Amino Acids

Several past studies with the perfused liver and isolated hepatocytes have indicated that the number of amino acids which contribute to the overall suppression of hepatic proteolysis is comparatively small, although it is larger than that in muscle where only leucine is active [49–51]. In general these studies have shown that phenylalanine and tryptophan are effective inhibitors [9, 27–29]. Proline, glutamine, and methionine have been suggested [9, 18, 27–29], and Seglen has recently reported requirements for leucine, tyrosine, and histidine in hepatocyte experiments [28]. Inhibition was also observed with large amounts of asparagine, but it differed from other amino acids in its failure to suppress autophagy [28, 52]. Pösö et al. [19] have demonstrated that at 4 times its normal plasma concentration, asparagine has no inhibitory activity when combined with other amino acids, although it does cause inhibition by itself.

More recently, experiments were undertaken by Pösö and Mortimore to define the inhibitory group under more physiological conditions, utilizing the single-pass perfusion technique as a means of controlling external amino acid concentrations. Results were evaluated largely on effects obtained at 4× plasma levels because they represent the lowest concentrations that consistently give full suppression [19] and also correspond to the upper limit of amino acid concentrations in portal vein plasma [21, 22]. Aiming first at establishing the largest number of ineffective amino acids, Pösö et al. [19] found that 12 of the 20 amino acids that normally turn over in liver protein were completely devoid of direct inhibitory activity when tested as a group at 1, 4, and 10× plasma concentrations (see Fig. 6 and Table 1). The complementary group of 8 amino

Table 1. Effects on hepatic protein degradation of 0.74 m*M* glutamine (1× normal plasma concentration) given alone and in combination with other amino acids. Livers from fed rats were perfused in the single-pass mode as described in Fig. 1 with the various amino additions below. Resident protein degradation was determined as in [16]. All amino acids were present at their "normal" plasma concentration as defined in [16]. See text for description of the noninhibitory and complete mixtures. Values are means ± one SE; numbers of experiments in parentheses. (A. R. Pösö and G. E. Mortimore, unpublished data)

Perfusate amino acids (normal plasma conc.)	protein degradation (nmoles valine · min^{-1}/ liver 100-g rat)	Δ (A vs B)	p
A. None (28)	339 ± 11	6 ± 36	n.s.
B. Glutamine (11)	333 ± 34		
A. Noninhibitory (6)	339 ± 17	56 ± 21	< 0.05
B. Noninhibitory plus glutamine (4)	282 ± 13		
A. Complete (10)	165 ± 8	62 ± 11	< 0.001
B. Complete minus glutamine (6)	227 ± 8		

acids (leucine, phenylalanine, tyrosine, glutamine, proline, methionine, tryptophan, and histidine) thus appeared to contain the active inhibitors.

But whether all of the 8 amino acids are true inhibitors was not easily answered since in some instances individual activity was too low to measure [19]. In the case of phenylalanine and tyrosine, though, a somewhat different problem arose. Because phenylalanine is rapidly hydroxylated to form tyrosine [53], we wondered whether the former, the latter, or perhaps both was the active regulator. Each amino acid was found to suppress the deprivation response by more than 30%, and a similar inhibition was observed when the two were combined [19]. It is also known that phenylalanine activates phenylalanine hydroxylase strongly under the same conditions used for testing proteolytic regulation [54]. Taken together, these and related findings suggested to us that tyrosine is the active inhibitor, and it was sufficient for our purposes to exclude phenylalanine from the inhibitory category.

As illustrated in Fig. 4, leucine was by far the most effective inhibitory regulator at the 4× plasma amino acid level, suppressing the maximal deprivation response by more than 60%. Tyrosine, glutamine, and proline also evoked significant effects, but inhibitory responses to histidine, tryptophan, and methionine were too small to evaluate separately [19]. When the four strongest inhibitors were individually deleted from the mixture (Fig. 4, I–L) only the omission of leucine and proline produced demonstrable effects, suggesting that they are required for inhibition at the 4× level. However, despite the fact that glutamine was dispensible, its effect was clearly additive to that of leucine when all combinations of leucine, tyrosine and glutamine were tested (Fig. 4, M–P). Tyrosine exhibited no tendency to produce additive effects in these experiments, but proline, histidine, tryptophan, and methionine as a group appeared to act cumulatively with leucine. Our findings are generally in agreement with those of

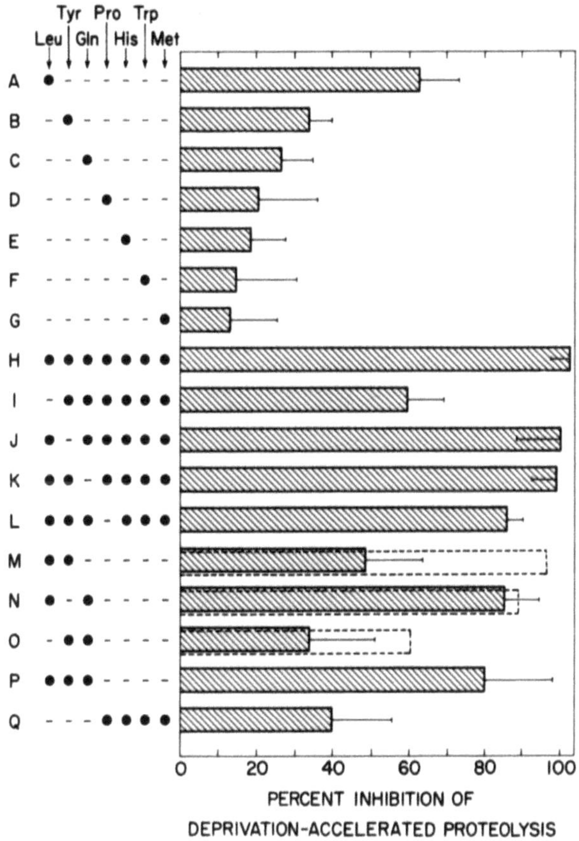

Fig. 4. Inhibition of deprivation-accelerated protein degradation by inhibitory (regulatory) amino acids. Individual and combined effects of inhibitory amino acids were assessed in livers perfused as in Fig. 1. All amino acids were given at 4× normal plasma concentrations, and the results are expressed as percentages of the difference between proteolytic rates at 0 and 4× (complete mixture). *Dashed bars* show calculated sums of individual effects from lines *A, B,* and *C* [19]

Seglen et al. [28] in suggesting that leucine plays a dominant role in providing numerous regulatory combinations with other inhibitory amino acids.

Proteolytic Responses to Graded Levels of Amino Acids

Inhibitory Amino Acids

In an earlier study, we reported that proteolytic responses to graded levels of the inhibitory (regulatory) amino acid mixture paralleled those of the complete mixture [19]. By contrast, the dose-response curve for leucine deviated strongly from this pattern in exhibiting two points of inflection between 0.5 and 2× together with a sharp, almost complete loss of inhibitory activity that peaked at about 1.25× [19]. Because our original conclusion regarding the shape of the curve for the inhibitory mixture was based largely on close agreement at 0.5 and 4× with only a limited number of experiments at 2×, we reinvestigated the question, adding data points at 1 and 2× (A. R. Pösö and G. E. Mortimore,

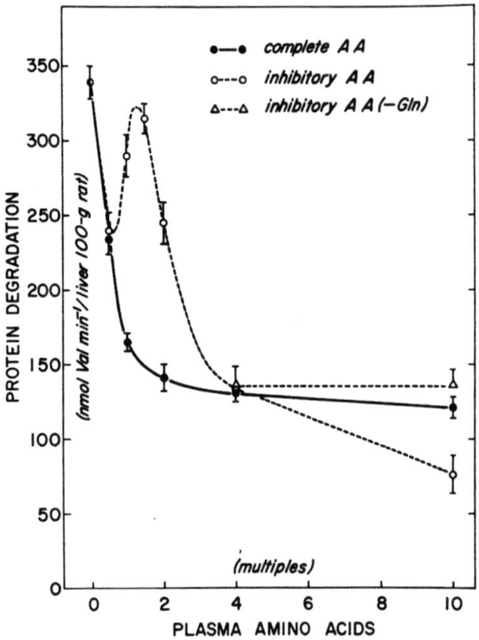

Fig. 5. Comparison of the effects of graded levels of complete and inhibitory (regulatory) amino acid mixtures on resident protein degradation. Except for the amino acid mixtures, the experiments were carried out as described in the legend to Fig. 1. The inhibitory mixture consisted of Leu, Tyr, Gln, Pro, His, Met, and Trp [19]. Values are means ± one S.E. of 4 to 20 determinations. [19] and A. R. Pösö and G. E. Mortimore, unpublished data

unpublished). The results in Fig. 5 show the same unique curve for the inhibitory mixture that Pösö et al. [19] observed with leucine. Although the scale of the curve for the inhibitory mixture is larger, both exhibit a sharply defined zone of inhibitory unresponsiveness that peaks slightly above the 1× level. The similarity in pattern suggests that responses to most or all of the individual inhibitory amino acids would follow the same course.

The significance of this zonal phenomenon to the physiological regulation of protein degradation will be discussed in the last section of this chapter. However, it should be noted here that the marked difference in effectiveness of the inhibitory and complete mixture at 1× clearly means that one or more noninhibitory amino acids are required for activity of the inhibitory amino acids at normal concentrations.

A second important difference between the curves of the complete and inhibitory mixtures in Fig. 5 is evident at concentrations above 4×. Whereas the course of the complete mixture was virtually flat between 4 and 10× (see also Fig. 1), responses to the inhibitory mixture deviated downward. Because rates at 10× were significantly lower than basal values, it is probable that intralysosomal proteolysis was directly inhibited. An increase in free ammonia would be the most likely cause of this effect [15], and since glutamine is a potential source of extra ammonia [55, 56] – particularly at high concentrations – we repeated proteolytic measurements at 4 and 10× with an inhibitory mixture that lacked glutamine. As shown in Figs. 5 and 6, the proteolytic suppression at 10× was totally eliminated.

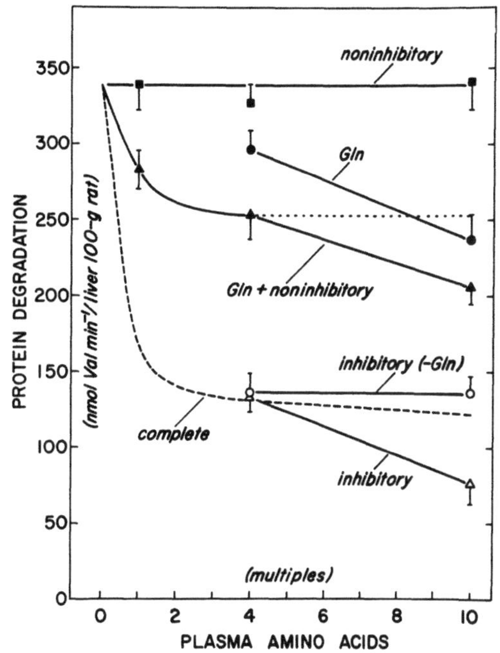

Fig. 6. Effects on resident protein degradation of varying concentrations of glutamine (Gln) alone and Gln in combination with inhibitory and noninhibitory amino acids. The experimental procedures were the same as those in Fig. 1, and the inhibitory amino acid mixture identical to that in Figs. 4 and 5; the noninhibitory mixture consisted of Ala, Arg, Asn, Asp, Cys, Glu, Gly, Ile, Lys, Ser, Thr, and Val [19, 20]. The experiments depicted in the lowest curve are the same as those in Fig. 5. Values are means ± S.E. of 5 to 10 observations. Modified from Pösö et al. [20] as described in text

Lysosomotropic and Regulatory Effects of Glutamine

Figure 6 depicts inhibitory effects of glutamine alone and glutamine plus the inhibitory amino acid mixture in comparison with responses to the inhibitory mixture (± glutamine) discussed in the preceding paragraph. With the exception of the complete amino acid mixture, all conditions in which glutamine was present were associated with comparable downward deviations in rates of proteolysis between 4 and 10×. The effect of glutamine alone at the 4× level (3 m*M*) is similar to that shown in Fig. 4 and, for purposes of this discussion, may be presumed to reflect a diminution in the rate of autophagy. Coadministration of the 12 noninhibitory amino acids significantly enhanced the effectiveness of glutamine at 4× (Fig. 6). Augmentation at this level was not observed with the full inhibitory mixture (Fig. 5), and its significance is not presently known.

Because the 10× inhibitory amino acid mixture decreased proteolytic rates to values below the level at which overt autophagy is virtually nonexistent (Fig. 2), the suppression would not be categorized as a regulatory response (see above) but rather a direct interference with basal protein degradation. Although the mechanism of this latter process in the hepatocyte is not fully established, the final step or steps almost certainly involves the microautophagic uptake of cytoplasmic protein by the lysosomal system [5, 16, 17] and thus would be subject to inhibition by ammonia [15], amines [15], and other lysosomotropic agents [14, 15, 47, 48]. In the present study, the most reasonable explanation is the accumulation of free ammonia generated from the hydrolysis of glutamine at

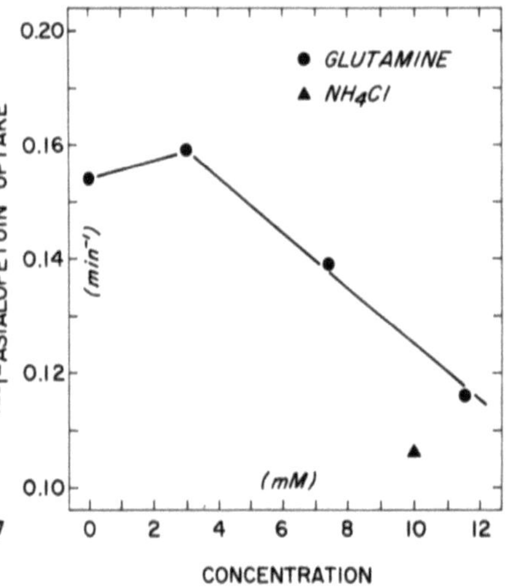

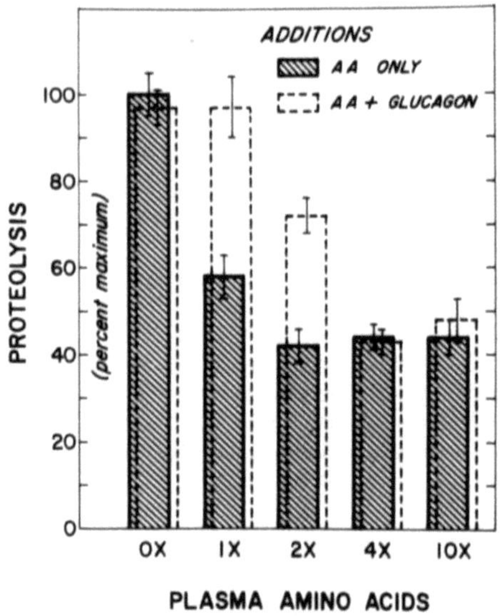

Fig. 7. Effects of glutamine and ammonium chloride on the uptake of ^{125}I-asialofetuin by the cyclically perfused rat liver. Single additions of glutamine or NH_4Cl, yielding the concentrations shown, were made at time zero. Ten minutes later, 10 μg of ^{125}I-asialofetuin were added. Perfusate samples were taken at intervals and the trichloroacetic acid-precipitable radioactivity determined as in [60]. The first-order rate constant of uptake for each experiment was calculated from a plot of the logarithm of the disappearance of label (see ref. 60) [20]

Fig. 8. Effects of graded levels of plasma amino acids with (dashed bars) and without (hatched bars) 1.5 n*M* glucagon on total protein degradation in perfused livers of fed rats. Experimental procedures were the same as in Fig. 1 except that the degradation rates were not corrected for the short lived release of valine [18]

unphysiologically high concentrations. Little is known of the requirements for ammonia accumulation under these conditions [57, 58], but recent evidence from perfusion studies suggests that ammonia could increase rapidly at glutamine concentrations in excess of 2–4 m*M* [56].

Because of known problems in measuring free intracellular ammonia [59], we indirectly monitored ammonia levels from alterations in the uptake of ^{125}I-asialofetuin [60]. Ammonia and the primary means have been shown to inhibit the internalization of receptor-bound ligands in fibroblasts [61, 62] and to interfere with lysosome-phagosome fusion in macrophages [63]. Although others have not observed effects of ammonium chloride on ^{125}I-asialofetuin uptake in isolated hepatocytes [64], we found that 10 m*M* NH_4Cl significantly decreased the rate constant of ^{125}I-asialofetuin removal by the cyclically perfused rat liver (control, 0.154 ± 0.01 min^{-1}; NH_4Cl, 0.106 ± 0.007 min^{-1}, $p < 0.01$ Fig. 7). Additional studies (Fig. 7) showed that glutamine was also effective. No response was observed with the addition of 3 m*M* glutamine, but from 3–12 m*M* the uptake rate constant decreased in direct proportion to the increase in glutamine concentration.

Since the above transition concentration, 3 m*M*, is equal to the 4× level of glutamine employed in our studies, results in Figs. 6 and 7 are subject to similar interpretations. Both, for example, display effects that can readily be attributed to increases in intracellular ammonia when the concentration of glutamine is appreciably greater than 3 m*M*, but in each case the effects are nonexistent at 3 m*M*. With regard to the proteolytic responses in Fig. 6, it should be emphasized that glutamine, when included in the 4× inhibitory amino acid mixture, exhibited no suppressive activity of any type since it could be deleted without effect. The possibility that expression of lysosomotropic inhibition was blocked by the remaining inhibitory amino acids seems unlikely since they failed to do so at the 10× level. It is thus reasonable to conclude that 3 m*M* (4×) glutamine has no lysosomotropic potential under these conditions.

The only circumstance in which unphysiologically high concentrations of glutamine failed to decrease protein degradation below the basal rate was that in which glutamine is a normal component of the complete amino acid mixture. This is evident from the leveling off of proteolytic rates between 4 and 10× in Figs. 1, 6, and 8. It is of interest that ^{125}I-asialofetuin uptake was similarly unaffected at these concentrations; the uptake rate constants at 4 and 10× were, respectively, $0.148 \pm 0.016\ \text{min}^{-1}$ ($n = 3$) and $0.146 \pm 0.009\ \text{min}^{-1}$ ($n = 4$) [20]. The reason for this presumed interference by the mixture with the lysosomotropic action of glutamine is unknown. The findings do suggest, though, that the net production of ammonia from glutamine can be modulated by amino acids under specific conditions.

Although glutamine is not required for the action of the inhibitory amino acid mixture at 4×, it is nevertheless an effective inhibitor by itself at this concentration (Figs. 4 and 6). On the other hand, the previous suggestion of Pösö and Mortimore [20] that glutamine alone at 1× (0.74 m*M*) was also effective is in doubt. More recent experiments in Table 1 reveal no inhibition by glutamine at 1×, a finding that is in keeping with the zonal loss of responsiveness of the full inhibitory amino acid mixture at normal plasma concentrations (Fig. 5).

The ability of amino acids that have no demonstrable inhibitory activity to augment the effectiveness of known inhibitors was initially observed by us during the course of defining the inhibitory and noninhibitory groups [19]. As shown in Fig. 6, we found that addition of the 12 noninhibitory amino acids greatly enhanced glutamine's inhibitory effect at 4× and induced a strong response at 1× also. The latter effect was particularly welcome because it provided direct evidence that glutamine has a potential physiological role as an inhibitor of proteolysis. Proof that glutamine is in fact required for the suppression of protein degradation at normal plasma concentrations is given in Table 1. It may be seen that its deletion from a complete 1× amino acid mixture enhanced protein degradation by an amount comparable to the effect of glutamine combined with the 12 noninhibitory amino acids.

Effects of Glucagon and Glucogenic Amino Acid Deletion on Protein Degradation

As was mentioned earlier, glucagon is also an important inducer of overt autophagy and protein degradation in liver [31, 39, 40, 65] and in hepatocytes

[66]. Because autophagy is stimulated by epinephrine [35] and cyclic AMP [66, 67], cyclic AMP undoubtedly serves as the initial intracellular agent for both hormonal actions. In view of the morphologic similarity between the autophagic vacuoles produced by glucagon and those induced by amino acid deprivation [18, 37, 39, 40], and the fact that glucagon decreases some glucogenic amino acid pools as a result of enhanced gluconeogenesis [18, 68], we considered two possible explanations for the hormonal stimulation: first, that it could arise from the depletion of certain free intracellular amino acids, and second, that it was the result of a direct action of cyclic AMP, independent of amino acids [18].

In subsequent studies, we found that proteolytic and autophagic responses to glucagon were strongly affected by the level of perfusate amino acids [18]. As depicted in Fig. 8, glucagon proved to be a potent catabolic agent, but the effect was noted only at 1× and 2× normal plasma levels; little or no stimulation was observed in the absence of amino acids, and the effects, both proteolytic and autophagic, were completely blocked by 4 and 10× additions. Because glucagon increased cyclic AMP and rates of glycogenolysis in the presence of 10× amino acids, the suppressive effect of the amino acids could not be attributed to a generalized interference with the actions of the hormone or the cyclic nucleotide [18].

The results in Fig. 8 are exactly what one would predict if the catabolic action of glucagon were mediated by amino acid deprivation. In agreement with Mallette et al. [68], we found that glucagon evoked significant decreases in intracellular alanine, glycine, glutamine, glutamate, and phenylalanine when livers were perfused with normal plasma amino acid concentrations [18]. Under these conditions, the addition of glutamine at 10 times its normal plasma concentration prevented the proteolytic response to glucagon; none of the other four amino acids, given singly or in combination, were effective [18]. We now recognize the possibility that part of the inhibition at this high concentration of glutamine was the result of ammonia accumulation (Figs. 6 and 7), but we cannot exclude the notion that another part was the consequence of glutamine replacement [18]. Data in Table 1 clearly show that glutamine is required for optimal proteolytic inhibition at normal amino acid concentrations. Thus a decrease in free glutamine would probably have contributed to the overall proteolytic effect of the hormone, although the extent of the contribution would be difficult to assess.

The above problem is made even more complicated by recent observations indicating that cellular sites of glutamine hydrolysis and synthesis are located at opposite ends of the hepatic sinusoids [69]. With much of the hydrolysis presumably occurring in the periportal region of the lobule, intrasinusoidal concentrations of glutamine might be lower and ammonia higher than conventional determinations have led us to believe. Because glucagon stimulates this hydrolytic step in hepatocytes [70], the uncertainty becomes even larger for conditions described in our glucagon studies [18].

The results in Fig. 5 show that suppression of protein degradation by inhibitory amino acids at normal concentrations requires one or more noninhibitory amino acids. The unusual sharpness of the effect defies ordinary metabolic explanation and has suggested to us that the cell might have special

Table 2. Effects of glucogenic amino acid deletion on intracellular protein degradation in the perfused rat liver. Livers were perfused and protein degradation measured as described in the experiments of Table 1. The deletion results were taken from [20]

Perfusate amino acid conc. (multiples of normal)	Protein degradation (nmol valine · min^{-1}/liver/100-g rat)	
	Complete mixture	Complete minus Gln, Glu, Ala, Gly
None (0×)	339 ± 11	339 ± 11
1×	165 ± 8	312 ± 13
2×	141 ± 9	241 ± 10
4×	131 ± 6	135 ± 15

sites for the recognition of the particular amino acid or acids involved. In view of the potential complexity of this regulation, it is possible that under certain circumstances the cell might respond as though a true deficiency existed because of an induced interference with or modulation of the recognition. In the case of glucagon, for example, it is important that we consider this possibility since if true, one would not necessarily be able to correct the "deficiency" by adding an excess of the amino acid in question.

In recent experiments, we have sought to identify the noninhibitory amino acid or acids whose lack is presumably responsible for the sharp peak in Fig. 5. The same amino acid might eventually prove to be important in mediating the proteolytic response to glucagon. Table 2 shows the effects of deleting from various levels of the complete mixture the four glucogenic amino acids whose pools are decreased by glucagon. Interestingly, their omission at 1× evoked an acceleration of protein degradation that nearly equalled responses to maximal deprivation or glucagon (Fig. 8). Moreover, the acceleration was modulated by the level of amino acids in a manner similar to that seen with glucagon; strong effects were evoked at 1 and 2× and none at 4×. As suggested from results in Table 1, a lack of glutamine would not likely have caused such a large effect, but deletion of one or more of the remaining three amino acids could have done so. When glutamate, alanine, and glycine were omitted, the proteolytic response was comparable in magnitude to deletion of the four (319 ± 14 nmol valine min^{-1}/100-g rat, A. R. Pösö and G. E. Mortimore, unpublished), representing 94% of the maximal rate of proteolysis. Further work on the amino acid requirement of this response is currently under way. Whatever the outcome, the answer should greatly increase our understanding of the regulatory link between the demand for gluconeogenic substrate and intracellular protein degradation in liver.

Conclusions

There is now general agreement that the striking acceleration in the breakdown of resident cellular proteins of the hepatocyte which follows amino acid and insulin deprivation is the result of overt autophagy. Because the turnover of the sequestered cytoplasm is rapid ($t_{1/2}$ = 8–9 min), the rate of cytoplasmic internalization responds quickly to regulatory agents. Virtually all of the cell protein is subject to removal by this mechanism, and increases in protein degradation above the basal or steady state rate of turnover can quantitatively explain net losses of total cell protein.

Protein degradation in the hepatocyte in situ is exquisitely responsive to plasma amino acid alterations within physiological limits of concentration; degradation is only slightly higher than basal at normal plasma levels and is fully suppressed at 4× normal concentrations. Recent studies indicate that amino acid control of proteolysis is mediated mainly by: (a) inhibitory actions of Leu, Tyr, Gln, Pro, Met, Trp, and His at plasma amino acid concentrations above and below normal and (b) by a strong requirement for one or more of the 12 noninhibitory amino acids to allow the foregoing inhibition to be expressed at normal amino acid concentrations. Preliminary experiments suggest that Glu, Ala, and/or Gly will satisfy this requirement. A third mechanism of inhibition is associated with the administration of glutamine at unphysiologically high concentrations. This is believed to occur by the direct suppression of lysosomal proteolysis following the accumulation of free ammonia. Neither the amino acid specificity nor the mechanism of the first two classes of regulation is currently understood, but it is likely that the mechanism or mechanisms of amino acid recognition is complex and contains multiple sites for both positive and negative modulation of autophagy.

Current evidence suggests that protein degradation could be linked to the demand for gluconeogenic substrate in two ways: first, by a decrease in free glutamine and/or proline, alterations that would reduce the normal restraint on degradation and second, by a decrease in one or more of the noninhibitory amino acids listed above. The fact that the inhibitory amino acids are virtually inactive as a group at normal concentrations in the absence of one or more noninhibitory amino acids suggests a way by which protein degradation could be linked to energy metabolism. The same link might also be involved in mediating the stimulatory action of glucagon and cyclic AMP on protein breakdown.

Acknowledgements. The work reported here was supported by a U.S. Public Health Service grant AM-21624. A.R.P is a recipient of a research fellowship from the Research Council for the Natural Sciences of the Academy of Finland. The authors thank Bonnie Merlino for typing the manuscript.

References

1. White A (1947) Protein metabolism. In: Duncan RT (ed) Diseases of metabolism. Saunders, Philadelphia, pp 83–157
2. Addis T, Poo LJ, Lew W (1936) The quantities of protein lost by the various organs and tissues of the body during a fast. J Biol Chem 115: 111–118
3. Soberón G, Sánchez QE (1961) Changes in effective enzyme concentration in the growing rat liver. J Biol Chem 236: 1602–1606

4. Hutson NJ, Mortimore GE (1982) Suppression of cytoplasmic protein uptake by lysosomes as the mechanism of protein regain in livers of starved-refed mice. J Biol Chem 257: 9548–9554
5. Mortimore GE, Hutson NJ, Surmacz CA (1983) Quantitative correlation between proteolysis and macro- and microautophagy in mouse hepatocytes during starvation and refeeding. Proc Natl Acad Sci USA 80: 2179–2183
6. Millward DJ, Nnanyelugo DO, James WPT, Garlick PJ (1974) Protein metabolism in skeletal muscle: The effect of feeding and fasting on muscle RNA, free amino acids and plasma insulin concentrations. Br J Nutr 32: 127–142
7. Millward DJ, Waterlow JC (1978) Effect of nutrition on protein turnover in skeletal muscle. Fed Proc 37: 2283–2289
8. Mortimore GE, Mondon CE (1970) Inhibition by insulin of valine turnover in liver. J Biol Chem 245: 2375–2383
9. Woodside KH, Mortimore GE (1972) Supression of protein turnover by amino acids in the perfused rat liver. J Biol Chem 247: 6474–6481
10. Garlick PJ, Millward DJ, James WPT (1973) The diurnal response of muscle and liver protein synthesis in vivo in meal-fed rats. Biochem J 136: 935–945
11. Garlick PJ, Millward DJ, James WPT, Waterlow J (1975) The effect of protein deprivation and starvation on the rate of protein synthesis in tissues of the rat. Biochim Biophys Acta 414: 71–84
12. Conde RD, Scornik OA (1976) Role of protein degradation in the growth of livers after a nutritional shift. Biochem J 158: 385–390
13. Ward WF, Cox JR, Mortimore GE (1977) Lysosomal sequestration of intracellular protein as a regulatory step in hepatic proteolysis. J Biol Chem 252: 6955–6961
14. Neff NT, DeMartino GN, Goldberg AL (1979) The effect of protease inhibitors and decreased temperature on the degradation of different classes of proteins in cultured hepatocytes. J Cell Physiol 101: 439–458
15. Grinde B, Seglen PO (1980) Differential effects of proteinase inhibitors and amines on the lysosomal and non-lysosomal pathways of protein degradation in isolated rat hepatocytes. Biochim Biophys Acta 632: 73–86
16. Schworer CM, Shiffer KA, Mortimore GE (1981) Quantitative relationship between autophagy and proteolysis during graded amino acid deprivation in perfused rat liver. J Biol Chem 256: 7652–7658
17. Mortimore GE, Ward WF (1981) Internalization of cytoplasmic protein by hepatic lysosomes in basal and deprivation-induced proteolytic states. J Biol Chem 256: 7659–7665
18. Schworer CM, Mortimore GE (1979) Glucagon-induced autophagy and proteolysis in rat liver: Mediation by selective deprivation of intracellular amino acids. Proc Natl Acad Sci USA 76: 3169–3173
19. Pösö AR, Wert JJ Jr, Mortimore GE (1982) Multifunctional control by amino acids of deprivation-induced proteolysis in liver. J Biol Chem 257: 12114–12120
20. Pösö AR, Schworer CM, Mortimore GE (1982) Acceleration of proteolysis in perfused rat liver by deletion of glucogenic amino acids: Regulatory role of glutamine. Biochem Biophys Res Commun 107: 1433–1439
21. Peraino C, Harper AE (1963) Observations on protein digestion in vivo: Free amino acids in blood plasma of rats force-fed zein, casein, or their respective hydrolyzates. J Nutr 80: 270–278
22. Elwyn DH, Parikh HC, Shoemaker WC (1968) Amino acid movements between gut, liver and periphery in unanesthesized dogs. Am J Physiol 215: 1260–1275
23. Khairallah EA, Mortimore GE (1976) Assessment of protein turnover in perfused rat liver. J Biol Chem 251: 1375–1384
24. Mortimore GE, Woodside KH, Henry JE (1972) Compartmentation of free valine and its relation to protein turnover in perfused rat liver. J Biol Chem 247: 2776–2784
25. Poole B, Wibo M (1973) Protein degradation in cultured cells. J Biol Chem 248: 6221–6226
26. Epstein D, Elias-Bishko S, Hershko A (1975) Requirement for protein synthesis in the regulation of protein breakdown in cultured hepatoma cells. Biochemistry 14: 5199–5204
27. Hopgood MF, Clark MG, Ballard FJ (1977) Inhibition of protein degradation in isolated rat hepatocytes. Biochem J 164: 399–407

28. Seglen PO, Gordon PB, Poli A (1980) Amino acid inhibition of the autophagic/lysosomal pathway of protein degradation in isolated rat hepatocytes. Biochim Biophys Acta 630: 103–118
29. Sommercorn JM, Swick RW (1981) Protein degradation in primary monolayer cultures of adult rat hepatocytes. J Biol Chem 256: 4816–4821
30. De Duve C, Wattiaux R (1966) Functions of lysosomes. Annu Rev Physiol 28: 435–492
31. Ashford TP, Porter KR (1962) Cytoplasmic components in hepatic cell lysosomes. J Cell Biol 12: 198–202
32. Izzo JL, Glasser SR (1961) Comparative effects of glucagon, hydrocortisone, and epinephrine on the protein metabolism of the fasting rat. Endocrinology 68: 189–198
33. Miller LL (1960) Glucagon: A protein catabolic hormone in the isolated perfused rat liver. Nature (London) 185: 248
34. Deter RL, De Duve C (1967) Influence of glucagon, an inducer of cellular autophagy, on some physical properties of rat lysosomes. J Cell Biol 33: 437–449
35. Neely AN, Nelson PB, Mortimore GE (1974) Osmotic alterations of the lysosomal system during rat liver perfusion: Reversible supression by insulin and amino acids. Biochim Biophys Acta 338: 458–472
36. Mitchener JS, Shelburne JD, Bradford WD, Hawkins HK (1976) Cellular autophagocytosis induced by deprivation of serum and amino acids in HeLa cells. Am J Pathol 83: 485–498
37. Mortimore GE, Schworer CM (1977) Induction of autophagy by amino acid deprivation in perfused rat liver. Nature (London) 270: 174–176
38. Jefferson LS, Rannels DE, Munger BL, Morgan HE (1974) Insulin in the regulation of protein turnover in heart and skeletal muscle. Fed Proc 33: 1098–1104
39. Arstila AU, Trump BF (1968) Studies on autophagocytosis: The formation of autophagic vacuoles in the liver following glucagon administration. Am J Pathol 53: 687–733
40. Deter RL (1971) Quantitative characterization of dense body, autophagic vacuole, and acid phosphatase-bearing particle populations during the early phases of glucagon-induced autophagy in rat liver. J Cell Biol 48: 473–489
41. Pfeifer U (1978) Inhibition by insulin of the formation of autophagic vacuoles in rat liver. J Cell Biol 78: 152–167
42. Blouin A, Bolender RP, Weibel EW (1977) Distribution of organelles and membranes between hepatocytes and nonhepatocytes in the rat liver parenchyma. J Cell Biol 72: 441–455
43. Knook DL, Sleyster EC (1980) Isolated parenchymal, Kupffer and endothelial rat liver cells characterized by their lysosomal enzyme content. Biochem Biophys Res Commun 96: 250–257
44. Munthe-Kaas AC, Berg T, Seljelid R (1976) Distribution of lysosomes in different types of rat liver cells. Exp Cell Res 99: 146–154
45. Pfeifer U (1973) Cellular autophagy and cell atrophy in the rat liver during long-term starvation. Virchows Arch Abt B Zellpathol 12: 195–211
46. Knowles SE, Ballard FJ (1976) Selective control of the degradation of normal and aberrant proteins in Reuber H35 hepatoma cells. Biochem J 156: 609–617
47. Ward WF, Chua BL, Li JB, Morgan HE, Mortimore GE (1979) Inhibition of basal and deprivation-induced proteolysis by leupeptin and pepstatin in perfused rat liver and heart. Biochem Biophys Res Commun 87: 92–98
48. Dean RT (1975) Direct evidence of importance of lysosomes in degradation of intracellular proteins. Nature (London) 257: 414–416
49. Fulks RM, Li JB, Goldberg AL (1975) Effects of insulin, glucose, and amino acids on protein turnover in rat diaphragm. J Biol Chem 250: 290–298
50. Buse MG, Reid SS (1975) Leucine: A possible regulation of protein turnover in muscle. J Clin Invest 56: 1250–1261
51. Chua B, Siehl DL, Morgan HE (1979) Effect of leucine and metabolites of branched chain amino acids on protein turnover in heart. J Biol Chem 254: 8358–8362
52. Grinde B, Seglen PO (1981) Leucine inhibition of autophagic vacuole formation in isolated rat hepatocytes. Exp Cell Res 134: 33–39
53. Shiman R, Mortimore GE, Schworer CM, Gray DW (1982) Regulation of phyenylalanine hydroxylase activity by phenylalanine in vivo, in vitro, and in perfused rat liver. J Biol Chem 257: 11213–11216

54. Shiman R, Gray DW (1980) Substrate activation of phenylalanine hydroxylase. J Biol Chem 255: 4793–4800
55. Chamalaun RAFM, Tager JM (1970) Nitrogen metabolism in the perfused rat liver. Biochim Biophys Acta 222: 119–134
56. Deaciuc IV, Petrescu I (1980) Regulation of glutamine catabolism in the perfused guinea-pig liver in relation to ureogenesis and gluconeogenesis. Int J Biochem 12: 605–618
57. Lund P (1980) Glutamine metabolism in the rat. FEBS Lett 117: K86–K92
58. Kovacevic Z, McGiven JD (1983) Mitochondrial metabolism of glutamine and glutamate and its physiological significance. Physiol Rev 63: 547–605
59. Wanders RJA, Hoek JB, Tager JM (1980) Origin of the ammonia found in protein-free extracts of rat liver mitochondria and rat hepatocytes. Eur J Biochem 110: 197–202
60. Surmacz CA, Ward WF, Mortimore GE (1982) Distribution of ^{125}I-Asialofetuin among liver particles separated on colloidal silica gradients. Biochem Biophys Res Commun 107: 1425–1432
61. Maxfield FR, Willingham MC, Davies PJA, Pastan I (1979) Amines inhibit the clustering of α_2-macroglobulin and EGF on the fibroblast cell surface. Nature (London) 277: 661–663
62. Sando G, Titus-Dillon P, Hall CW, Neufeld EF (1979) Inhibition of receptor-mediated uptake of a lysosomal enzyme into fibroblasts by chloroquine, procaine, and ammonia. Exp Cell Res 119: 359–364
63. Gordon AH, D'Arcy Hart P, Young MR (1980) Ammonia inhibits phagosome-lysosome fusion in macrophages. Nature (London) 286: 79–80
64. Berg T, Tolleshaug H (1980) The effects of ammonium ions and chloroquine on uptake and degradation of ^{125}I-labeled asialo-fetuin in isolated rat hepatocytes. Biochem Pharmacol 29: 917–925
65. Woodside KH, Ward WF, Mortimore GE (1974) Effects of glucagon on general protein degradation and synthesis in perfused rat liver. J Biol Chem 249: 5458–5463
66. Hopgood MF, Clark MG, Ballard FJ (1980) Effects of glucagon, 3':5'-cyclic monophosphate and insulin. Biochem J 186: 71–79
67. Rosa F (1971) Ultrastructural changes produced by glucagon, cyclic 3'-5'AMP and epinephrine on perfused rat livers. J Ultrastruct Res 34: 205–213
68. Mallette LE, Exton JH, Park CR (1969) Effects of glucagon on amino acid transport and utilization in the perfused liver. J Biol Chem 244: 5724–5728
69. Häussinger D (1983) Hepatocyte heterogeneity in glutamine and ammonia metabolism and the role of an intracellular glutamine cycle during ureogenesis in perfused rat liver. Eur J Biochem 133: 269–275
70. Joseph SK, McGiven JD (1978) The effect of ammonium chloride and glucagon on the metabolism of glutamine in isolated liver cells from starved rats. Biochim Biophys Acta 543: 16–28

Kidney

Renal Glutamine Metabolism and Hydrogen Ion Homeostasis

T. C. WELBOURNE and V. PHROMPHETCHARAT[1]

Renal Base Generation in Cellular H^+ Homeostasis

Cellular H^+ concentration is rigorously maintained within very narrow limits. If not, acidity collapses proton gradients, disrupts energy generation, and accelerates cellular autophagy. Most animals, including man, are faced with a surfeit of acid. The daily metabolic CO_2 production obligates the formation of some 15 mol of carbonic acid; normally expiration matches cellular CO_2 production so the H^+ homeostais is not threatened by this volatile acid. However, with pulmonary dysfunction carbonic acid accumulates and the resulting rise in H^+ concentration is termed respiratory acidosis. The daily oxidation of sulfur containing amino acids and phospholipids generates some 40 to 60 mmol of nonvolatile acids, sulfuric, and phosphoric acid. The excretion of sulfate and phosphate is accompanied by sodium, potassium, magnesium, calcium, or other cations leaving protons to reduce the alkaline reserves; acidosis due to nonvolatile acid production is termed metabolic acidosis. During uncontrolled diabetes the incomplete oxidation of fats accelerates nonvolatile acid production inducing life threatening metabolic acidosis. In both physiological and clinical circumstances cellular H^+ homeostasis is ultimately dependent upon renal acid excretion and bicarbonate generation.

Renal Glutamine Metabolism in Acid Excretion

The ability of the kidneys to generate bicarbonate in defense against acidosis is dependent upon the production of ammonia derived from glutamine; this dependency is illustrated in Fig. 1. Proximal tubule cells are functionally asymmetrical, an inherent property of their luminal and antiluminal membranes. The brush border membrane exchanges intracellular protons for filtered sodium; the basolateral membrane releases sodium bicarbonate. Whether this base release represents generation of new bicarbonate or the reclamation of the filtered moiety depends upon intraluminal reactions. In the normal condition over 98% of the secreted protons react with the filtered bicarbonate (Table 1). The reabsorption of this bicarbonate represents reclamation of the filtered moiety; no new base is formed in the process. Additional acid reacts with avialable proton acceptors, largely filtered phosphate. Accordingly, H^+ are

1 Department of Physiology, LSU Medical Center, Shreveport, LA 71130, USA

Glutamine Metabolism in Mammalian Tissues
Edited by D. Häussinger and H. Sies

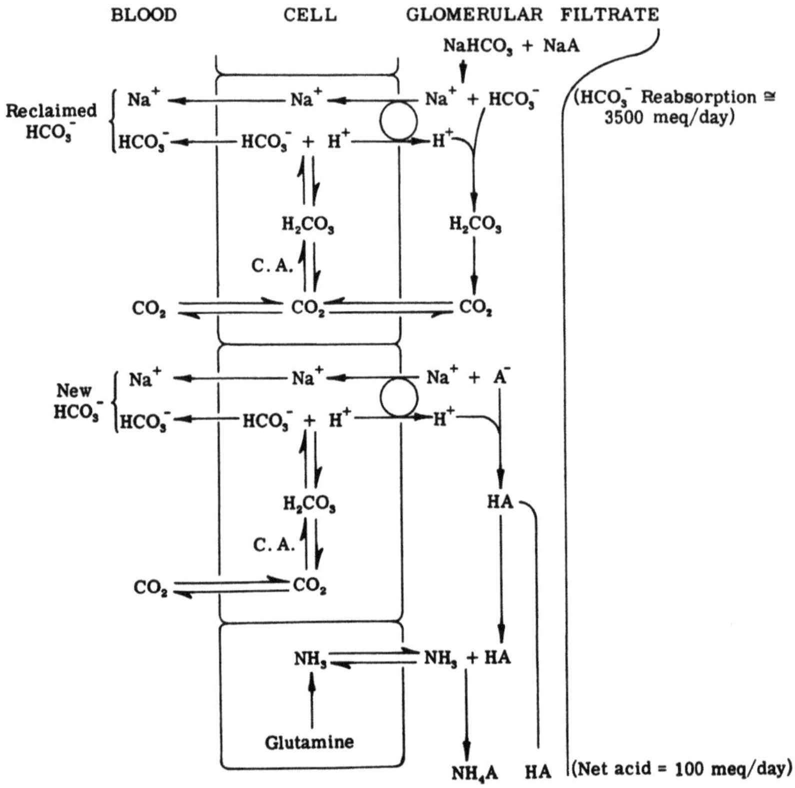

Fig. 1. Generation of bicarbonate as a consequence of proton secretion. [1]

exchanged for Na^+ of the filtered $Na_2H\ PO_4$ resulting in acid excreted as Na H_2PO_4; for each mole of acid so eliminated one mole of new bicarbonate appears in the renal vein. However, the filtered load and tubular phosphate reabsorption limits the amount of base generated in this fashion. Continued acid secretion rapidly achieves the limiting urine to plasma proton gradient, 1,000 : 1, and further base generation is nullified by back diffusion of H^+. The amount of protons excreted in the free form some 60 μmol, is only 0.001 of the daily nonvolatile acid production. Therefore, to preserve cellular H^+ homeostasis the kidney must generate its own intraluminal proton acceptor: ammonia produced from glutamine. Ammonia formed in tubular cells diffuses down its partial pressure gradient into the lumen where the dissolved molecule avidly reacts with the free protons forming ammonium. In contrast to the membrane permeable pNH_3, which moves through water filled pores, the ammonium is relatively impermeable and remains "trapped" in the urine. By effectively lowering the proton gradient, glutamine-dependent ammonia production allows the generation of bicarbonate, mole for mole. In the normal state (Table 1), renal glutamine metabolism supplies 50% of the new bicarbonate constituting 0.7% of the bicarbonate released across the basolateral membrane; in respiratory

Table 1. Quantitative aspects of bicarbonate generation from acid excretion[a]

	Normal	Respiratory acidosis	Metabolic acidosis
GFR (L/24 h)	180	180	180
$[HCO_3^-]$p (m*M*/L)	26	30	10
HCO_3^- filtered (m*M*/24 h)	4,680	5,400	1,800
HCO_3^- excreted (m*M*/24 h)	0	0	0
HCO_3^- readsorbed (m*M*/24 h)	4,680	5,400	1,800
T.A. excreted (m*M*/24 h)	35	70	70
NH_4^+ excreted (m*M*/24 h)	35	70	230
Total H^+ secreted (m*M*/24 h)	4,750	5,540	2,100
% Reclaimed HCO_3^-	98.5	97.4	85.8
% New HCO_3^-	1.5	2.6	14.2

[a] After [2]; data on T.A. and NH_4^+ [15]

acidosis this contribution doubles; in metabolic acidosis the glutamine dependent contribution jumps nearly 16-fold. In the absence of glutamine or the ammonia produced therefrom, secreted H^+ filters back into the blood titrating the alkaline reserves and threatening cellular homeostasis.

Conceptual Development of Renal Glutamine-Dependent Base Generation

Historically, progress towards the realization of the above concept came not in an evolutionary progression, but rather in quantum jumps. A brief historical digression provides insight into this process as well as reveals the cornerstones upon which this concept was built. An awareness of acid excretion and the role of ammcnium excretion preceeds that of acid secretion and for our purposes can be said to have begun with Salkowski's study in 1873 [3]. He administered a mineral acid load to dogs and observed a brisk ammonium excretion response; this led to the generalization that carnivora disposed of their nonvolatile acid load by neutralization with ammonia, formed in the periphery, and excreted as ammonium; herbivora, in contrast, lacked this base generating mechanism and consequently depleted their alkaline reserves. Despite numerous subsequent studies supporting the primary role of ammonium excretion, acid eliminated as titratable acid was given supremacy by Henderson [4]. His studies were based on excretion rates of normal and nephritic subjects in which urine acidity was allowed to vary spontaneously. Although he viewed acid excretion as the sum of titratable acid plus ammonia, the fact that the latter varied little led him to conclude that it does not play a regulatory role in acid base homeostasis. Thus, nephritic subjects develop metabolic acidosis because of reduced titratable acid excretion with retention of phosphoric acid in the blood [5] and not due to the absence of the ammonia response described by Salkowski [3]. Regardless of the actual role of ammonia in acid elimination, it was "a commonly accepted view that neutralization of acids by ammonia is a function of the organism in general, or of the liver" [6]. In 1921 this view was challenged and vanquished by Nash and Benedict [6]. In its place came a cornerstone of the present paradigm: acid excretion in the form of ammonium derives from renal ammoniagenesis. Their starting point was a not so simple technical point, blood ammonia levels. They reasoned, given the extraordinary levels of ammonium in even the normal urine, blood ammonium levels must be very high indeed; or, if not, the kidney itself must synthesize the excreted ammonium. The test was to measure the arterial and renal venous ammonia levels. The result, renal venous concentration greater than arterial, allowed them to conclude that the kidney synthesizes ammonia. Armed with this they advanced beyond their experimental base postulating depletion of alkaline reserves: (1) due to "introduction of acids into the blood more rapidly than the normal kidney can make ammonia to combine with them" and (2) "results should the kidney become defective in its power of ammonia formation (nephritic)."

By 1921 the concept of acid excretion was essentially in place lacking only the connection with tubular acid secretion and concomitant renal vein bicarbonate release. Unfortunately, tubular acid secretion was mired in the much broader controversy concerning the processes involved in the elaboration of urine. Although Ludwig had proposed his monumental filtration-reabsorption hypothesis in 1842, renal physiology was in fact dominated by the tubular secretion hypothesis of Bowman, 1842, and Heidenhain, 1874 [7]. In 1917 Cushny published his classic with the intriguing title[2], *The Secretion of the Urine;* this "modern theory" in essence resurrected Ludwig's view while strenuously opposing all tubular secretion, including acid [8]. He puts forward the view, consistent with his thesis, that urine acidity derives from the filtration of conjugate acid base pairs Na H_2PO_4/Na_2HPO_4 and $H_2CO_3/NaHCO_3$ followed by selective reabsorption of the base forms relinquishing the acid moiety to acidify the urine. Cushny concludes: "Followers of Bowman and Heidenhain assert that acid is actively excreted, those of the modern theory that alkali is absorbed." Subsequent studies of subjects in the spontaneous acid base state seemingly confirmed this hypothesis [9]. Nor did studies of subjects given an NH_4Cl acid load reveal its limitation; rather, quite the contrary. Believing chloride anions represents acid which obligates the excretion of a base cation, sodium, the rise in ammonium excretion was seen as an adaptive consequence of tubular $NH_4^+ \gtrless Na^+$ exchange [10]; this misconception has in turn led to the use of NH_4Cl as diuretic, defining in one instance [11] the limits of H^+ homeostasis.

In 1944 came the unequivocal demonstration that tubules secrete protons[3]. Working in the seemingly unrelated area of tubular phosphate reabsorption, Robert Pitts presented his findings of a fixed transport maximum, Tm, at the Cornell Research Forum; his conclusion was quickly challenged from the perspective of the filtration-reabsorption hypothesis since Tm must decrease in acidosis to accommodate rising titratable acid excretion. Pitts then carefully

2 Cushny's usage of the terms secretion and excretion was extraordinary; the former being what is excreted and the latter being what is secreted, i.e., transferred from tubule cell to urine

3 A fascinating first hand account of this event has recently been published, see Alexander, R. S. (1983) The Physiologist 26: 364

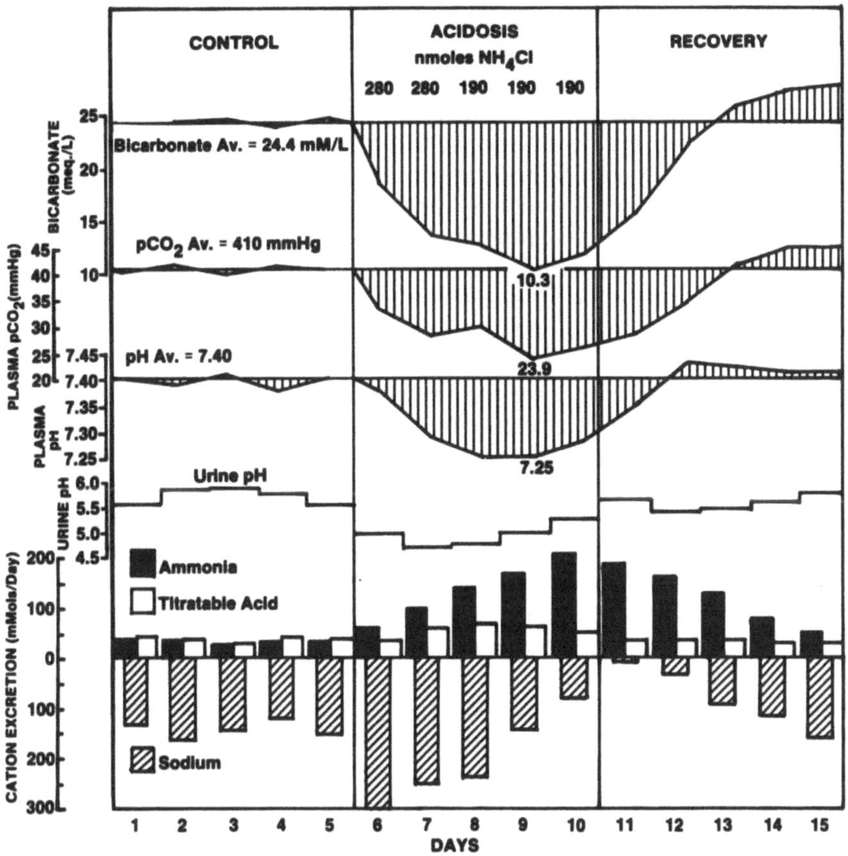

Fig. 2. Plasma bicarbonate levels and ammonium excretion. [15]

designed experiments from which tubular secretion of acid could be inferred; these were carried out by raising the plasma phosphate concentration and thereby the filtered phosphate load, exceeding the Tm, and demonstrating that titrable acid excretion exceeded the sum of filtered H_2CO_3 and NaH_2PO_4 [12]. (Ironically, Cushny himself performed this experiment in 1904 [13], also observing a remarkable acid excretion but unlike Pitts in 1944 was unable to quantify the tubular interactions.) Pitts then went far beyond the data proposing that tubular cells secret protons in exchange for filtered sodium and that the source of this proton was carbonic acid; a $H^+ \gtrless Na^+$ exchange rather than the gastric HCl model was proposed since tubular chloride secretion was not observed; H_2CO_3 was inferred to be the source of the secreted H^+ since carbonic anhydrase inhibition reduced acid secretion. [Sulfonamides, forerunners of the modern line of diuretics, were recently shown to deplete alkaline reserves by a direct renal action (14)]. Ammonium excretion was seen as following passively from the diffusion of ammonia down its concentration gradient established by active proton secretion; the generation of bicarbonate released into the renal vein was inferred to follow. Although this hypothesis could not, strictly speaking, be proven its power was quickly appreciated and deployed, Fig. 2 [15]. Now ammonium chloride ingestion was seen as an acid load by merit of the generation of hydrochloric acid equivalents titrating the alkaline reserves. Initially urine pH falls as a consequence of reduced filtered bicarbonate and laggard ammonia production and, hence, limited bicarbonate generation. In the chronic phase plasma bicarbonate stabilizes as a consequence of the adaptive ammonia production dependent generation of bicarbonate. Production of bicarbonate, generated by ammonium excretion, continues despite cessation of acid loading inducing a metabolic alkalosis overshoot. Thus, these investigators clearly associated by 1948 the adaptive increase in renal ammonia production, however regulated, with generation of bicarbonate; indeed, both were seen as an essential couplet in H^+ homeostasis.

Overlooked, however, was the precursor of the ammonia produced in tubule cells. Nash and Benedict [6] considered amino acids, but opted for urea based on the reciprocal urea/ammonia excretion rates. Lotspeich and Pitts [16] suggested ammonia generated from amino acid oxidase activity. However, in a remarkable two page report to the editors of the *Journal of Biological Chemistry,* Van Slyke and associates [17] had, in fact, already shown that the amide nitrogen of glutamine extracted by the kidney "sufficed to provide the ammonia removed from the kidney via the renal vein and 60% or more of the ammonia excreted in the urine." Like the other two cornerstones, renal ammoniagenesis and tubular H^+ secretion, this finding has never been challenged. In fact, glutamine extraction provides the precursor nitrogen for ammonia production during acidosis in man [18] and rat [19] as well as dog [20]. Van Slyke et al. concluded, appropriately enough: "The data provided the first direct evidence by experiments in vivo of a physiological function of glutamine in animals."

Recently, the concept that renal glutamine metabolism does in fact generate base has come under attack [21, 22]. The argument goes like this: hydrolysis of glutamine within tubule cells produces glutamate and ammonium, NH_4^+; the latter dissociates to a free proton subsequently pumped into the urine with ammonia diffusing to react with that proton. Thus, no net acid elimination occurs as ammonium excretion nor can bicarbonate be generated. Instead ammonium excretion is seen as a means of eliminating nitrogen rather than acid and hepatic urea production by consuming bicarbonate is seen as regulating acid base homeostasis. Accordingly in metabolic acidosis, when the alkaline reserves are limited, nitrogen is shifted from bicarbonate-consuming-urea production to glutamine synthesis as an adaptation to the acid load. In turn, this glutamine is extracted by the kidneys in increased amounts during acidosis and the amide nitrogen is excreted as ammonium, while glutamate is recycled back to the liver. Thus, the liver and its rate of urea production is seen as the regulator of H^+ homeostasis: not the kidney.

In order for renal glutamine metabolism to generate bicarbonate it must undergo the reactions shown in Fig. 3. The conversion of 1 mol of glutamine to 2 mol of ammonium will result in 2 mol of bicarbonate generated if, and only if, the carbon skeleton, α-ketoglutarate is converted to the neutral products CO_2 and glucose. If only 1 mol of ammonium is formed per glutamine, $NH_3/Gln = 1.0$, and glutamate is released into the renal vein, no net acid excretion nor base

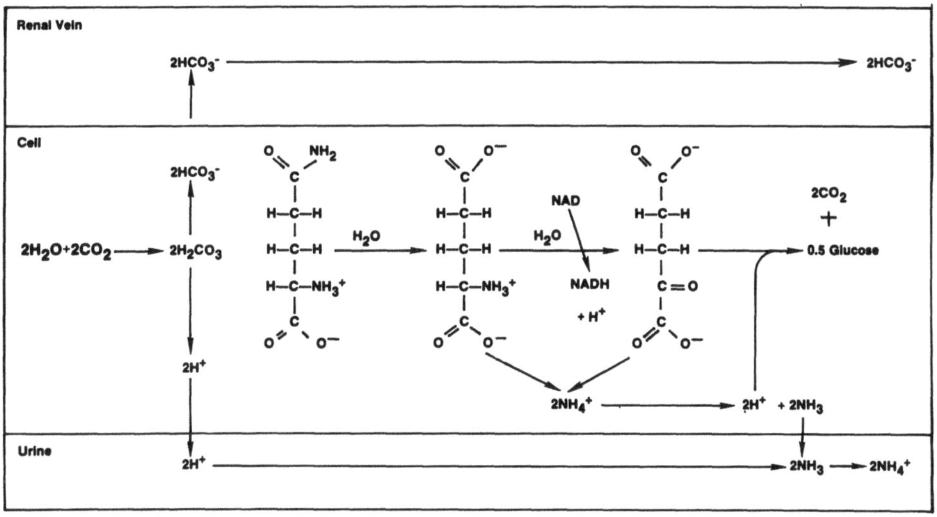

Fig. 3. Renal glutamine metabolism and bicarbonate generation

generation occurs; if α-ketoglutarate is released, despite an $NH_3/Gln = 2.0$, no new bicarbonate is generated. Incorporation of glutamine carbon into the neutral product glucose has been shown in vitro using rat kidney cortical slices [23]; incorporation increased in slices from chronically acidotic rats. In vivo studies performed on chronically acidotic and alkalotic dogs demonstrated predominant conversion of glutamine carbon to CO_2 [24]; conversion rates increased in chronic acidosis. Although the relative role of the oxidative versus gluconeogenic pathway is unclear, for the purpose of bicarbonate generation conversion of α-ketoglutarate to either product suffices for the formation of new base.

Proof That Renal Glutamine Metabolism Generates Bicarbonate

Historically, net acid excretion rather than bicarbonate generation has been the approach of studies relating renal function to acid base homeostasis. An obvious reason was the ease in measuring acid excretion in contrast to renal vein bicarbonate. Not only is renal vein sampling demanding, but owing to the huge disparity in urine and blood volumes, the small amount added to the latter would be difficult to accurately measure. The truly impassable obstacle, however, is that presented by the preformed renal bicarbonate load; even in severe metabolic acidosis the newly formed bicarbonate is less than 3% of the total exiting the kidney. Consequently, renal glutamine metabolism dependent bicarbonate is lost in a virtual sea of preformed bicarbonate; yet proof of the glutamine conversion shown in Fig. 3 requires simultaneous measurement of both urine acid excretion and plasma bicarbonate release. The solution to this problem can be had by perfusing the isolated kidney [25] with a bicarbonate less plasma, Fig. 4. Closing the system as shown ensures the retention of gaseous

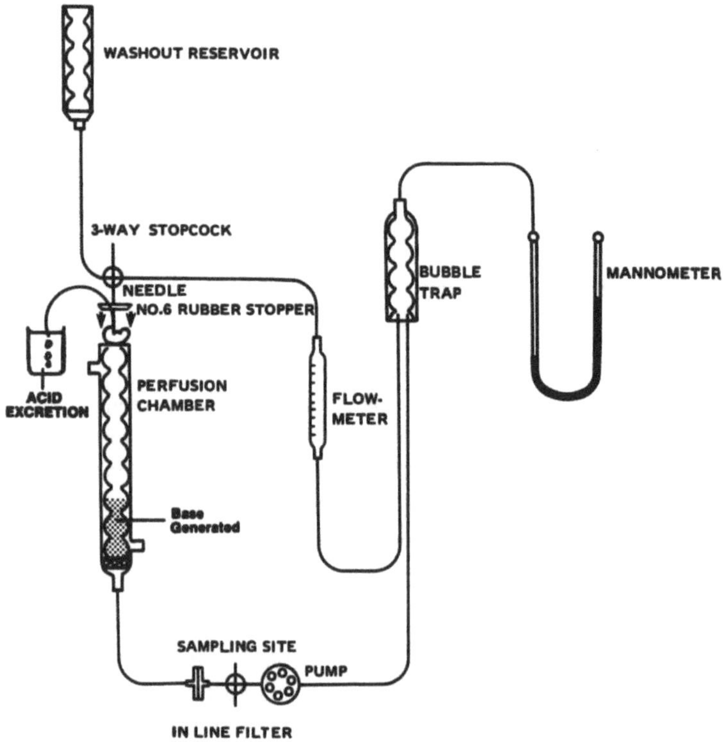

Fig. 4. Isolated perfused rat kidney: acid excretion and plasma bicarbonate accumulation

CO_2 and NH_3 and, therefore, accurate accounting of both carbon and nitrogen products. Using ^{14}C labelled glutamine allows the tracing of glutamine's carbon disposal along either the oxidative or gluconeogenic pathway. The actual experiments were performed as follows: kidneys from control and chronically acidotic rats were isolated, mounted enclosed in the perfusion chamber, and perfused with 1 m*M* L-glutamine over a 40 min time course; plasma samples were taken at the beginning and end of the time course; urine was collected via a ureter cannula exiting the system through an air tight portal. Despite the absence of exogenous CO_2 both the kidneys from nonacidotic and chronically acidotic rats secreted protons in maintaining a urine to plasma gradient of 10 : 1; bicarbonate generation was equivalent to acid excretion (Fig. 5). Adding 60 μmol of glutamine to the nonacidotic kidney resulted in the uptake of 14 μmol and production of 17 μmol of ammonia, NH_3/Gln = 1.21, Table 2; glutamine carbon recovered in neutral products, CO_2 plus glucose, averaged 6.5 μmol equal to the accumulation of 6 μmol bicarbonate in the plasma. Of the 14 μmol of glutamine taken up only 6 μmol of bicarbonate, or some 21% of the maximal base generating capacity, were released. The presentation of glutamine to the chronically acidotic rat kidney resulted in a strikingly different response. Glutamine uptake jumped fourfold to 55 μmol with an 80% efficiency in generating bicarbonate; plasma bicarbonate rose from 6 to 88 μmol in acidosis

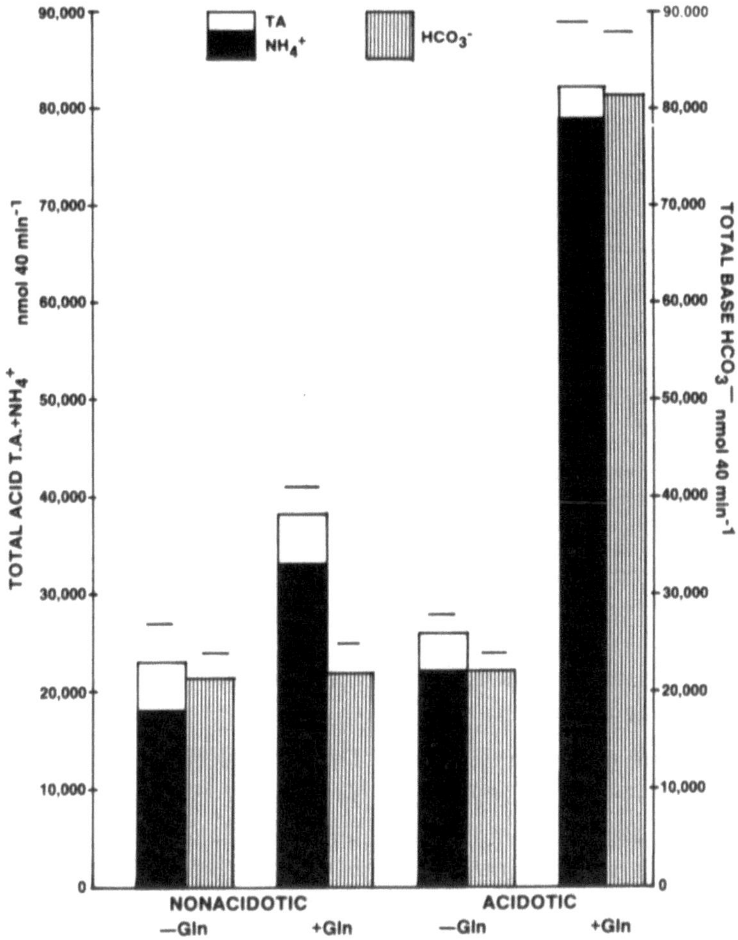

Fig. 5. Effect of glutamine on bicarbonate generation and acid excretion

(Fig. 5 and Table 2). Consistent with this is the rise in the NH_3/Gln ratio to 1.87 and sixfold rise in ammonia production. Furthermore, of the 55 μmol of glutamine extracted 41 μmol could be accounted for as neutral products CO_2 and glucose in good agreement with the rise in bicarbonate, 88 μmol (approx. 41 μmol glutamine times 2 mol bicarbonate per glutamine); the ratio of glutamine carbon atoms incorporated into CO_2 to those incorporated into glucose, 85 : 144, is consistent with disposal of α-ketoglutarate via the gluconeogenic pathway. Therefore, in chronic metabolic acidosis the renal glutamine metabolism approximates the following stoichiometry:

$$1 \text{ mol glutamine} \rightarrow 2 \text{ mol } NH_3 + 2 \text{ mol } CO_2 + 0.5 \text{ mol glucose.}$$

Utilization of glutamine by the neutral γ-glutamyltransferase with an NH_3/Gln ratio of 1.0 is relatively more significant in the nonacidotic kidney [26]. An unexpected but obviously necessary correlate of the tremendous increase in

Table 2. Renal glutamine metabolism and bicarbonate generation

	Glutamine uptake	Ammonia production	Ammonia/Glutamine	Neutral CO_2
NA $\bar{X}$ SEM	14[a] ± 4	17 ± 3	1.21 ± 0.04	3.0[b] ± 0.3 (15)
A $\bar{X}$ SEM	55 ± 5	103 ± 10	1.87 ± 0.06	17 ± 3 (85)

	Products Glucose	Total bicarbonate generated	Bicarbonate from glutamine
NA $\bar{X}$ SEM	3.5[c] ± 0.2 (21)	25[d] ± 5	6[e]
A $\bar{X}$ SEM	24 ± 3 (144)	109 ± 14	88

[a] μMol per 40 min from six kidneys in each group
[b] μMol of glutamine converted to CO_2
[c] μMol glucose formed from glutamine see [38] for details; numbers in parentheses are glutamine carbon atoms in CO_2 or glucose
[d] μMol of bicarbonate in plasma after 40 min
[e] Total bicarbonate generated in presence of glutamine minus that formed in absence of glutamine

bicarbonate is the increase in proton secretion, indicating acid secretion is also glutamine dependent:

$$1 \text{ mol glutamine} \rightarrow 2 \text{ mol } CO_2 + 2 \text{ mol } H_2O$$
$$\rightarrow 2 \text{ mol } H^+ + 2 \text{ mol } HCO_3^-.$$

The asymetrical tubule cells partition protons in the urine where they are held by ammonia releasing new bicarbonate, mole for mole, into the renal vein as predicted (Fig. 1). The new bicarbonate elevates plasma bicarbonate, accompanied by sodium [27].

Regulation of Bicarbonate Generation from Glutamine

Although bicarbonate generation occurs in both respiratory and metabolic acidosis [28], the magnitude of the response in the latter gives this approach a particular vantage point for studies of regulation. Bicarbonate generation occurs in three distinct phases (Fig. 2): (1) acute, day 6; (2) chronic, day 10; and (3) recovery, day 13. *Acute phase:* Acid excretion is only slightly enhanced due to small increases in titratable acid and ammonium excretion. Bicarbonate generation derives from a rise in the NH_3/Gln ratio 1.2 to 2.4 [29] and a shift of ammonium from renal vein to urine; glutamine uptake remains essentially unchanged (Fig. 6). *Chronic phase:* The NH_3/Gln ratio remains near maximal and now increased ammonia production is provided by enhanced glutamine extraction, Fig. 6 [18]. Increased glutamine uptake is all the more remarkable since the plasma levels are reduced some 40% [18, 30]; increasing glutamine concentration drives ammonium excretion [17, 31] and elevates plasma

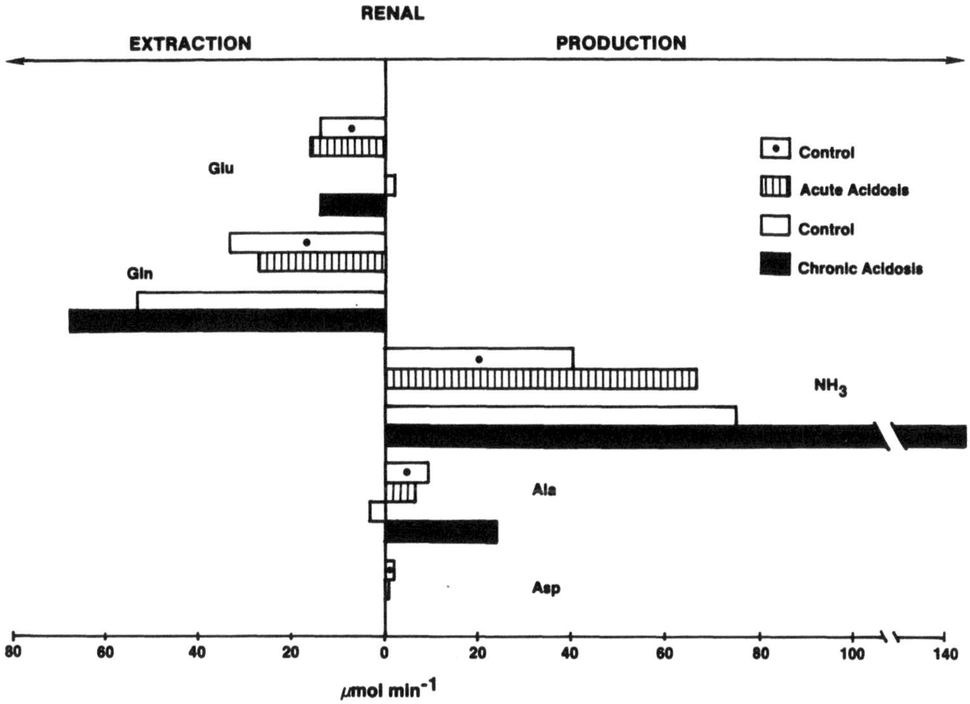

Fig. 6. Renal glutamine metabolism and nitrogen release in vivo by acutely acidotic and chronically acidotic humans. Acutely acidotic subjects from Tizianello et al. [29] after 24 h NH_4Cl ingestion, 159 mmol; chronically acidotic subject G. J. served as own control followed by 6 days of NH_4Cl, maximum 150 mmol per day [18]

bicarbonate in healthy subjects [30], but interestingly not in nephritic patients [30]. In this phase bicarbonate generation is near maximal despite rising plasma bicarbonate concentration and decreased urine acidity.

Recovery Phase. The excessive and inappropriate bicarbonate generation from sustained glutamine metabolism characterizes this period. Acid excretion as ammonium remains high despite normal urine pH obligating large amounts of ammonium to the renal vein; nevertheless metabolic alkalosis appears in virtually every study [15, 30, 32]. Pitts in reflecting upon this response concluded: "It is obvious that some other factor than plasma bicarbonate must condition the rate of excretion of ammonia" [15].

That this factor derives from the adrenal gland was suggested by the reduction in alkaline reserves of adrenalectomized subjects [33] and supported by the reduction in net acid excretion [34]. Moreover, adrenalectomized animals exhibited a blunted response to an acid load, largely, but not entirely due to impaired ammonium excretion [35]; evidence implicating the adrenals in the response to acidosis was also provided [35]. Following this path, synthetic glucocorticoids were shown to be potent stimulators of ammonium excretion and ammonia production from glutamine by kidneys from adrenalectomized rats [36,

Table 3. Renal glutamine metabolism in acute and chronic acidosis

	In vivo		
	Total CO_2 (m*M*)	Urinary ammonium ($\mu mol \cdot 4\ h^{-1}$)	Plasma corticosterone ($\mu g/100\ ml$)
Nonacidotic	27.1 ± 0.4	93 ± 11	15.8 ± 0.4
Acidotic (4 h)	22.5 ± 0.7	344 ± 11	31.4 ± 2.7
Acidotic (7 days)	22.6 ± 0.5	546 ± 30	36.3 ± 2.1

	In vitro[a]		
	Ammonia production	Glutamine (Gln) uptake	
	($\mu mol \cdot h^{-1} \cdot g^{-1}$)	($\mu mol \cdot h^{-1} \cdot g^{-1}$)	NH_3/Gln
Nonacidotic	20.6 ± 2.3	16.4 ± 0.8	1.26 ± 0.02
Acidotic (4 h)	45.2 ± 2.7	25.2 ± 2.1	1.82 ± 0.13
Acidotic (7 days)	79.5 ± 3.3	43.2 ± 1.2	1.84 ± 0.04

[a] Isolated kidneys perfused with 1 m*M* L-glutamine. Means ± SE ($n = 5-7$) significantly different from control [37]

37]; acute exposure to triamcinolone, 20 min, enhanced the NH_3/Gln ratio [36]. Chronic exposure to triamcinolone stimulated glutamine uptake at the sustained NH_3/Gln ratio and generated bicarbonate in excess leading to metabolic alkalosis reminiscent of the recovery phase [38]. Detailed balance studies in dogs administered dexamethasone clearly demonstrated ammonia production dependent acid excretion in that species [39]. The link between natural glucocorticoids and ammonium excretion [40] and ammonia production was subsequently provided. Within 4 h of an acute acid load both plasma corticosterone and the NH_3/Gln ratio are elevated (Table 3) with little increase in glutamine extraction; chronically corticosterone levels remain elevated as glutamine uptake is enhanced (Table 3). It should be noted that these in vivo responses are preserved in vitro in the perfused kidney despite normal pH and bicarbonate levels. Thus, the renal glutamine metabolism response to acidosis, acute, chronic, and recovery phases, is identical to that after glucocorticoid administration regardless of the acid base status. In the absence of glucocorticoids an acid load results in a precipitous fall in plasma bicarbonate, quite remarkable in view of the small, but significant rise in ammonium excretion [41]. The ineffectiveness of this response appears to reflect the latency of the base generating mitochondrial pathway and acidosis enhanced activity of the neutral γ-glutamyltransferase pathway [42]. In addition to glucocorticoid regulation of renal glutamine metabolism, mineralocorticoids are also elevated in acid loading [41] and appear to play their primary role in distal tubule proton secretion [40, 43] and not in ammonia production [44]. Nevertheless, the adrenal gland and specifically glucocorticoids unquestionably play a major role in regulating renal bicarbonate generation.

Table 4. Enlarged mitochondria and reduced renal energy levels in metabolic acidosis[a]

	Mitochondrial fractions (%)			
Density (g/cm^3)	1.165	1.175	1.188	1.200
Nonacidotic (5)	35 ± 5	51 ± 6	9 ± 2	2 ± 1
Metabolic acidotic (5)	18^b ± 4	46 ± 8	29^b ± 3	5^b ± 2

	Energy levels (nmol/g)		
	ATP	ADP	AMP
Nonacidotic (5)	1,593 ± 153	830 ± 73	210 ± 33
Metabolic acidotic (5)	$1,303^b$ ± 156	917 ± 57	310^b ± 37

[a] Five control and five 10-day chronically acidotic rats were anesthesized with inactin and the left kidney freeze clamped in situ and processed for adenine nucleotide analysis on HPLC; right kidney was homogenized in 0.44 *M* sucrose and subjected to sucrose density gradient analysis; mitochondrial distribution was determined by cytochrome oxidase assay with results expressed as percent of starting homogenate population in each fraction; from Norwood, Henderson, and Welbourne, unpublished

[b] Significantly different from nonacidotic, $p < 0.05$

Glucocorticoids appear to act by "conditioning" the mitochondria of the proximal convoluted S_2 segment, for it is the mitochondria of this segment which undergo striking enlargement during chronic acidosis [45]. Since the enzymes of glutamine metabolism, glutaminase-glutamate dehydrogenase, are localized within the inner mitochondrial membrane [46] and since permeability of this barrier to glutamine increases in acidosis [45], it is likely that glucocorticoids act to facilitate glutamine's penetrance. Inasmuch as the key gluconeogenic enzymes are also localized in the proximal convoluted tubule [47], the increase in mitochondrial permeability would virtually ensure bicarbonate generation [48, 49]. The conditioning effects appears to reflect recruitment of proximal convoluted mitochondria into the permeable and, thus, activated subpopulation; accordingly, aged and, therefore, more susceptible organelles may respond initially giving the acute response and as more are recruited a near maximal response is achieved. Consistent with this is the increased mitochondrial turnover and phospholipid synthesis [50] in this population as well as accelerated autophagy of these mitochondria [45]. Uncouplers of oxidative phosphorylation, for example 2,4-dinitrophenol, are potent stimulators of ammonia formation while swelling the in situ mitochondria [51] suggesting both glutamine and proton gradients are abolished in these conditioned mitochondria. Indeed, Table 4 shows that renal ATP levels are reduced in chronic metabolic acidosis roughly in proportion to the shift in mitochondria to the enlarged, more dense, mitochondrial population. As a consequence energy dependent proximal tubule functions, such as reabsorption of the glomerular filtrate, may be impaired. If so, the chronically acidotic kidney may be viewed as compromising its own functional integrity in achieving maximal base generating capacity thereby preserving H^+ homeostasis of extrarenal cells at the expense of extracellular fluid volume homeostasis.

Regulation of Glutamine Homeostasis to Supply Renal Bicarbonate Generation in Metabolic Acidosis

Renal bicarbonate generation as well as numerous extrarenal biosynthetic reactions depend upon plasma glutamine for delivery of nitrogen and carbon moieties [52]. Consequently, it is not surprising that the body has a means for directing the flow of glutamine according to prevailing metabolic pressures. A reflection of this mechanism is the glutamine dependent rise in ammonium excretion and fall in urea excretion during metabolic acidosis; acute reversibility of this mechanism is shown by the return of urea production following elimination of ammonium excretion [53]. Undoubtedly, the flow of glutamine between muscle, kidney, intestine, and liver is closely regulated since conversion of glutamine nitrogen into urea is, in terms of ATP, much more expensive than conversion to ammonia and excretion as ammonium. The direction of flow may be dependent in large part on plasma glutamine concentration itself. In metabolic acidosis plasma glutamine concentration is reduced about 40% [18, 30]. Despite the reduced level renal glutamine uptake is highly effective [54], driven by the wide open proximal tubule mitochondria; other organ uptake is not. As a consequence, the kidneys siphon off a larger portion of the reduced pool. A more intricate mechanism yet may involve the gut and liver operating in a feed-forward manner. Intestinal glutamine extraction appears to be maintained in chronic acidosis [53, 55], but in place of alanine released in nonacidosis [55] now releases ammonia [53, 55]. Consequently, the portal vein profile normally seen, high alanine and glutamine and low ammonia, is skewed to high ammonia and low alanine and glutamine [53]; the former mixture supports ureagenesis, the latter glutamine synthesis. This relationship between portal vein NH_3/Gln appears to be critically important in determining net release or uptake of glutamine by the liver [57]. Thus, in metabolic acidosis hepatic glutamine formation is enhanced at the expense of urea and glucose formation, while renal bicarbonate and glucose formation is enhanced at the expense of glutamine. In the recovery phase renal vein ammonia loads deliver precursor to muscle, the major manufacturer [58] and plasma glutamine levels rise. Faced with elevated glutamine the gut reverts to releasing alanine; alanine in combination with higher portal vein glutamine and reduced ammonia fuels ureagenesis and gluconeogenesis. Although hypothetical, a sparing of glutamine would leave unaltered glutamine turnover [59] and thereby preventing a spill over of acidosis induced compensations into nitrogen homeostasis. Plasma glutamine concentration is set by the prevailing pCO_2 [60] and conceivably by way of intracellular pH which, if so, provides a link between extrarenal supply of the precursor of renal base generation and cellular hydrogen ion homeostasis.

Acknowledgements. We are grateful to Jennifer Hadley for her excellent secretarial assistance. Dedicated to Emanuel Bresler, 1918–1981.

References

1. Rector FC Jr (1971) Renal secretion of hydrogen. In: Rouiller C, Muller A (eds) The kidney, vol III. Academic Press, London New York, p 245
2. Winters RW, Bell RB (1965) Regulation of acid base equilibrium. In: Yamamoto WS, Brobeck JR (eds) Physiological controls and regulations. Saunders, Philadelphia London, p 234
3. Salkowski E (1873) Über die Entstehung der Schwefelsäure und das Verhalten des Taurins im tierischen Organismus. Arch Pathol Anat Physiol 58: 460–508
4. Henderson LJ (1911) A critical study of the process of acid excretion. J Biol Chem 9: 403–424
5. Henderson LJ, Palmer WW (1915) On the several factors of acid excretion. J Biol Chem 21: 37–55
6. Nash TP Jr, Benedict SR (1921) The ammonia content of the blood, and its bearing on the mechanism of acid neutralization in the animal organism. J Biol Chem 48: 463–488
7. Smith HW (1964) Renal physiology. In: Fishman AP, Richards DW (eds) Circulation of the blood: Men and ideas. Oxford Univ Press, New York, pp 562–582
8. Cushny AR (1917) The secretion of the urine. Longman, Green and Co, London, pp 165–173
9. Senroy J Jr, Seelig S, Van Slyke DD (1934) Studies of acidosis: the carbon dioxide tension and acid base balance of human urine. J Biol Chem 106: 479–500
10. Gamble JL, Blackfan KD, Hamilton B (1924) A study of the diuretic action of acid producing salts. J Clin Invest 1: 359–387
11. Relman AS, Shelburne PF, Talman A (1961) Profound acidosis resulting from excessive ammonium chloride in previously healthy subjects. New Engl J Med 264: 848–852
12. Pitts RF, Alexander RS (1945) The nature of the tubular mechanism for acidifying the urine. Am J Physiol 144: 239–245
13. Cushny AR (1904) On the secretion of acid by the kidney. J Physiol 31: 188–203
14. Southworth R (1937) Acidosis associated with the administration of para-amino-benzene sulfonamide. Proc Soc Exp Biol Med 36: 58–61
15. Sartorius OW, Roemmelt JC, Pitts RF (1949) The renal regulation of acid base balance in man. IV. The nature of the renal compensations in ammonium chloride acidosis. J Clin Invest 28: 423–439
16. Lotspeich WD, Pitts RF (1947) The role of amino acids in the renal tubular secretion of ammonia. J Biol Chem 168: 611–622
17. Van Slyke DD, Phillips RA, Hamilton PB, Archibald RM, Futcher PH, Hiller A (1943) Glutamine as source material of urinary ammonia. J Biol Chem 150: 481–482
18. Owen EE, Robinson RR (1963) Amino acid extraction and ammonia metabolism by the human kidney during the prolonged administration of ammonium chloride. J Clin Invest 42: 263–276
19. Squires EJ, Hall DE, Brosnan JT (1976) Arteriovenous differences for amino acids and lactate across kidneys of normal and acidotic rats. Biochem J 160: 125–128
20. Pitts RF, Pilkington LA, De Haas J (1965) N^{15} tracer studies on the origin of urinary ammonia in the acidotic dog, with notes on the enzymatic synthesis of labeled glutamic acid and glutamine. J Clin Invest 44: 731–745
21. Atkinson DE, Camien MN (1982) The role of urea synthesis in the removal of metabolic bicarbonate and the regulation of blood pH. Curr Top Cell Regul 21: 261–302
22. McGilvery RW (1979) Biochemistry: A functional approach. Saunders, Philadelphia pp 764–767
23. Goodman AD, Fuizy RE, Cahill GF (1966) Renal gluconeogenesis in acidosis, alkalosis and potassium deficiency: Its possible role in regulation of renal ammonia production. J Clin Invest 45: 612–619
24. Pitts RF, Pilkington LA, MacLeod MB, Leal-Pinto E (1972) Metabolism of glutamine by the intact functioning kidney of the dog. J Clin Invest 51: 557–565
25. Bowman R (1975) The perfused rat kidney. In: Hardman JG, O'Malley BW (eds) Methods in enzymology, vol 39. Academic Press, London New York, pp 3–10
26. Welbourne T (1975) Mechanism of renal ammonia production adaptation to chronic acidosis. Med Clin N Am 59: 629–648

27. Rector FC Jr, Martinez-Maldonado M, Brunner FP, Seldin DW (1966) Evidence for passive reabsorption of NaCl in proximal tubule of rat kidney. J Clin Invest 45: 1060
28. Gurreri G, Bertocchi I, DeFerrari G, Acquarone N, Tizanello A (1975) Utilizzazione renale di glutamina e produzione totale di ammoniaca nel ratto in condizioni normali, di acidosi metabolica e respiratoria croniche. Bull Soc Ital Biol Sper 51: 196–202
29. Tizianello A, Deferrai G, Garibotto G, Robaudo G, Acquarone N, Griggeri GM (1982) Renal ammoniagenesis in an early stage of metabolic acidosis in man. J Clin Invest 69: 240–250
30. Welbourne T, Weber M, Bank N (1972) The effect of glutamine administration on urinary ammonium excretion in normal subjects and patients with renal disease. J Clin Invest 51: 1852–1860
31. Dies F, Telley R, Navarete R, Ordoney A (1972) Estimulo de la excretion urinaria de hidrogeniones mediante la injeccion de glutamina en perros. Rev Invest Clin 22: 291–302
32. Wood FJY (1955) Ammonium chloride acidosis. Clin Sci 14: 81–89
33. Jimenez-Diaz C (1936) Death in Addison's disease. Lancet 2: 1135–1139
34. Sartorius OW, Calhoon D, Pitts RF (1952) The capacity of the adrenalectomized rat to secrete hydrogen and ammonium ions. Endocrinology 51: 444–450
35. Sartorius OW, Calhoon D, Pitts RF (1953) Studies on the interrelationship of the adrenal cortex and renal ammonia excretion by the rat. Endocrinology 52: 256–265
36. Welbourne T (1974) Influence of adrenal glands on pathways of renal glutamine utilization and ammonia production. Am J Physiol 226: 555–559
37. Welbourne T (1976) Acidosis activation of the pituitary adrenal-renal glutaminase I axis. Endocrinology 99: 1071–1079
38. Welbourne T, Phenix P, Thornley-Brown G, Welbourne CJ (1976) Triamcinolone activation of renal ammonia production. Proc Soc Exp Biol Med 153: 539–542
39. Hulter H, Sigala J, Sebastian A (1981) Effects of dexamethasone on renal and systemic acid-base metabolism. Kidney Int 20: 43–49
40. Damasco MC, Diaz F, Cenal JP, Lantos CP (1979) Acute effects of three natural corticosteroids on the acid-base and electrolyte composition of urine in adrenalectomized rats. Acta Physiol Lat 29: 305–314
41. Dubrovsky AHE, Nair RC, Byers MK, Levine DZ (1981) Renal net acid excretion in the adrenalectomized rat. Kidney Int 19: 516–528
42. McFarlane-Anderson N, Allayne GAO (1981) The effect of maleate on glutamine metabolism by rat kidney. Biochem Med 25: 149–159
43. Wilcox CS, Cemerikic DA, Giebisch G (1982) Differential effects of acute mineralo- and glucocorticosteroid administration on renal acid excretion. Kidney Int 21: 546–556
44. Welbourne TC, Francoeur D (1977) Influence of aldosterone on renal ammonia production. Am J Physiol 233: E56–E60
45. Welbourne TC, Bazer G (1980) Mitochondrial glutamine permeability and renal ammonia production in metabolic acidosis. Am J Physiol 239: E51–E56
46. Curthoys NP, Weiss RF (1974) Regulation of renal ammoniagenesis. Subcellular localization of rat kidney glutaminase isoenzymes. J Biol Chem 249: 3261–3266
47. Guder WG, Schmidt U (1974) The localization of gluconeogenesis in rat nephron. Z Physiol Chem 3355: 273–278
48. Pitts RF (1972) Control of renal production of ammonia. Kidney Int 1: 297–305
49. Lund P (1980) Glutamine metabolism in the rat. FEBS Lett 117: K86–K92
50. Dass PD (1983) Functional changes in metabolic acidosis. Fed Proc 42: 477
51. Barnett D, Welbourne TC (1982) 2,4-Dinitrophenol stimulation of renal ammoniagenesis. Biochem Biophys Acta 715: 215–221
52. Meister A (1956) Metabolism of glutamine. Physiol Rev 36: 103–127
53. Phromphetcharat V, Jackson A, Dass PD, Welbourne TC (1981) Ammonia partitioning between glutamine and urea: Interorgan participation in metabolic acidosis. Kidney Int 20: 598–605
54. Lemieux G, Vinay P, Cartier P (1974) Renal hemodynamics and ammoniagenesis: characteristics of the antiluminal site for glutamine extraction. J Clin Invest 53: 884–894
55. Addae SK, Lotspeich WD (1968) Relation between glutamine utilization and production in metabolic acidosis. Am J Physiol 215: 269–277

56. Windmueller HG, Spaeth AE (1974) Uptake and metabolism of plasma glutamine by the small intestine. J Biol Chem 249: 5070–5079
57. Häussinger D, Sies H (1979) Hepatic glutamine metabolism under the influence of the portal ammonia concentration in the perfused rat liver. Eur J Biochem 101: 179–184
58. Garber AJ, Karl IE, Kipnis DM (1976) Alanine and glutamine synthesis and release from skeletal muscle. J Biol Chem 251: 836–843
59. Squires EJ, Brosnan JT (1983) Measurement of the turnover rate of glutamine in normal and acidotic rats. Biochem J 210: 272–280
60. Phromphetcharat V, Welbourne TC (1982) Glutamine homeostasis: role of pCO_2 in regulating plasma Gln levels in metabolic acidosis. Fed Proc 41: 1114

Effects of 2-Oxoglutarate and Glutamate on Glutamine Metabolism by Rat Kidney Mitochondria*

A. C. Schoolwerth and K. F. LaNoue[1]
(With the technical assistance of W. J. Hoover)

In order to provide further insight into the control of renal ammoniagenesis, recent interest has focused on the initiating events in acute metabolic acidosis. A perplexing problem has been the observation that despite an increase in renal ammoniagenesis from glutamine known to occur in acidosis in vivo, glutamine metabolism in vitro was depressed by acid pH [1–3], probably by inhibition of phosphate-dependent glutaminase [4]. Although glutamine is the primary precursor for renal ammonia formation, other metabolites present in the cell in vivo may influence its metabolism. Their absence from in vitro incubation media could explain the anomalous pH response. The observation that 2-oxoglutarate metabolism was augmented in acute acidosis by stimulation of 2-oxoglutarate dehydrogenase provided a clue. Subsequently, Lowry and Ross [5] and Schoolwerth et al. [6] demonstrated that the K_m of the enzyme for 2-oxoglutarate was decreased at acid pH values. Moreover, additional studies from our laboratory demonstrated that renal glutamine metabolism was accelerated by hydrogen ions in the presence, but not in the absence, of 2-oxoglutarate, suggesting an important role for this metabolite in the augmented ammonia synthesis in acute metabolic acidosis [7]. Subsequent work has focused on the role of glutamate metabolism in the initiation of enhanced ammoniagenesis in acute acidosis.

The experiments reviewed below suggest that enhanced glutamate oxidative deamination leads to the initial increase in ammoniagenesis observed acutely in acidosis. The H^+ induced increase in glutamate oxidation is mediated by alterations in mitochondrial glutamate transport and 2-oxoglutarate metabolism.

Methods

Renal cortical mitochondria from Sprague-Dawley rats were isolated and purified as previously described [3, 8]. For the measurement of total mitochondrial metabolites, mitochondria (2–3 mg/ml) were incubated at 28° C in a medium containing (in m*M*) 135 KCl, 5 $MgCl_2$, 5 KH_2PO_4, 20 MOPS,

* This work was supported by National Institutes of Health Grant AM-19714

1 Renal and Electrolyte Division, Departments of Medicine and Physiology, The Milton S, Hershey Medical Center, The Pennsylvania State University, P.O. Box 850, Hershey, PA 17033, USA

Glutamine Metabolism in Mammalian Tissues
Edited by D. Häussinger and H. Sies

10 Tris-HCl, 30 glucose, 0.1 malate, 0.2 ADP, hexokinase (6 U/ml), and 1 glutamine plus 0.5 2-oxoglutarate. In several experiments, 0.5 m*M* 2-[1-^{14}C] oxoglutarate was used to determine the fate of the 2-oxoglutarate added. Samples were taken at intervals following the onset of incubation and transferred to 14% perchloric acid. The acidified samples were then centrifuged to remove the acid precipitate and neutralized. Changes in medium ammonia, glutamate, aspartate, and 2-oxoglutarate were measured enzymatically, allowing calculation of enzyme fluxes in the intact mitochondria [3, 9].

Measurement of mitochondrial glutamate transport was performed as previously described [10]. In these studies mitochondria were incubated at 28° C in a medium containing (final concentration, in m*M*): 130 KCl, 20 MOPS, 10 succinate 0.5–10 [U-^{14}C]glutamate, [1-^{3}H]sucrose, 6 μg/ml rotenone, 7% dextran. The pH was varied as indicated in the text. When glutamate efflux was evaluated, mitochondria were loaded with 10–100 m*M* [U-^{14}C]glutamate [10, 11]. Following loading, the mitochondria were washed several times and the efflux of glutamate from the matrix space was measured [10, 11].

Intramitochondrial metabolites and spaces were evaluated utilizing silicone oil centrifugation [3, 7]. Mitochondrial $^{3}H_2O$ and [U-^{14}C]sucrose spaces were determined in parallel incubations. Intramitochondrial pH was estimated by the distribution of [2-^{14}C]5,5′dimethyloxazolidine-2,4-dione (DMO), as described previously [3, 7].

Results and Discussion

With 1 m*M* glutamine as sole substrate, ammonia formation was decreased when medium pH was reduced from 7.4 to 6.8 (data not shown, see [7]). However, when a physiological level (0.5 m*M*) of 2-oxoglutarate was added to the incubation medium, the opposite pH response was observed (Fig. 1). With the addition of 2-oxoglutarate to glutamine, ammonia formation was lower at pH 7.4 than with glutamine alone (data not shown). However, as medium pH was decreased to 6.8 in the presence of 2-oxoglutarate, a 66% increase in ammonia production was observed compared with pH 7.4 (11.8 ± 0.8 vs 7.1 ± 0.6 nmol · mg^{-1} · min^{-1}, $p < 0.05$). This increase resulted entirely from an increased flux through glutamate dehydrogenase (+ 2.5 ± 0.7 vs − 0.8 ± 0.5 nmol · mg^{-1} · min^{-1}, $p < 0.05$). Glutaminase flux was unchanged at acid pH under these conditions. There was no definite effect of pH on glutamate oxalacetate transaminase flux (data not shown) in these series of experiments. Additionally, it was noted that accumulation of glutamate in the incubation medium was reduced significantly at pH 6.8 compared to 7.4 (+ 3.4 ± 0.9 vs + 5.4 ± 0.9 nmol · mg^{-1} · min^{-1}, $p < 0.05$).

In order to provide further information with respect to the stimulation of ammonia formation from glutamine when medium pH was decreased in the presence of 2-oxoglutarate, intramitochondrial pH was estimated by the distribution of [2-^{14}C]DMO under experimental conditions identical to those of Fig. 1. As shown in Table 1, at more alkaline medium pH values, intramitochondrial pH was approx. 0.6 pH units higher than the medium pH. When

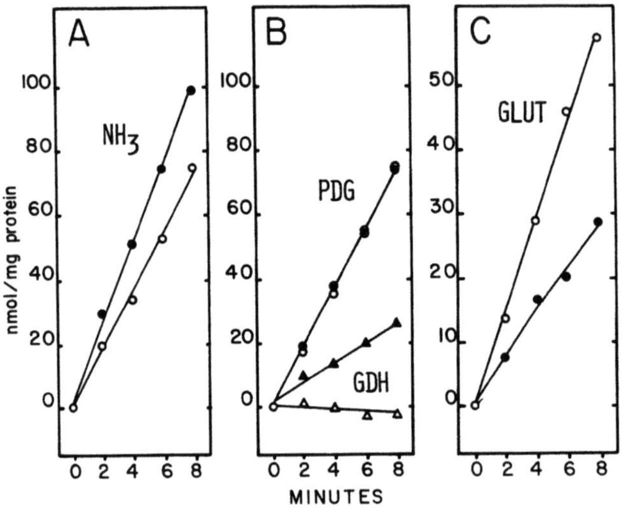

Fig. 1A–C. Effect of pH on glutamine and glutamate metabolism by rat kidney mitochondria. The data are presented as the means of 3 experiments in which mitochondria were incubated with 1 m*M* glutamine and 0.5 m*M* 2-oxoglutarate at pH 6.8 *(closed symbols)* and 7.4 *(open symbols)*. Samples were taken at 2 min intervals from 0 to 8 min. Medium ammonia (NH_3), glutamate (Glut), and aspartate were measured and glutaminase (PDG) and glutamate dehydrogenase (GDH) fluxes were calculated [3]. **A** NH_3 production; **B** PDG *(circles)* and GDH *(triangles)* flux; **C** total glut accumulation. Adapted from [7]

Table 1. Effect of medium pH on mitochondrial matrix pH and ΔpH[a]

Substrates	Medium pH	Matrix pH	ΔpH
1 m*M* Glutamine	7.31 ± 0.01	7.94 ± 0.02	+ 0.64 ± 0.02
+ 0.5 m*M* OG	6.75 ± 0.02	7.67 ± 0.02	+ 0.92 ± 0.02[b]

[a] The data are means ± SE from 5 experiments in which mitochondria (2–5 mg/ml) were incubated with (in m*M*) 135 KCl, 5 $MgCl_2$, 5 KH_2PO_4, 20 MOPS, 10 Tris-HCl, 30 glucose, 0.1 malate, 0.2 ADP, hexokinase (6 U/ml), 7% Dextran, ^3H_2O, [2-^{14}C]DMO, and 1 glutamine and 0.5 2-oxoglutarate, at pH 7.4 and 6.8. Samples were taken between 2 and 6 min of incubation. The medium was separated from the mitochondrial pellet by silicone oil centrifugation. Medium pH was measured with a pH electrode. Matrix pH was calculated by the distribution of [2-^{14}C]DMO, using mitochondrial spaces determined in parallel incubations with ^3H_2O and [U-^{14}C]sucrose. ΔpH, the pH gradient across the inner mitochondrial membrane was calculated as the difference between medium and matrix pH

[b] $p < 0.005$ vs higher pH value

Data were taken from reference [7]

medium pH was reduced to approx. 6.8, matrix pH decreased by a smaller amount such that the pH gradient across the inner mitochondrial membrane (ΔpH) rose to 0.9 pH units. In contrast when glutamine was the only substrate, the ΔpH did not increase as medium pH was reduced. Consequently, matrix pH was considerably more acid in the presence of glutamine alone than with glutamine plus 2-oxoglutarate as substrates. The observation that the addition of

Table 2. Effect of medium pH on 2-oxoglutarate metabolism[a]

Measured parameter ($nmol \cdot mg^{-1} \cdot min^{-1}$)	*n*	Medium pH	
		7.4	6.8
Decrease medium OG	10	25.5 ± 2.2	34.5 ± 3.3[b]
[1-^{14}C]OG oxidation	3	25.8 ± 4.5	42.1 ± 1.0[b]
2-OGDH flux	10	28.6 ± 3.1	42.0 ± 4.1[b]

[a] Mitochondria were incubated with 1 m*M* glutamine + 0.5 m*M* 2-oxoglutarate (OG) or 1 m*M* glutamine + 0.5 m*M* 2-[1-^{14}C]oxoglutarate. The change in medium 2-oxoglutarate, labeled 2-oxoglutarate oxidation and 2-oxoglutarate dehydrogenase (2-OGDH) flux are shown in $nmol \cdot mg^{-1} \cdot min^{-1}$. The data are taken from reference [7] and are expressed as mean ± SE

[b] $p < 0.05$ vs pH 7.4

arsenite, an inhibitor of 2-oxoglutarate dehydrogenase, resulted in a decrease in matrix pH and ΔpH suggests that the metabolism of 2-oxoglutarate helps to maintain matrix pH levels at the more alkaline level. The increased ΔpH observed with glutamine and 2-oxoglutarate as substrates might explain the increased level of glutamate measured in the matrix space when medium pH was reduced. However, the increased ΔpH cannot explain the increase in ammonia formation. Glutaminase and glutamate dehydrogenase are known to have alkaline pH optimum, and, although ΔpH increased, the matrix pH did decrease somewhat as medium pH was reduced. For this reason additional studies were performed testing the effects of pH on the metabolism of 2-oxoglutarate by intact mitochondria.

The results of studies performed to evaluate hydrogen ion induced 2-oxoglutarate metabolism are shown in Table 2. This table summarizes experiments performed with 1 m*M* glutamine and 0.5 m*M* labeled or unlabeled 2-oxoglutarate as substrates [7]. Compared to pH 7.4, medium 2-oxoglutarate levels decreased more rapidly, more $^{14}CO_2$ was formed from 2-[1-^{14}C]oxoglutarate, and calculated 2-oxoglutarate dehydrogenase flux was accelerated at pH 6.8. The data provide experimental support for augmented 2-oxoglutarate oxidation by rat kidney mitochondria at acid medium pH.

2-Oxoglutarate has been shown previously to inhibit renal ammoniagenesis [12, 13], probably by competitively inhibiting glutamate deamination [8, 14, 15]. Thus, to provide additional insight into the augmented flux through glutamate dehydrogenase as medium pH was lowered with glutamine and 2-oxoglutarate as substrates, the matrix concentrations of the reactants and products of the reaction were determined. The results are summarized in Table 3. As medium pH was decreased, calculated matrix 2-oxoglutarate and ammonia concentrations decreased, whereas matrix glutamate concentration rose. The decrease in 2-oxoglutarate levels were compatible with enhanced 2-oxoglutarate oxidation observed at acid pH due to stimulation of 2-oxoglutarate dehydrogenase (Table 2). The decrease in matrix ammonia and increase in matrix glutamate concentrations were ascribed to an increase in the mitochondrial pH gradient (ΔpH) which occurred when medium pH was decreased (Table 1). A decrease in

Table 3. Effect of pH on matrix concentrations of glutamate, 2-oxoglutarate, and ammonia[a]

Medium pH	Matrix (m*M*)		
	[Glut]	[OG]	[NH_3]
7.40	3.80 ± 0.40 (8)	1.83 ± 0.23 (12)	0.055 ± 0.002 (8)
6.80	4.93 ± 0.30[b] (8)	1.30 ± 0.21[c] (12)	0.038 ± 0.001[d] (8)

[a] The data represent means ± SE of 8 or 12 determinations. Mitochondria (4–6 mg/ml) were incubated with 1 m*M* glutamine, 0.5 m*M* 2-oxoglutarate, [2-^{14}C]DMO, and [1-^{3}H]sucrose, pH 7.4 or 6.8, in a medium similar to that described in Table 1. Matrix metabolite levels were determined on samples obtained at 6 min following silicone oil centrifugation. Matrix ammonia was determined using measured medium ammonia and extra- and intramitochondrial pH. The data are presented as the means ± SE from 10 experiments

[b] $p < 0.002$

[c] $p < 0.025$

[d] $p < 0.01$ vs pH 7.4

Taken from reference [7] with permission from The American Physiological Society

the gradient of ammonia and, hence, a decrease in matrix ammonia occurs when ΔpH increases because ammonia is transported as neutral NH_3 and trapped as NH_4^+. The increase in ΔpH was also expected to cause an increase in the mitochondrial glutamate gradient because glutamate is cotransported with a proton.

Additional studies were subsequently performed to test the effect of pH on glutamate hydroxyl transport. It was anticipated that glutamate uptake would be accelerated at acid medium pH levels. However, surprisingly, no increase in glutamate uptake was observed over the pH range 6.6–7.4. Kinetic studies were performed at both pH 6.6 and 7.0 and indicated that the K_m and V_{max} of glutamate transport were identical (K_m 1.5 m*M*, V_{max} 3.3 nmol · mg^{-1} · min^{-1}). Thus, the increase in matrix glutamate concentration observed at acid medium pH could not be ascribed to augmented uptake of glutamate into the matrix space. In order to test the possibility that pH affects the efflux rather than the uptake of glutamate across the inner mitochondria membrane, additional studies were performed which are summarized in Fig. 2. As described previously [10, 11], glutamate efflux was found to be first-order with respect to matrix glutamate. The data demonstrate clearly that the rate of glutamate efflux decreased progressively as medium pH was reduced from 7.9 to 6.4. The rate constant (k) of glutamate efflux was calculated from these data. The mean rate constant at pH 7.9 was 0.99 and decreased to 0.36 at pH 6.4. Rate constants were 0.72, 0.57, and 0.44 at pH 7.6, 7.4, and 6.9, respectively. These studies suggest that in kidney acidification of the medium results in a significant decrease in the rate of glutamate efflux from the mitochondrial matrix space. Thus, the rise in matrix glutamate observed previously upon medium acidification of mitochondria incubated with glutamine and 2-oxoglutarate can be explained by a reduction in the rate of glutamate efflux.

The accelerated rate of 2-oxoglutarate oxidation at acid pH can explain the well-described decrease in tissue oxoglutarate levels in vivo [16–18] as well as

Fig. 2. Effect of medium pH on glutamate efflux. Mitochondria were loaded with 10–100 m*M* [U-^{14}C]glutamate at pH 7.0. Efflux was measured at medium pH 7.9 (▲——▲), 7.4 (△——△), 6.9 (●——●), and 6.4 (○——○). In individual experiments, matrix glutamate was plotted against time. In order to normalize the data from different experiments, the matrix glutamate concentrations at different times were expressed as fractions of zero time concentration, which was set at 1.0 on the y intercept. The data are taken from [11] and represent the mean of 2–5 experiments

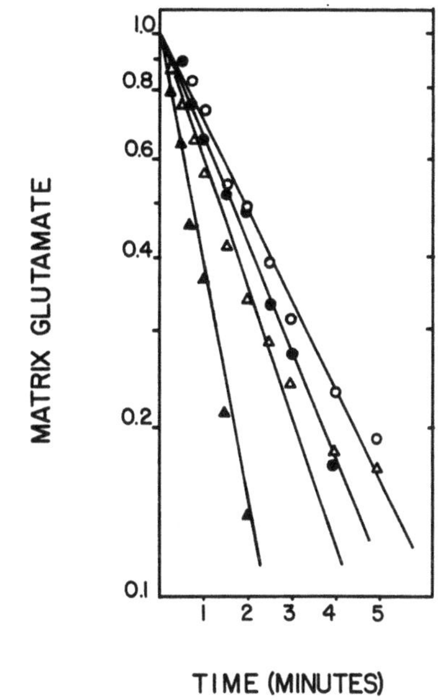

the observed decrease in matrix 2-oxoglutarate levels which we have noted (Table 2). In order to estimate whether the increase in glutamate oxidation and the alteration in glutamate efflux can explain the observed fall in tissue glutamate noted in acute acidosis in vivo, the calculations shown in Table 4 were made. For this calculation it was assumed that under normal acid-base conditions, cytosolic and mitochondrial glutamate concentrations are stable, the latter being presumably higher because of the ΔpH across the inner mitochondrial membrane. These steady state levels are maintained since the rates of glutamate entry and exit from each compartment are equal. Glutamate is formed in the mitochondria from glutamine which is deamidated to glutamate by glutaminase, accounting for the basal level of ammonia synthesis. The resulting glutamate leaves the mitochondria and enters the cytosol on the glutamate hydroxyl carrier. No significant glutamate oxidation via the glutamate dehydrogenase reaction occurs (Fig. 1). Glutamate is also continuously recycled across the inner mitochondrial membrane by the concerted action of the glutamate-aspartate transporter and the cytosolic and mitochondrial glutamate oxalacetate transaminases (see Fig. 3).

In acute acidosis the flux through 2-oxoglutarate dehydrogenase increases, resulting in a decrease in matrix 2-oxoglutarate concentrations. Also, glutamate efflux from the matrix space is reduced, resulting in an increase in matrix glutamate concentration despite the fact that glutamate formation from glutamine is unchanged. The rise in the matrix glutamate/2-oxoglutarate ratio serves to drive the glutamate dehydrogenase reaction with the resultant increase in ammonia formation. The augmented rate of glutamate oxidation without a change in its production results in a decrease in cortical glutamate levels (largely

Table 4. Renal cortical glutamate concentration in acute acidosis adapted from [11]

Mitochondrial glutamate generation unchanged
Glutamate oxidation increased 2.5 nmol/mg mito protein/min at 28° C
Assume: Doubling of rate at 37° C
Mito protein ≃ 64 mg/g wet wt kidney
Rat kidney 1 g wet wt, 0.85 of which cortex
Therefore, cortical glutamate could decrease by
$(2.5 \times 2) \times (64 \times 0.85) = 272$ nmol/min, and
in 3 min = 816 nmol/min
As glutaminase flux increases, glutamate levels may stabilize at higher level

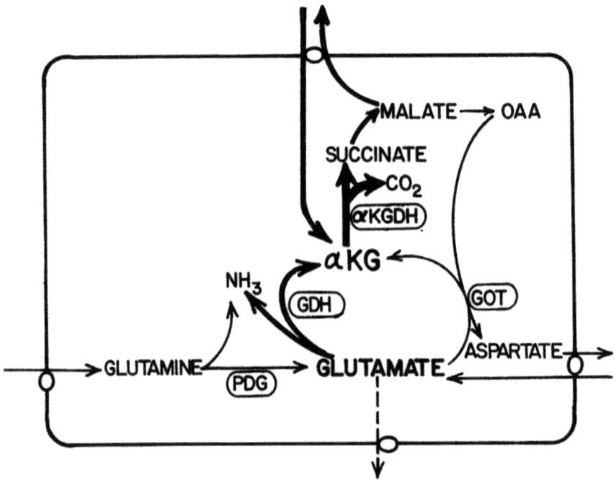

Fig. 3. Alterations in glutamine metabolism in acute acidosis. Abbreviated terms enclosed in *boxes* refer to the enyzmes glutaminase (PDG), glutamate dehydrogenase (GDH), glutamate oxaloacetate transaminase (GOT), and 2-oxoglutarate dehydrogenase (αKGDH). Other abbreviated terms are 2-oxoglutarate (αKG) and oxaloacetate (OAA). Mitochondrial transporters involved are clockwise from the top: the 2-oxoglutarate : malate carrier, the glutamate : aspartate carrier, the glutamate : hydroxyl carrier, and the glutamine carrier. In this schematized mitochondrion, events accelerated in acidosis are depicted by *heavy lines*. Thus, 2-oxoglutarate oxidation is augmented directly by H^+, tending to decrease matrix levels of this metabolite. Matrix glutamate is increased as a result of a decrease in glutamate efflux, indicated by the *dashed line* at the bottom of the figure. Increased glutamate : 2-oxoglutarate ratio results in increased glutamate deamination and ammonia formation. The figure also shows that 2-oxoglutarate influx is augmented in acidosis. It is not known whether this is the result of a pH effect on the 2-oxoglutarate : malate carrier or to the increased rate of malate production consequent to enhanced 2-oxoglutarate dehydrogenase flux. From [7] with permission from the American Physiological Society

cytosolic). Indeed, as shown in Table 4, the observed effects on glutamate efflux and glutamate oxidation are sufficient to account for the alterations in cortical glutamate concentrations reported in acute acidosis in vivo. For this calculation, as mentioned previously, mitochondrial glutamate generation was initially assumed to be unchanged. Glutamate oxidation increased approx. 2.5 nmol/mg mitochondrial protein/min at 28° C. If we assume an approximate doubling of this rate at the in vivo temperature of 37° C and that mitochondrial protein is approximately 64 mg/g weight of kidney tissue (Idell-Wenger JA and Schoolwerth AC, unpublished observations) and finally that an average rat kidney weights 1 g, it is possible to calculate that cortical glutamate could decrease by 272 nmol/min/kidney such that within 3 min over 800 nmol had been utilized. This value is sufficient to account for the reported decreases in tissue glutamate which represent values 60%–80% of control values [19].

As acute acidosis progresses toward a more chronic state glutaminase flux is augmented resulting in an increase in glutamate generation within the matrix space. Total tissue glutamate levels may then stabilize at a slightly higher level intermediate between the values which occur acutely and those found more chronically in metabolic acidosis [19].

In conclusion, the experiments summarized here indicate that at acid pH, mitochondrial ammonia formation is accelerated by enhanced glutamate dehydrogenase flux. The increased glutamate oxidation is in turn driven by an increase in the glutamate/2-oxoglutarate ratio in the matrix space. This increased ratio is brought about on the one hand by an increase in 2-oxoglutarate oxidation by H^+ induced 2-oxoglutarate dehydrogenase flux and by an increase in matrix glutamate consequent to a reduction in glutamate efflux from the matrix space at acid medium pH. Enhancement in glutamate deamination contributes to the early rise in ammonia formation and can explain the observed changes in tissue glutamate. Increased 2-oxoglutarate oxidation accounts for the decrease in tissue 2-oxoglutarate levels. As acidosis progresses, it is postulated that glutaminase flux increases with a consequent rise in glutamate generation within the matrix space such that glutamate levels stabilize at a value intermediate between those that have been observed under normal acid-base conditions and in acute acidosis.

Acknowledgements. The expert secretarial assistance of Mrs Janet M. True is appreciated.

References

1. Relman AS, Narins RG (1975) The control of ammonia production in the rat. Med Clin North Am 59: 583–593
2. Tannen RL, Kunin AS (1976) Effect of pH on ammonia production by renal mitochondria. Am J Physiol 231: 1631–1637
3. Schoolwerth AC, Nazar BL, LaNoue KF (1978) Glutamate dehydrogenase activation and ammonia formation by rat kidney mitochondria. J Biol Chem 253: 6177–6183
4. Preuss HG, Vavatsi-Manos O, Vertuno L, Baird K (1974) The effects of pH change on renal ammoniagenesis in vitro. Proc Soc Exp Biol Med 146: 803–808
5. Lowry M, Ross BD (1980) Activation of oxoglutarate dehydrogenase in the kidney in response to acute acidosis. Biochem J 190: 771–780
6. Schoolwerth AC, Strzelecki T, LaNoue KF, Hoover WJ (1982) Effect of pH and α-ketoglutarate on mitochondrial ammonia production. Contrib Nephrol 31: 127–133
7. Schoolwerth AC, LaNoue KF (1983) Control of ammoniagenesis by α-ketoglutarate in rat kidney mitochondria. Am J Physiol 244 (Renal Fluid Electrolyte Physiol 13): F399–F408
8. Schoolwerth AC, LaNoue KF (1980) The role of microcompartmentation in the regulation of glutamate metabolism by rat kidney mitochondria. J Biol Chem 255: 3403–3411
9. Nazar BL, Schoolwerth AC (1979) An improved microfluorometric enzymatic assay for the determination of ammonia. Anal Biochem 95: 507–511

10. Schoolwerth AC, LaNoue KF, Hoover WJ (1983) Glutamate transport in rat kidney mitochondria. J Biol Chem 258: 1735–1739
11. Schoolwerth AC, LaNoue KF, Hoover WJ (1984) Effect of pH on glutamate efflux from rat kidney mitochondria. Am J Physiol 246: F 266–F 271
12. Balagura-Baruch S, Shurland LM, Welbourne TC (1970) Effects of α-ketoglutarate on renal ammonia release in the intact dog. Am J Physiol 218: 1070–1075
13. Welbourne TC, Balagura-Baruch S (1972) Renal metabolism of glutamine in dogs during infusion of α-ketoglutaric acid. Am J Physiol 223: 663–666
14. Engel PC, Chen SS (1975) A product-inhibition study of bovine liver glutamate dehydrogenase. Biochem J 151: 305–318
15. Schoolwerth AC, Hoover WJ, Daniel CH, LaNoue KF (1980) Effect of aminooxyacetate and α-ketoglutarate on glutamate deamination by rat kidney mitochondria. Int J Biochem 12: 145–149
16. Alleyne GAO (1970) Renal metabolic response to acid-base changes II. The early effects of metabolic acidosis on renal metabolism in the rat. J Clin Invest 49: 943–951
17. Hems DA, Brosnan JT (1971) Effects of metabolic acidosis and starvation on the content of intermediary metabolites in rat kidney. Biochem J 123: 391–397
18. Narins RG, Relman AS (1974) Acute effects of acidosis on ammoniagenic pathways in the kidneys of the intact rat. Am J Physiol 227: 946–949
19. Vinay P, Allignet E, Pichette C, Watford M, Lemieux G, Gougoux A (1980) Changes in renal metabolite profile and ammoniagenesis during acute and chronic metabolic acidosis in dog and rat. Kidney Int 17: 312–325

Role of Fatty Acids in Simultaneous Regulation of Flux Through Glutaminase and Glutamine Synthetase in Rat Kidney Cortex

G. Baverel, C. Michoudet, and G. Martin[1]

Introduction

Although its activity may be low in herbivorous animals, phosphate-dependent glutaminase (EC 3.5.1.2; hereafter referred to as glutaminase), which is the only important enzyme thought to initiate renal degradation of glutamine (Tannen 1978), the main precursor of the urinary ammonia needed to serve as a buffer for excess protons (Van Slyke et al. 1943, Pitts et al. 1965), has been shown to be present in the kidney of all the species that have been studied (Krebs 1935, Richterich and Goldstein 1958, Mattenheimer and De Bruin 1964). In contrast, glutamine synthetase (EC 6.3.1.2.), which catalyzes the condensation of glutamate and ammonia, is present in the kidney of rat, sheep, guinea pig, rabbit, mouse and hamster, but absent from that of man, dog, cat, and pig (Krebs 1935, Richterich and Goldstein 1958, Wu 1963a, Mattenheimer and DeBruin 1964, Janicki and Goldstein 1969, Lemieux et al. 1976, Burch et al. 1978a). Thus, in the species whose kidney contains both glutaminase and glutamine synthetase, the possibility exists that renal degradation and synthesis of glutamine occur simultaneously.

Arteriovenous difference measurements have shown that unlike the kidney of man, dog and monkey, which extracts significant amounts of circulating glutamine (Owen and Robinson 1963, Hills et al. 1967, Tizianello et al. 1980, Silverman et al. 1981, Fine 1982), the kidney of the rat, which contains a high activity of both glutamine synthetase and glutaminase, takes up or releases very little glutamine under normal conditions in vivo (Lund and Watford 1976, Squires et al. 1976, Relman and Yablon 1978, Hughey et al. 1980, Schröck et al. 1980, Schröck and Goldstein 1981); whereas large amounts of glutamine are extracted by the kidney of rats rendered acidotic or diabetic or fed a casein diet (Yamamoto et al. 1974, Squires et al. 1976, Relman and Yablon 1978, Parry and Brosnan 1978, Hughey et al. 1980, Schröck et al. 1980, Schröck and Goldstein 1981, Brosnan et al. 1983). Conversely, the rat kidney releases significant amounts of glutamine after an ammonia load (Vinay et al. 1978a). Thus, depending on the experimental condition, the rat kidney expresses its capacity to either degrade or synthesize glutamine.

However, arteriovenous difference measurements across the rat kidney give no information on the possible existence of concomitant fluxes through glutaminase and glutamine synthetase. By means of isotopic tracer studies in

1 Laboratoire de Physiologie Rénale et Métabolique and Centre National de la Recherche Scientifique GRECO no 24, Faculté de Médecine Alexis Carrel, 69008 Lyon, France

Glutamine Metabolism in Mammalian Tissues
Edited by D. Häussinger and H. Sies

vivo, Damian and Pitts (1970) measured flux through both enzymes in the rat kidney. They found that under normal acid-base balance conditions, flux through glutaminase was virtually equal to that through glutamine synthetase, which explains that no net renal uptake or release of glutamine is observed. In the latter experiments, flux through each enzyme was measured in separate experiments. In the present study, we show that in the same experiment and under appropriate conditions, simultaneous synthesis and degradation of glutamine can be demonstrated in isolated rat kidney-cortex tubules.

Methodology

Kidney-cortex tubules from fed male rats (200–300 g) were prepared by collagenase treatment as described by Baverel et al. (1978).

Incubations were carried out at 37° C in a shaking water bath, in 25-ml Erlenmeyer flasks, each with a centre well, in an atmosphere of 95% O_2–5% CO_2. Tubules were incubated for 30 or 60 min in 4 ml of Krebs-Henseleit buffer (Krebs and Henseleit 1932) with 1 m*M* [1-^{14}C] glutamine or 1 m*M* [1-^{14}C] glutamate in the absence of albumin and in the presence of dialysed bovine serum albumin (2.5% w/v). Albumin solutions were dialysed as described by Krebs et al. (1974).

In Table 4, the [1-^{14}C]glutamine + [1-^{14}C]glutamate found was measured by the combined action of glutaminase and glutamic decarboxylase as described by Squires and Brosnan (1978) ([1-^{14}C]glutamine and [1-^{14}C]glutamate could not be measured separately because the commercial glutamic decarboxylase used was found to be significantly contaminated by glutaminase activity). The unlabelled glutamine + glutamate found was calculated as the total glutamine + glutamate found in each flask minus the [1-^{14}C]glutamine + [1-^{14}C]glutamate found. The net unlabelled glutamine + glutamate released was calculated as the difference between the unlabelled glutamine + glutamate found and the glutamate found in the zero-time flask (Table 4).

In Table 5, the specific radioactivity of the glutamine + glutamate found after incubation was calculated by dividing the radioactivity left in the medium (after collection of the $^{14}CO_2$ formed) by the glutamine + glutamate found, except in the zero-time flask where only the glutamine found was taken into account. It was ascertained that all the radioactivity left in the medium was due to the sum of [1-^{14}C]glutamine and [1-^{14}C]glutamate, except in the presence of amino-oxyacetate, which was found to be a potent inhibitor of the glutamic decarboxylase used. The unlabelled glutamine + glutamate found was calculated by multiplying the percent decrease in specific radioactivity of glutamine + glutamate by the glutamine + glutamate found.

Results and Discussion

Effect of Fatty Acid Contaminants of Dialysed Albumin on Glutamine Metabolism in Rat Kidney Tubules

Table 1 shows the results of experiments performed with 1 m*M* [1-^{14}C]glutamine as substrate in the course of our studies on the usefulness of albumin as a

Table 1. Metabolism of 1 m*M* L-[1-^{14}C]glutamine in rat kidney-cortex tubules[a]

Experimental condition	Incubation (min)	Metabolite removal or production				
		Glutamine	Glutamate	Ammonia	Glucose	$^{14}CO_2$
[1-^{14}C] Glutamine (1 m*M*)	30	− 209.6 ± 14.6	+ 30.6 ± 4.7	+ 449.8 ± 19.4	+ 75.3 ± 7.4	+ 198.9 ± 13.8
[1-^{14}C] Glutamine (1 m*M*) + 2.5% albumin	30	− 145.6 ± 13.7[c]	+ 41.1 ± 7.4[b]	+ 388.0 ± 10.7[d]	+ 72.2 ± 4.8	+ 167.1 ± 6.3[b]
[1-^{14}C] Glutamine (1 m*M*) + 2.5% defatted albumin	30	− 205.8 ± 18.2	+ 34.5 ± 7.1	+498.9 ± 28.3[c]	+ 80.9 ± 7.5[b]	+ 214.7 ± 11.2[c]
[1-^{14}C] Glutamine (1 m*M*)	60	− 369.8 ± 23.9	+ 31.3 ± 6.4	+ 793.8 ± 55.9	+ 116.4 ± 15.2	+ 337.5 ± 25.9
[1-^{14}C] Glutamine (1 m*M*) + 2.5% albumin	60	− 240.9 ± 15.6[c]	+ 41.1 ± 7.5[b]	+ 622.8 ± 33.3[c]	+ 109.1 ± 10.8	+ 277.7 ± 12.8[b]
[1-^{14}C] Glutamine (1 m*M*) + 2.5% defatted albumin	60	− 359.3 ± 22.4	+ 29.7 ± 5.1	+ 882.1 ± 26.8[b]	+ 125.8 ± 16.1[b]	+ 358.8 ± 24.2
No added substrate	60	− 2.6 ± 5.3	− 8.7 ± 1.9	+ 126.1 ± 7.4	+ 18.1 ± 1.1	−

[a] Kidney tubules (6.4 ± 0.7 mg dry wt per flask) were incubated as described in the Methodology Section. Results (μmol/g dry wt) for metabolite removal (−) or production (+) are reported as means ± SEM for five experiments, each performed in duplicate. Statistical difference was measured by the paired Student's *t*-test against the control

[b] $p < 0.05$

[c] $p < 0.01$

[d] $p < 0.001$

medium component for incubation of isolated rat kidney-cortex tubules. Under all conditions, the only products of the metabolism of [1-^{14}C]glutamine were glutamate, ammonia, $^{14}CO_2$ and glucose; carbon balance calculations indicate that in the absence of albumin or in the presence of defatted albumin, a significant portion of the glutamine carbon skeleton used was oxidized in the tricarboxylic acid cycle. Glutamate represented only a small fraction of the glutamine removed indicating that most of the glutamate formed from glutamine was further metabolized via glutamate dehydrogenase.

Addition of 2.5% dialysed albumin inhibited glutamine removal and the formation of ammonia and $^{14}CO_2$, but did not change glucose synthesis. At the same time, accumulation of glutamate was increased by the presence of dialysed albumin, which means that flux through glutamate dehydrogenase was inhibited by some contaminant in the dialysed albumin–probably fatty acids. This is borne out by results obtained when the dialysed albumin was replaced by the same albumin defatted as described by Chen (1967) (see Table 1); under this condition, the values obtained returned towards those found in the absence of albumin. These results confirm the inhibitory effects of fatty acids on renal glutamine metabolism already reported by Vinay et al. (1976, 1978b). They also indicate that albumin should be used with the utmost caution in experiments with kidney tubules.

Thanks to the use of [1-^{14}C]glutamine a new observation is made. If one assumes that glutamine was the sole source of carbon in the present experiments, then glutamine removal should be equal to the sum of the glutamate released and the $^{14}CO_2$ formed, because no accumulation of α-oxoglutarate occurred. Table 1 shows that after 60 min of incubation, this was the case only in the absence of albumin; in the presence of albumin and to a much lesser extent in the presence of defatted albumin, glutamine removal was less than glutamate accumulation plus $^{14}CO_2$ formation, suggesting that de novo synthesis of glutamate and/or glutamine occurred. As the difference between the $^{14}CO_2$ found and the glutamine removed was only of borderline significance, these experiments do not firmly establish that glutamine synthesis occurred.

Effect of Fatty Acid Contaminants of Dialysed Albumin on Glutamate Metabolism in Rat Kidney Tubules

The capacity of the rat kidney-cortex tubules used in our experiments to synthesize glutamine is demonstrated in experiments where [1-^{14}C]glutamate was used as substrate (Table 2). This means that our tubule suspension contained not only proximal and distal convoluted tubules (Vinay et al. 1981), but also a significant amount of proximal straight tubules, which have been shown to be the only cortical segments of the rat nephrons to possess glutamine synthetase activity (Burch et al. 1978a). The data of Table 2 also show that in the presence of fatty acid contaminants of dialysed albumin, glutamate removal was not altered, but glutamine synthesis was greatly increased, whereas ammonia release was inhibited. Flux through glutamate dehydrogenase, taken as the $^{14}CO_2$ found or as the sum of the glutamine + ammonia found, was slightly

Table 2. Metabolism of 1 m*M* L-[1-^{14}C] glutamate in rat kidney-cortex tubules[a]

Experimental condition	Incubation (min)	Metabolite removal or production: Glutamate	Glutamine	Ammonia	Glucose	$^{14}CO_2$
[1-^{14}C] Glutamate (1 m*M*)	30	− 214.1 ± 15.8	+ 71.2 ± 4.1	+ 63.9 ± 4.5	+ 31.3 ± 0.7	+ 136.1 ± 8.7
[1-^{14}C] Glutamate (1 m*M*) + 2.5% albumin	30	− 188.0 ± 12.3	+ 101.0 ± 8.3[b]	+ 27.0 ± 6.7[c]	+ 35.5 ± 1.5[b]	+ 126.2 ± 10.3
[1-^{14}C] Glutamate (1 m*M*)	60	− 341.5 ± 23.4	+ 98.3 ± 10.2	+ 158.3 ± 9.6	+ 48.7 ± 2.6	+ 243.8 ± 10.9
[1-^{14}C] Glutamate (1 m*M*) + 2.5% albumin	60	− 340.6 ± 19.9	+ 169.8 ± 10.3[d]	+ 69.3 ± 14.7[c]	+ 61.5 ± 3.6[c]	+ 230.9 ± 10.2[c]
No added substrate	60	− 6.9 ± 3.2	+ 2.0 ± 2.7	+106.8 ± 8.6	+ 3.6 ± 0.5	−
No added substrate + 2.5% albumin	60	− 4.2 ± 3.6	+ 9.9 ± 7.0	+115.2 ± 10.2	+ 17.0 ± 1.3[d]	−

[a] Kidney tubules (8.3 ± 1.8 mg dry wt per flask) were incubated for 60 min as described in the Methodology Section. Results (μmol/g dry wt) for metabolite removal (−) or production (+) are reported as means ± SEM for four experiments, each performed in duplicate. Statistical difference was measured by the paired Student's *t*-test against the control

[b] $p < 0.05$

[c] $p < 0.01$

[d] $p < 0.001$

inhibited. Calculation of the amount of [1-^{14}C]glutamate carbon converted into glutamine as the difference between the glutamate removed and the $^{14}CO_2$ found indicates that in the absence of albumin, all the glutamine found arose from [1-^{14}C]glutamate, whereas in the presence of dialysed albumin, the contaminating fatty acids promoted the synthesis of glutamine mainly from endogenous sources and not from added [1-^{14}C]glutamate. This means that in the cells which synthesize glutamine, there are two different pools of glutamate. In this respect, it should be mentioned that Schoolwerth and La Noue (1980) have recently proposed a mitochondrial compartmentation of glutamate metabolism.

Effect of Oleate on Glutamine Metabolism in Rat Kidney Tubules; Evidence for Simultaneous Synthesis and Degradation of Glutamine

Evidence for glutamine synthesis in rat kidney tubules metabolizing a near physiological concentration of [1-^{14}C)glutamine was obtained by addition of 0.5 m*M* oleate (bound to 2.5% dialysed fatty acid-free albumin) to the incubation medium (Table 3). Under this condition, the $^{14}CO_2$ formed was significantly higher than the glutamine removed, clearly indicating that de novo synthesis of glutamine occurred from endogenous sources. Under this condition, flux of [1-^{14}C]glutamine through glutaminase cannot be considered as the glutamine apparently removed which represents the difference between actual flux through glutaminase and flux through glutamine synthetase. Taking the $^{14}CO_2$ found as a minimal value for the actual flux through glutaminase, the difference between the $^{14}CO_2$ found and the glutamine removed (see Table 3) provides an estimation of the rate of the de novo synthesis of glutamine; in the presence of oleate flux through glutamine synthetase (72.9 μmol/g dry wt × h) was about one-third that through glutaminase (233.5 μmol/g dry wt × h).

Thus, [1-^{14}C]glutamine appears to be a very useful tool to demonstrate that concomitant degradation and synthesis of glutamine occurred under this condition because only the $^{14}CO_2$ data allow calculation of the minimal flux through glutaminase. This indicates that in experiments with rat kidney tubules, the use of unlabelled glutamine in the presence of fatty acids leads to an underestimation of the true flux through glutaminase and does not allow flux through glutamine synthetase to be detected.

Our data also demonstrate that oleate inhibited net (or apparent) glutamine removal not only by inhibiting flux through glutaminase, but also by promoting flux through glutamine synthetase.

As expected, addition of methionine sulfoximine, a potent inhibitor of glutamine synthetase (Pace and McDermott 1952), which completely suppressed glutamine synthesis from 1 m*M* [1-^{14}C]glutamate (results not shown), abolished the oleate-induced glutamine synthesis in tubules metabolizing [1-^{14}C]glutamine (Table 3); under this condition, the $^{14}CO_2$ found was less than the glutamine removed.

A novel finding is that methionine sulfoximine itself was found to be a potent stimulator of renal glutamine utilization (Table 3); as methionine sulfoximine has been reported to be without effect on the assayed activity of glutaminase

Table 3. Effect of oleate, L-methionine DL-sulfoximine (MSO) and amino-oxyacetate (AoA) on the metabolism of 1 m*M* L-[1-^{14}C] glutamine in rat kidney-cortex tubules[a]

Experimental condition	Metabolite removal or production				
	Glutamine	Glutamate	Ammonia	Glucose	$^{14}CO_2$
[1-^{14}C] Glutamine (1 m*M*)	− 342.7 ± 48.3	+ 11.7 ± 2.6	+ 737.4 ± 120.8	+ 88.9 ± 16.0	+ 337.5 ± 52.6
[1-^{14}C] Glutamine (1 m*M*) + oleate (0.5 m*M*)	− 160.6 ± 21.5[c]	+ 35.3 ± 5.9[c]	+ 449.8 ± 73.4[c]	+ 94.3 ± 15.5	+ 233.5 ± 36.6[c]
[1-^{14}C] Glutamine (1 m*M*) + AoA (1 m*M*)	− 438.6 ± 71.4[b]	+ 45.0 ± 8.0[c]	+ 964.1 ± 185.3[b]	+ 101.8 ± 19.1[b]	+ 342.9 ± 61.9
[1-^{14}C] Glutamine (1 m*M*) + MSO (5 m*M*)	− 538.1 ± 98.8[b]	+ 76.8 ± 7.6[c]	+ 1,098.0 ± 199.0[b]	+ 134.1 ± 24.8[b]	+ 435.6 ± 77.1[b]
[1-^{14}C] Glutamine (1 m*M*) + oleate (0.5 m*M*) + AoA (1 m*M*)	− 205.0 ± 31.4[c]	+ 77.7 ± 10.8[c]	+ 543.0 ± 94.2[c]	+ 80.5 ± 13.7[b]	+ 167.6 ± 25.9[c]
[1-^{14}C] Glutamine (1 m*M*) + oleate (0.5 m*M*) + MSO (5 m*M*)	− 356.6 ± 71.8	+ 137.6 ± 18.0[c]	+ 761.2 ± 135.2	+ 128.2 ± 22.9[b]	+ 276.5 ± 49.3[d]
No added substrate	+ 1.7 ± 1.9	− 1.9 ± 1.2	+ 136.0 ± 4.8	+ 5.1 ± 0.8	–
No added substrate + AoA (1 m*M*)	+ 1.3 ± 1.4	+ 2.7 ± 1.7	+ 112.7 ± 5.8[c]	− 4.1 ± 1.1[d]	–
Oleate (0.5 m*M*)	+ 13.4 ± 3.0[c]	+ 4.8 ± 0.6[c]	+ 125.4 ± 4.6	+ 22.8 ± 2.5[c]	–

[a] Kidney tubules (7.5 ± 0.2 mg dry wt per flask) were incubated for 60 min as described in the Methodology Section. Results (μmol/g dry wt) for metabolite removal (−) or production (+) are reported as means ± SEM for four experiments, each performed in duplicate. Statistical difference was measured by the paired Student's *t*-test against the control

[b] $p < 0.05$

[c] $p < 0.01$

[d] $p < 0.001$

Table 4. Release of unlabelled glutamine and glutamate during the metabolism of 1 m*M* [1-^{14}C] glutamine in rat kidney-cortex tubules[a]

Experimental condition	Incubation (min)	Glutamine found	Glutamate found	Glutamine + glutamate found	[1-^{14}C] Glutamine + [1-^{14}C] glutamate found	Unlabelled glutamine + glutamate found	Net unlabelled glutamine + glutamate released
[1-^{14}C] Glutamine (1 m*M*)	0	3.89	0.15	4.04	3.89	0.15	–
[1-^{14}C] Glutamine (1 m*M*)	60	1.58	0.21	1.79	1.29	0.50	0.35
[1-^{14}C] Glutamine (1 m*M*) + oleate (0.5 m*M*)	60	2.82	0.35	3.17	2.15	1.02	0.87
[1-^{14}C] Glutamine (1 m*M*) + MSO (5 m*M*)	60	0.52	0.66	1.18	0.84	0.34	0.19
[1-^{14}C] Glutamine (1 m*M*) + oleate (0.5 m*M*) + MSO (5 m*M*)	60	1.68	1.05	2.73	1.99	0.74	0.59

[a] Kidney tubules (7.4 mg dry wt per flask) were incubated as described in the Methodology Section. Results (µmol/flask) are reported as means of duplicate conditions of one out of four experiments performed with the same protocol. The [1-^{14}C] glutamine + [1-^{14}C] glutamate found or released were calculated as described in the Methodology Section

Table 5. Decrease in specific radioactivity of glutamine and glutamate during the metabolism of 1 m*M* [1-^{14}C] glutamine in rat kidney-cortex tubules[a]

Experimental condition	Incubation (min)	Glutamine found	Glutamate found	Glutamine + glutamate found	Specific radioactivity of glutamine + glutamate	Unlabelled glutamine + glutamate found	Net unlabelled glutamine + glutamate released
[1-^{14}C] Glutamine (1 m*M*)	0	3.89	0.15	4.04	238,937	0.15	–
[1-^{14}C] Glutamine (1 m*M*)	60	1.58	0.21	1.79	188,684	0.38	0.23
[1-^{14}C] Glutamine (1 m*M*) + oleate (0.5 m*M*)	60	2.82	0.35	3.17	165,445	0.98	0.83
[1-^{14}C] Glutamine (1 m*M*) + AoA (1 m*M*)	60	1.06	0.39	1.45	216,786	0.13	0
[1-^{14}C] Glutamine (1 m*M*) + MSO (5 m*M*)	60	0.52	0.66	1.18	184,506	0.27	0.12
[1-^{14}C] Glutamine (1 m*M*) + oleate (0.5 m*M*) + AoA (1 m*M*)	60	2.59	0.63	3.22	194,039	0.61	0.46
[1-^{14}C] Glutamine (1 m*M*) + oleate (0.5 m*M*) + MSO (5 m*M*)	60	1.68	1.05	2.73	174,272	0.74	0.59

[a] Values (µmol/flask or cpm/µmol) are reported as means of duplicate conditions of the same experiment whose results are given in Table 4. The specific activity of glutamine + glutamate and the unlabelled glutamine + glutamate found or released were calculated as described in the Methodology Section

(Lamar 1968), stimulation by this compound of flux through renal glutaminase, which deserves further study, was probably the result of an indirect rather than a direct effect on the enzyme.

Promotion by oleate of flux through glutamine synthetase raises the question of the origin of the carbon skeleton and nitrogens of the newly synthesized glutamine. Suppression by methionine sulfoximine of the oleate-induced de novo synthesis of glutamine clearly indicates that glutamate synthesized from endogenous substrates was the source of the glutamine carbon skeleton and amino nitrogen; since our tubule preparation was in a protein-catabolic state, as indicated by large production of ammonia in the absence (and even in the presence) of exogenous substrate (see Table 3), it is likely that the glutamate used for glutamine synthesis was formed by transamination with α-oxoglutarate of amino acids released by proteolysis. The ammonia incorporated as the amide nitrogen of glutamine was probably provided by oxidative deamination of glutamate via glutamate dehydrogenase or by operation of the purine nucleotide cycle (Bogusky et al. 1976), whose role in renal ammonia production is however still under debate (Strzelecki et al. 1983). Evidence for the proteolytic origin of the newly synthesized glutamine was obtained by addition of amino-oxyacetate, an inhibitor of transaminases (Rognstad and Katz 1970), which both reduced the ammonia released from endogenous substrates (in the absence of exogenous substrate) and abolished the glutamine synthesis observed in the presence of [1-^{14}C]glutamine plus oleate (see Table 3).

Synthesis of glutamine from endogenous substrates implies that the added [1-^{14}C]glutamine was diluted by unlabelled glutamine. Tables 4 and 5 show that unlabelled glutamate + glutamine entered the [1-^{14}C]glutamate + [1-^{14}C]glutamine pool under all experimental conditions, except in the presence of [1-^{14}C]glutamine plus amino-oxyacetate; this means that the latter compound efficiently prevented the dilution of the pool of radioactive glutamine and glutamate by unlabelled glutamate arising from endogenous sources. The rate of entry of unlabelled glutamate + glutamine into the [1-^{14}C]glutamate + [1-^{14}C]glutamine pool was minimal with [1-^{14}C]glutamine as sole substrate, but especially high in the presence of [1-^{14}C]glutamine plus oleate; it was, however, significantly decreased by further addition of either amino-oxyacetate (Table 5) or methionine sulfoximine (Tables 4 and 5).

All these results indicate that the calculations based on the assumption that the specific radioactivity of added [1-^{14}C] glutamine remained constant during the entire incubation period are not necessarily valid; a large decrease in the specific radioactivity of the [1-^{14}C]glutamate + [1-^{14}C]glutamine pool cannot be neglected in the presence of fatty acids.

Possible Existence of a Glutamine-Glutamate Cycle

The present study clearly demonstrates the capacity of isolated rat kidney-cortex tubules to synthesize and degrade glutamine simultaneously. A similar finding has been recently reported in the isolated perfused rat liver by Häussinger and Sies (1979), Häussinger et al. (1983) and Häussinger (1983), who referred to the concomitant occurrence of both processes as a glutamine-glutamate cycle. It should be noted that in the present study, we are not dealing with a cycle stricto sensu, because the glutamate used for de novo synthesis of glutamine is not the [1-^{14}C]glutamate produced from [1-^{14}C]glutamine by the glutaminase reaction. A futile cycle between glutamine and glutamate cannot be demonstrated by using ^{14}C-labelled glutamine as substrate, because such a cycle would not be accompanied by any $^{14}CO_2$ release.

This study, however, strongly suggests that a futile cycle may occur between glutamine and glutamate in the rat renal cortex, but methods other than the one employed here are needed to demonstrate the existence of such a cycle.

The large stimulation of [1-^{14}C]glutamine removal by methionine sulfoximine (see Table 3), which was thought at first sight to result from inhibition of the glutamine synthetase component of a futile cycle between glutamine and glutamate, cannot be taken as evidence for the existence of such a cycle because glutamine utilization was also strikingly accelerated in isolated dog kidney-cortex tubules, which as already mentioned, are devoid of glutamine synthetase activity (results not shown).

Compartmentation of Glutamine Synthesis and Degradation in Rat Kidney-Cortex Tubules

In the rat renal cortex, phosphate-dependent glutaminase has been shown to be an intramitochondrial enzyme (Kovacevic 1974, Curthoys and Weiss 1974, Kalra and Brosnan 1974), whereas by analogy with rat liver (Wu 1963b), renal glutamine synthetase is considered to be bound to the microsomes.

Simultaneous synthesis and degradation of glutamine, which involve a cytosolic and a mitochondrial process, are very unlikely to occur in the same renal cell because distributions of glutamine synthetase and phosphate-dependent glutaminase do not overlap along the rat nephron; glutamine synthetase activity has been found only in the proximal straight tubule (Burch et al. 1978a), whereas this part of the nephron contains an extremely low phosphate-dependent glutaminase activity (Curthoys and Lowry 1973a). High phosphate-independent glutaminase activity found in the proximal straight tubule (Curthoys and Lowry 1973a) has been identified as a side-activity of γ-glutamyltranspeptidase (EC 2.3.2.2.) (Tate and Meister 1974, Curthoys and Kuhlenschmidt 1975) with a very low affinity for glutamine. Thus, the available information points to the rat proximal straight tubule as a site of synthesis, but not of degradation of glutamine.

Consistent with the distribution of phosphate-dependent glutaminase along the rat nephron (Curthoys and Lowry 1973a) is the recent demonstration by Vinay et al. (1981) that both the proximal (convoluted) and distal tubular fragments of a tubule suspension prepared from the rat renal cortex metabolize glutamine.

It should be mentioned that a spatial segregation of glutamine synthesis and degradation in the rat kidney has already been proposed by Burch et al. (1978b).

It is likely that the glutamate needed for de novo synthesis of glutamine, which seems to be of proteolytic origin, may be formed in the proximal convoluted and distal tubules as well as in the proximal straight tubules of our tubule suspension. It is also conceivable that in vivo the glutathione degraded by renal γ-glutamyltranspeptidase (Silbernagl et al. 1982), whose activity is very high in the proximal straight tubule (Heinle et al. 1977), provides some glutamate for glutamine synthesis.

Glutamate dehydrogenase which together with glutaminase may provide the glutamine amide nitrogen, is present in all cortical segments of the rat nephron (Schmidt and Guder 1976).

It should be noted that substantial amounts of ATP must be synthesized in the cells of the proximal straight tubules to meet the energy requirements of the glutamine synthetase reaction (Speck 1947); in this respect, it is noteworthy that enzymes of both mitochondrial and peroxisomal oxidation of fatty acids are present in these nephron segments (Le Hir and Dubach 1982).

Possible Mechanisms for the Regulation by Fatty Acids of Flux Through Glutaminase and Glutamine Synthetase

An obvious drawback in interpreting the effects of fatty acids on glutamine metabolism is the heterogeneity of our preparation of renal tubular fragments as well as our lack of information on the metabolite concentrations in the cytosolic and mitochondrial compartments of different nephron segments as opposed to total suspension measurements.

It has been proposed and some evidence presented (Vinay et al. 1976, Lemieux et al. 1979, 1980) that inhibition of renal glutamine removal by fatty acids or ketone bodies is due to a shift of the mitochondrial $NAD^+/NADH$ ratio towards reduction secondary to fatty acid oxidation; this results in a diminished flux through glutamate dehydrogenase which leads to accumulation of glutamate, a known end-product inhibitor of renal glutaminase (Goldstein 1966).

The observation by Kvamme (1982) that high concentrations of long chain fatty acyl-CoA inhibit purified pig kidney glutaminase might also apply to the present study.

The mechanism by which fatty acids promote the synthesis of glutamine in the cells of proximal straight tubules is not apparent from the present data. As the K_m of rat renal glutamine synthetase for glutamate (Deuel et al. 1973) is higher than the glutamate concentration measured in proximal straight tubules (Curthoys and Lowry 1973b), it is likely that small variations in glutamate availability play an important role in the control of glutamine synthesis. Increased availability of glutamate in the presence of fatty acids is indicated by an enhanced accumulation of glutamate (Tables 1 and 3). Nitrogen balance calculations from the data of Table 3 also indicate that addition of oleate was accompanied by an increased formation of nitrogenous products which exceeded by far the nitrogen removed as glutamine; this means that glutamate and ammonia, presumably derived from proteolysis, were made available in increased amounts to the glutamine synthetase reaction when oleate was present.

It is also conceivable that in the cells of proximal straight tubules, glutamine synthesis, which is an ATP-consuming and an ADP-providing process, was tightly coupled to fatty acid oxidation which requires ADP and provides ATP.

Physiological Significance

Effect of Fatty Acids on Renal Glutamine Metabolism

Our results indicate that fatty acids reduce net glutamine removal by rat renal cortical tissue by controlling flux through glutaminase and glutamine synthetase in a coordinated and opposite way. Since the fatty acids taken up by the rat kidney from the circulating blood (Hohenegger and Schuh 1980) or derived from renal lipid stores (Fonteles et al. 1983) are at least in part oxidized by the renal cortex (Weidemann and Krebs 1969), these effects may well be of physiological importance. They may account, at least in part, for the apparent absence already mentioned of significant net uptake or release of glutamine by the normal rat kidney in vivo.

Our in vitro results (see Table 3) indicate that complete oxidation (calculated as the $^{14}CO_2$ formed minus 2 glucose formed), which represented a major fate of renal glutamine metabolism under control conditions, was abolished in the presence of oleate; it appears therefore that fatty acids may reduce the renal consumption of glutamine when the kidney is not in need of its carbon skeleton as a fuel.

As the kidney is sensitive to the antiproteolytic action of insulin (Pfeifer and Warmuth-Metz 1983), it is also conceivable that under conditions of insulin deficiency (such as diabetes or starvation) which are accompanied by high plasma levels (Guder et al. 1975) and presumably by high renal consumption of fatty acids, a stimulated glutamine synthesis contributes to salvage part of the amino acid carbon and nitrogen arising from renal proteolysis.

Functional Significance of Renal Glutamine Synthesis in Other Situations

Renal glutaminase is closely connected with the control of the renal production of the ammonia needed to participate in the urinary elimination of hydrogen ions. The factors controlling flux through renal glutaminase during metabolic acidosis or alkalosis have been the subject of extensive studies during the last 4 decades (for reviews see Tannen 1978, Ross and Lowry 1981) and will not be covered in this chapter (see also Welbourne, this volume).

Adaptation of flux through renal glutamine synthetase may also be connected with the control of renal ammonia production and excretion. Although renal glutamine synthetase activity did not fall (Burch et al. 1978a) during chronic metabolic acidosis, a diminished flux through glutamine synthetase has been reported in the rat kidney in vivo (Damian and Pitts 1970), and with glutamate as substrate, less glutamine and more ammonia were synthesized by the isolated perfused kidney (Hems 1972) or by renal cortical slices (Kamm and Strope 1972) of rats rendered acidotic. The reverse was found to be true in renal cortical slices from chronically alkalotic rats (Kamm and Strope 1972). Thus, flux through glutamine synthetase appears to be a factor which adapts the availability of urinary ammonia to the needs during changes in the acid/base status of the body.

On the other hand, since elevated levels of ammonia are toxic to the animal cells, renal-synthesized glutamine may play an appreciable role as a nontoxic transporter of ammonia from kidney to other tissues. In herbivorous species with alkaline urine (or in alkalotic rats), glutamine synthesis may prevent the release of excessive amounts of renal-derived ammonia which diffuses freely into the renal venous blood rather than into the more alkaline urine. In agreement with such a role, already proposed by Janicki and Goldstein (1969) and Burch et al. (1978a), are the observations that the kidney of fed pregnant or nonpregnant ewe (Bergman et al. 1974, Heitmann and Bergman 1978) and fed guinea pig (Baverel et al. 1980) adds glutamine to the circulating blood. A role of renal glutamine synthesis in detoxicating ammonia has recently been demonstrated by Vinay et al. (1978a), who showed that an ammonia load resulted in a large glutamine release by the rat kidney. Colombo et al. (1977) have also found a significant increase in renal glutamine synthetase activity in rats after portocaval shunt which induces hyperammonemia.

Conclusions

By using [1-^{14}C]glutamine as substrate it can be demonstrated that simultaneous synthesis and degradation of glutamine may occur in isolated rat kidney-cortex tubules. This suggests but does not prove the existence of a futile cycle between glutamine and glutamate in rat renal tissue. The available information indicates that glutamine synthesis and degradation occur in different segments of the rat nephron. Fatty acids control renal glutamine metabolism in the rat by a dual mechanism: they inhibit flux through glutaminase and promote flux through glutamine synthetase. Although it is an ATP-consuming reaction, renal glutamine synthesis appears to be advantageous to the organism; it may help (a) to control net renal utilization of glutamine as a precursor of ammonia; (b) to salvage the carbon skeleton and nitrogen of amino acids released by renal proteolysis, and (c) to maintain low levels of blood ammonia.

Acknowledgements. This work was supported by grants from the INSERM (ATP 62-78-94, grant No. 002), the Université Lyon I and the CNRS.

References

Baverel G, Lund P (1979) A role for bicarbonate in the regulation of mammalian glutamine metabolism. Biochem J 184: 599–606

Baverel G, Bonnard M, d'Armagnac de Castanet E, Pellet M (1978) Lactate and pyruvate metabolism in isolated renal tubules of normal dogs. Kidney Int 14: 567–575

Baverel G, Genoux C, Forissier M, Pellet M (1980) Fate of glutamate carbon and nitrogen in isolated guinea pig kidney-cortex tubules. Biochem J 188: 873–880
Bergman EN, Kaufman CF, Wolff JE, Williams HH (1974) Renal metabolism of amino acids and ammonia in fed and fasted pregnant sheep. Am J Physiol 226: 833–837
Bogusky RT, Lowenstein LM, Lowenstein JM (1976) The purine nucleotide cycle. A pathway for ammonia production in the rat kidney. J Clin Invest 58: 326–335
Brosnan JT, Man KC, Hall DE, Colbourne SA, Brosnan ME (1983) Interorgan metabolism of amino acids in streptozotocin-diabetic ketoacidotic rat. Am J Physiol 244: E151–E158
Burch HB, Choi S, McCarthy WZ, Wong PY, Lowry OH (1978a) The location of glutamine synthetase within the rat and rabbit nephron. Biochem Biophys Res Commun 82: 498–505
Burch HB, Chan AWK, Alvey TR, Lowry OH (1978b) Localization of glutamine accumulation and tubular reabsorption in rat nephron. Kidney Int 14: 406–413
Chen RF (1967) Removal of fatty acids from serum albumin by charcoal treatment. J Biol Chem 242: 173–181
Colombo JP, Bachmann C, Peheim E, Berüter J (1977) Enzymes of ammonia detoxication after portocaval shunt in the rat II. Enzymes of glutamate metabolism. Enzyme 22: 399–406
Curthoys NP, Kuhlenschmit T (1975) Phosphate-independent glutaminase from rat kidney. Partial purification and identity with γ-glutamyltranspeptidase. J Biol Chem 250: 2099–2105
Curthoys NP, Lowry OH (1973a) The distribution of glutaminase isoenzymes in the various structures of the nephron in normal, acidotic, and alkalotic, rat kidney. J Biol Chem 248: 162–168
Curthoys NP, Lowry OH (1973b) Glutamate and glutamine distribution in the rat nephron in acidosis and alkalosis. Am J Physiol 224: 884–889
Curthoys NP, Weiss RF (1974) Regulation of renal ammoniagenesis. Subcellular localization of rat kidney glutaminase isoenzymes. J Biol Chem 249: 3261–3266
Damian AC, Pitts RF (1970) Rates of glutaminase I and glutamine synthetase reactions in rat kidney in vivo. Am J Physiol 218: 1249–1255
Deuel TF, Lerner A, Albrycht D (1973) Regulation of glutamine synthetase from rat liver and rat kidney. In: Prusiner S, Stadtman ER (eds) The enzymes of glutamine metabolism. Academic Press, London New York, pp 129–144
Fine A (1982) Effects of acute metabolic acidosis on renal, gut, liver and muscle metabolism of glutamine and ammonia in the dog. Kidney Int 21: 439–444
Fonteles MC, Cohen JJ, Blach AJ, Wertheim SJ (1983) Support of kidney function by long-chain fatty acids derived from renal tissue. Am J Physiol 244: F235–F246
Goldstein L (1966) Relation of glutamate to ammonia production in the rat kidney. Am J Physiol 210: 661–666
Guder WG, Stukowski B, Siess EA, Wieland OH (1975) Pyruvate dehydrogenase in rat kidney metabolism. In: Angielski S, Dubach UC (eds) Biochemical aspects of renal function. Huber, Bern Stuttgart Vienna, pp 70–75
Häussinger D (1983) Hepatocyte heterogeneity in glutamine and ammonia metabolism and the role of an intercellular glutamine cycle during ureogenesis in perfused rat liver. Eur J Biochem 133: 269–275
Häussinger D, Sies H (1979) Hepatic glutamine metabolism under the influence of the portal ammonia concentration in the perfused rat liver. Eur J Biochem 101: 179–184
Häussinger D, Gerok W, Sies H (1983) Regulation of flux through glutaminase and glutamine synthetase in isolated perfused rat liver. Biochim Biophys Acta 755: 272–278
Heinle H, Wendel A, Schmidt U (1977) The activities of the key enzymes of the γ-glutamyl cycle in microdissected segments of the rat nephron. FEBS Lett 73: 220–224
Heitmann RN, Bergman EN (1978) Glutamine metabolism, interorgan transport and glucogenicity in the sheep. Am J Physiol 234: E197–E203
Hems DA (1972) Metabolism of glutamine and glutamic acid by isolated perfused kidneys of normal and acidotic rats. Biochem J 130: 671–680
Hills AG, Reid EL, Kerr WD (1967) Circulatory transport of L-glutamine in fasted mammals: cellular sources of urine ammonia. Am J Physiol 223: 1470–1476
Hohenegger M, Schuh H (1980) Uptake and fatty acid synthesis by the rat kidney. Int J Biochem 12: 169–172
Hughey RB, Rankin BB, Curthoys NP (1980) Acute acidosis and renal arteriovenous differences of glutamine in normal and adrenalectomized rats. Am J Physiol 238: F199–F204

Janicki RH, Goldstein L (1969) Glutamine synthetase and renal ammonia metabolism. Am J Physiol 216: 1107–1110
Kalra J, Brosnan JT (1974) The subcellular localization of glutaminase isoenzymes in rat kidney cortex. J Biol Chem 249: 3255–3260
Kamm DE, Strope GL (1972) The effects of acidosis and alkalosis on the metabolism of glutamine and glutamate in renal cortex slices. J Clin Invest 51: 1251–1263
Kovacevic Z (1974) Distribution of glutaminase isoenzymes in kidney cells. Biochim Biophys Acta 334: 199–207
Krebs HA (1935) The synthesis of glutamine from glutamic acid and ammonia and the enzymatic hydrolysis of glutamine in animal tissues. Biochem J 29: 1951–1969
Krebs HA, Henseleit K (1932) Untersuchungen über die Harnstoffbildung im Tierkörper. Hoppe-Seyler's Z Physiol Chem 210: 33–66
Krebs HA, Cornell NW, Lund P, Hems R (1974) Isolated liver cells as experimental material. In: Lundquist F, Tygstrup N (eds) Regulation of hepatic metabolism. Mungsgaard, Copenhagen, pp 726–750 (Alfred Benzon Symposium, vol 6)
Kvamme E (1982) Regulation of pig kidney phosphate-activited glutaminase. In: Tannen RL, Goldstein L, Lemieux G, Simpson D, Vinay P (eds) Renal ammonia metabolism. Karger, Basel, pp 60–70
Lamar C (1968) The duration of the inhibition of glutamine synthetase by methionine sulfoximine. Biochem Pharmacol 17: 636–640
Le Hir M, Dubach UC (1982) Distribution of two enzymes of beta-oxidation of fatty acids along the rat nephron. In: Morel F (ed) Biochemistry of kidney function. Elsevier/North Holland, Biomedical Amsterdam New York Oxford, pp 87–94
Lemieux G, Baverel G, Vinay P, Wadoux P (1976) Glutamine synthetase and glutamyltransferase in the kidney of man, dog and rat. Am J Physiol 231: 1068–1073
Lemieux G, Baverel G, Vinay P, Gougoux A (1979) Effect of fluoroacetate on the inhibitory action of ketone bodies and fatty acids on renal ammoniagenesis. Am J Physiol 237: F7–F13
Lemieux G, Pichette C, Vinay P, Gougoux A (1980) Cellular mechanisms of the antiammoniagenic effect of ketone bodies in the dog. Am J Physiol 239: F420–F426
Lund P, Watford M (1976) Glutamine as a precursor of urea. In: Grisolia C, Baguena R, Mayor F (eds) The urea cycle. Wiley, New York London, pp 479–488
Mattenheimer H, De Bruin H (1964) Glutaminases. Enzymol Biol Clin 4: 65–83
Owen EE, Robinson RR (1963) Amino acid extraction and ammonia metabolism by the human kidney during the prolonged administration of ammonium chloride. J Clin Invest 42: 263–276
Pace J, McDermott EE (1952) Methionine sulfoximine and some enzyme systems involving glutamine. Nature (London) 169: 415–416
Parry DM, Brosnan JT (1978) Glutamine metabolism in the kidney during induction of, and recovery from, metabolic acidosis in the rat. Biochem J 174: 387–396
Pfeifer U, Warmuth-Metz M (1983) Inhibition by insulin of cellular autophagy in proximal tubular cells of rat kidney. Am J Physiol 244: E109–E114
Pitts RF, Pilkington LA, DeHaas J (1965) N^{15} tracer studies on the origin of urinary ammonia in the acidotic dog. J Clin Invest 44: 731–745
Relman AS, Yablon S (1978) The regulation of ammonia production in the rat. In: Guder WG, Schmidt U (eds) Biochemical nephrology. Huber, Bern Stuttgart Vienna, pp 198–200
Richterich RW, Goldstein L (1958) Distribution of glutamine metabolizing enzymes and production of urinary ammonia in the mammalian kidney. Am J Physiol 195: 316–320
Rognstad R, Katz J (1970) Gluconeogenesis in the kidney cortex. Effects of D-malate and amino-oxyacetate. Biochem J 116: 483–491
Ross B, Lowry M (1981) Recent developments in the renal handling of glutamine and ammonia. In: Greger R, Lang F, Silbernagl S (eds) Renal transport of organic substances. Springer, Berlin Heidelberg Hew York, pp 78–92
Schmidt U, Guder WG (1976) Sites of enzyme activity along the nephron. Kidney Int 9: 233–242
Schoolwerth AC, La Noue KF (1980) The role of microcompartmentation in the regulation of glutamate metabolism by rat kidney mitochondria. J Biol Chem 255: 3403–3411
Schröck H, Goldstein L (1981) Interorgan relationships for glutamine metabolism in normal and acidotic rats. Am J Physiol 240: E519–E525

Schröck H, Cha CJM, Goldstein L (1980) Glutamine release from hindlimb and uptake by kidney in the acutely acidotic rat. Biochem J 188: 557–560

Silbernagl S, Curthoys NP, Shapiro R (1982) The role of renal tubular brushborder enzymes in metabolism and absorption of glutathione and cysteinyl-glycine. In: Morel F (ed) Biochemistry of kidney function. Elsevier/North Holland, Biomedical, Amsterdam New York Oxford, pp 429–437

Silverman M, Vinay P, Shinobu L, Gougoux A, Lemieux G (1981) Luminal and antiluminal transport of glutamine in dog kidney: Effect of metabolic acidosis. Kidney Int 20: 359–365

Speck JF (1947) The enzymic synthesis of glutamine. J Biol Chem 168: 403–404

Squires EJ, Brosnan JT (1978) Determination of the specific radioactivity of ^{14}C-labeled glutamic acid and glutamine. Anal Biochem 84: 473–478

Squires J, Hall DE, Brosnan JT (1976) Arteriovenous differences for amino acids and lactate across kidneys of normal and acidotic rats. Biochem J 160: 125–128

Strzelecki T, Rogulski J, Angielski S (1983) The purine nucleotide cycle and ammonia formation from glutamine by rat kidney slices. Biochem J 212: 705–711

Tannen RL (1978) Ammonia metabolism. Am J Physiol 235: F265–F277

Tate SS, Meister A (1974) Stimulation of the hydrolytic activity and decrease of the transpeptidase activity of γ-glutamyl transpeptidase by maleate; identity of a rat kidney maleate-stimulated glutaminase and γ-glutamyl transpeptidase. Proc Natl Acad Sci USA 71: 3329–3333

Tizianello A, De Ferrari G, Garibotto G, Gurreri G, Robaudo C (1980) Renal metabolism of amino acids and ammonia in subjects with normal renal function and in patients with chronic renal insufficiency. J Clin Invest 65: 1162–1173

Van Slyke DD, Phillips RA, Hamilton PB, Archibald RM, Futcher PH, Hiller A (1943) Glutamine as a source material of urinary ammonia. J Biol Chem 150: 481–482

Vinay P, Lemieux G, Cartier P, Ahmad M, Baverel G (1976) Effect of fatty acids on renal ammoniagenesis in in vivo and in vitro studies. Am J Physiol 231: 880–887

Vinay P, Lemieux G, Gougoux A (1978a) Ammonia detoxification by the rat kidney in vivo. Can J Biochem 56: 305–314

Vinay P, Lemieux G, Gougoux A, Watford M (1978b) The metabolic fate of glutamine in the renal cell. Proc VIIth Int Cong Nephrol. Karger, Basel, pp 199–207

Vinay P, Gougoux A, Lemieux G (1981) Isolation of a pure suspension of rat proximal tubules. Am J Physiol 241: F403–F411

Weidemann MJ, Krebs HA (1969) The fuel of respiration of rat kidney cortex. Biochem J 112: 149–166

Wu C (1963a) Glutamine synthetase I. A comparative study of its distribution in animals and its inhibition by DL-allo-δ-hydroxylysine. Comp Biochem Physiol 8: 335–351

Wu C (1963b) Glutamine synthetase II. The intracellular localization in the rat liver. Biochim Biophys Acta 77: 482–493

Yamamoto H, Aikawa T, Matsutaka H, Okuda T, Ishikawa E (1974) Inter-organal relationships of amino acid metabolism in fed rats. Am J Physiol 226: 1428–1433

Other Tissues

Cyclic Nucleotide Regulation of Glutamine Metabolism in Skeletal Muscle

A. J. Garber, S. J. Allen, I. Moretti-Rojas, and C. M. Maillet[1]

Introduction

In the intact organism, glucose homeostasis in the postprandial state is maintained by a combination of both hepatic glycogenolysis as well as gluconeogenesis. After an overnight fast, the quantitative importance of glycogenolysis decreases substantially owing to depletion of glycogen stores [1]. Compensatory increases in hepatic gluconeogenesis therefore result such that the obligate daily requirement of approx. 180 g of glucose production per day is met by an increasing proportion of glucose production from 3-carbon precursors, such as lactate, alanine, and glycerol. Lactate, quantitatively the most important substrate for gluconeogenesis, is derived largely from anaerobic glycolysis in such peripheral tissues as the red blood cell mass. Glycerol is derived from fatty acid mobilization in adipose tissue. Glucose is used for re-esterification of such fatty acids in an ongoing cycle in adipose tissue. Thus, glycerol may also be viewed as being, in part, a product of glycolysis of glucose. Amino acids, principally alanine and perhaps glutamine, represent the most important sources of de novo carbon available for glucose production. Skeletal muscle is the predominant site of amino acid production and release in mammalian species. This tissue releases extremely large quantities of glutamine and to a lesser extent, alanine, as compared to the release of most other amino acids. Together, the release of glutamine plus alanine provides more than half of the α-amino nitrogen released from skeletal muscle under basal circumstances [2, 3]. Despite this, both glutamine and alanine together account for less than 10% of total amino acid residues in mammalian skeletal muscle protein [4]. In view of these observations, experimentation by a variety of authors has demonstrated that the preferential release of glutamine and alanine do not derive from the selective proteolysis of skeletal muscle proteins rich in glutamine and alanine residues. Instead, both amino acids seem to be produced selectively at the expense of the metabolism of other amino acids, particularly those released to a much lesser extent than would have been expected based on their content in skeletal muscle protein. Chief amongst these precursors for glutamine and alanine synthesis in muscle are branched-chain amino acids, although other amino acids, such as aspartate, also appear to make substantial contributions based on the profile of release of amino acids from muscle.

1 Departments of Medicine, Biochemistry, and Cell Biology, Baylor College of Medicine and the General Clinical Research Center, The Methodist Hospital, 1200 Moursund, Houston, TX 77030, USA

Glutamine Metabolism in Mammalian Tissues
Edited by D. Häussinger and H. Sies

Amino Acid Metabolism in Muscle

In our studies, we have used as an in vitro skeletal muscle preparation, the rat epitrochlaris skeletal muscle preparation, which maintains excellent viability in vitro for periods of at least 6 h duration. As noted in Fig. 1, the release of amino acids from this skeletal muscle preparation parallels to a substantial extent the release of amino acids from most other mammalian skeletal muscle preparations or from in vivo muscle studies reported. Glutamine release exceeds by 50% the release of alanine, the latter being the only other amino acid released to a substantial extent. Other amino acids are released at a negligible rate compared to release of alanine or glutamine. The principal differences between rat skeletal muscle preparations and human in vivo studies, such as forearm studies as performed using arteriovenous catheterizations of the human extremity, lie in the release of small quantities of glutamate from rat skeletal muscle preparations, whereas in humans, a small net uptake of glutamate has been observed [2, 3].

Studies with a variety of inhibitors of alanine aminotransferase, such as aminooxyacetate or cycloserine, have demonstrated that inhibition of alanine aminotransferase reduces alanine release from skeletal muscle by approx. 80% [5, 6]. Concomitant with such inhibition of alanine aminotransferase, the release of other, ordinarily poorly released amino acids, such as aspartate is increased. Similar results have also been obtained for inhibition of glutamine synthetase using methionine sulfoximine [6]. Inhibition of glutamine synthetase resulted in markedly reduced glutamine synthesis and release from skeletal muscle and a corresponding increase in alanine release.

A number of hypotheses have been advanced to explain the selective synthesis and release of alanine and glutamine from skeletal muscle. A

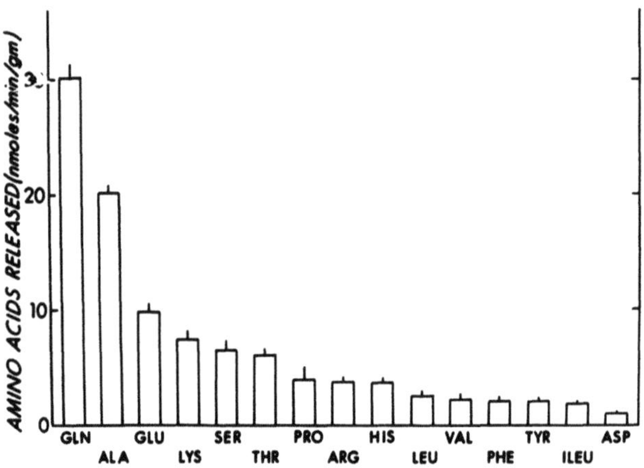

Fig. 1. Profile of amino acid release from epitrochlaris preparations of rat skeletal muscle. Epitrochlaris preparations of rat skeletal muscle were incubated for 1 h at 37° C in Krebs-Henseleit bicarbonate buffer (pH 7.4). After incubation muscles were rapidly removed from the media, rinsed, blotted, and frozen in liquid nitrogen. Amino acids in the media were assessed using amino acid analyzer techniques. Values are expressed as the means (± SE) for at least six experiments and are expressed as nmol/min/g wet wt of muscle [6]

glucose-alanine cycle has been proposed whereby alanine released from skeletal muscle is transported to the liver and converted by gluconeogenesis into glucose. This glucose is subsequently released and taken up by skeletal muscle where it is glycolyzed to pyruvate and subsequently transaminated to form alanine. The latter assigns the role to alanine released from skeletal muscle as primarily that of a nitrogen carrier for urea synthesis. Other investigators have found that a high proportion of the alanine released from skeletal muscle does not derive from anaerobic glycolysis of glucose, but derives instead from the net catabolism of other amino acids to form not only the amino group for net alanine synthesis, but also a significant fraction of the carbon skeleton as well [2, 5–7]. Estimates of the contribution of amino acids to alanine synthesis have suggested that approx. 75% of the alanine released from skeletal muscle derives from the catabolism of other amino acids and that not more than 20%–25% of the alanine derived from muscle reflects a carbon contribution from anaerobic glycolysis [8]. On the other hand, a comprehensive basis for an understanding of the metabolic utilization and disposal of glutamine released from skeletal muscle in the intact organism has not been well defined. Certainly glutamine release from skeletal muscle accounts for virtually half of the total amino acid output from skeletal muscle under basal circumstances [3]. Yet the metabolic fate of all of this carbon remains somewhat unclear. During periods of metabolic acidosis, glutamine utilization by the renal mass for ammonia generation with the subsequent formation of other amino acids, such as serine, as well as its utilization for renal gluconeogenesis has been demonstrated [9]. On the other hand, in the basal, nonacidotic state, the metabolic fate of glutamine has been somewhat unclear. The data of Windmueller and co-workers [10] clearly demonstrated that substantial oxidation of glutamine in the intestinal mucosa occurs and that in addition to oxidation to carbon dioxide, glutamine may also be metabolized to other 3-carbon precursors, such as alanine and lactate. It is possible that the latter are taken up by the liver and indeed support hepatic gluconeogenesis by an indirect mechanism using glutamine as substrate. This observation may also account for the higher alanine levels found in portal venous blood that have been observed in arterial blood in man [11].

Cyclic Nucleotide Control of Amino Acid Formation in Muscle

Two principal mechanisms controlling both alanine and glutamine synthesis and release from skeletal muscle have been demonstrated. First, regulation by the availability of amino acid substrates for conversion to alanine and glutamine is well documented. Such substrates may include excess free ammonium ion as produced by the addition of ammonium chloride to muscle incubation systems producing increased glutamine synthesis, in part at the expense of a concomitant decrease in alanine synthesis. On the other hand, excess pyruvate availability, as might be produced by the direct addition of pyruvate per se, results in increased alanine synthesis, in part at the expense of a diminished glutamine synthesis and release (Table 1). In contrast, the direct addition of glutamate increases the synthesis and release of both alanine and glutamine, the

Table 1. Effect of pyruvate, ammonium chloride, and glutamate addition on amino acid and pyruvate release from rat skeletal muscle[a]

Addition	Glutamine	Glutamate	Alanine	Pyruvate
	Release (nmol/min/g muscle)			
Control	29.9 ± 2.0	10.1 ± 1.0	19.8 ± 1.6	20.2 ± 1.8
Pyruvate	20.3 ± 1.8	5.8 ± 0.8	28.8 ± 2.3	
Glutamate	52.1 ± 3.6		29.5 ± 2.1	13.1 ± 2.0
Ammonium chloride	40.1 ± 2.4	15.3 ± 0.8	3ß.2 ± 2.4	11.9 ± 0.8

[a] Intact rat epitrochlaris preparations were incubated for 1 h at 37° C in Krebs-Henseleit buffer, pH 7.4, containing glucose (5 m*M*), insulin (100 mU/ml), and HEPES (5 m*M*). Neutral solutions of ammonium chloride, glutamate, and pyruvate were added to the concentrations indicated. After incubation, the release of amino acids and pyruvate was determined enzymatically in the media. Values given are means ± SE for at least four experiments and are expressed as nmol/min/g of muscle, wet wt

latter being accompanied by a concomitant decrease in lactate output. Glutamate is the precursor of both glutamine and alanine in skeletal muscle since it is the necessary substrate for glutamine synthetase as well as for alanine aminotransferase. These manipulations did not demonstrate net appearance of new carbon into the overall profile of amino acids and lactate released from skeletal muscle, but did demonstrate that redirection of enzymes operating at or near equilibrium may occur such that the relative proportion of glutamine or alanine or both may be manipulated under a variety of physiologic or metabolic circumstances.

On the other hand, there is also a broad range of hormonal influences on the pathways of alanine and glutamine synthesis and release in skeletal muscle. Insulin, cortisol, thyroid hormone, and testosterone have all been demonstrated to influence amino acid metabolism in muscle [12]. These effects are modest in magnitude and require prolonged times for the full expression of their action on skeletal muscle protein and amino acid metabolism. More recently, we have observed that hormones associated with cyclic nucleotide metabolism may also regulate protein and amino acid in skeletal muscle. In the course of our studies on potential relationships between glucose metabolism and alanine and glutamine synthesis in skeletal muscle, we found that physiologic concentrations of epinephrine, a glycogenolytic hormone for skeletal muscle, produced an inhibition rather than an augmentation of alanine and glutamine output from muscle [13]. This effect was found with physiologic concentrations of epinephrine (10^{-9}–10^{-7} *M*). These concentrations of epinephrine using skeletal muscle preparations in vitro, produced an activation of glycogenolysis as judged by increased lactate and pyruvate production from muscle, but also inhibited alanine and glutamine output in a concentration-dependent fashion (Fig. 2). This effect of epinephrine was reproduced by isoproterenol at concentrations approximately one order of magnitude less than those required for epinephrine to produce equal action on skeletal muscle amino acid and glycogen metabolism. On the other hand, phenylephrine (10^{-6} *M*) had no effect on glutamine and

Fig. 2. Effect of epinephrine on glutamine, alanine, and glutamate release from skeletal muscle preparations in vitro. Epitrochlaris preparations from control animals were obtained and incubated in Krebs-Henseleit bicarbonate buffer for 1 h. Varying concentrations of epinephrine were added to the media as indicated. Following incubation, alanine, glutamine, and glutamate released to the media were determined enzymatically. Values shown are the means (± SE) for at least six experiments and are expressed as nmol/min/g wet wt of muscle

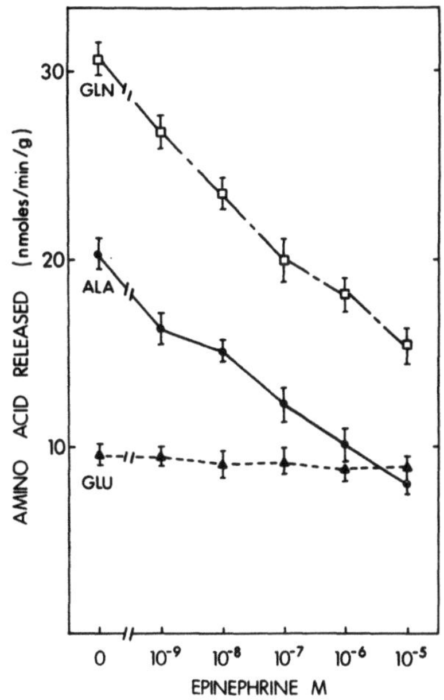

alanine synthesis and release. Furthermore, α-adrenergic antagonists, such as phentolamine, were found not to block the effects of epinephrine or of isoproterenol, but β-adrenergic antagonists, such as propranolol and oxprenolol, completely inhibited the inhibition of glutamine formation and release produced by agonists, such as epinephrine and isoproterenol. These findings suggest that the mechanism of catecholamine inhibition of amino acid formation and release in skeletal muscle depends upon the involvement of the β-adrenergic receptor. This receptor requirement is similar to the receptor requirement for adrenergic regulation of muscle glycogenolysis and a similar β-adrenergic effect on skeletal muscle amino acid metabolism as well as on cardiac muscle metabolism has been observed [14]. In view of the receptor participating in both metabolic processes and the demonstration that similar concentrations of hormone equally affect glycogenolysis as well as amino acid metabolism, it seems possible that the effects of catecholamines may be mediated by β-adrenergic stimulation of skeletal muscle adenylyl cylase activity and increased cAMP levels in skeletal muscle. Support of this concept is provided by the findings that phosphodiesterase inhibitors, such as theophylline or methylisobutylxanthine at low concentrations, potentiate the effects of catecholamines on amino acid metabolism, on muscle glycogenolysis, and on muscle cAMP levels. Furthermore, these effects of catecholamines are reproduced by the exogenous addition of cAMP derivatives, such as dibutyryl cAMP or 8-bromo cAMP.

Other neurohumoral agonists, such as serotonin, also appear to influence skeletal muscle protein and amino acid metabolism. Serotonin is an important autacoid influencing smooth muscle function, and also appears to have direct effects on skeletal muscle metabolism. This tryptamine derivative produces a concentration-dependent inhibition of glutamine as well as alanine output from

Table 2. Effects of serotonin and serotonergic antagonists on amino acid release from skeletal muscle[a]

Addition	Inhibitor	Alanine	Glutamine
		Release (nmol/min/g muscle)	
Control		20.1 ± 1.2	31.4 ± 2.3
Serotonin	10^{-8}	17.9 ± 1.6	27.1 ± 2.3
	10^{-6}	14.8 ± 2.0	23.0 ± 2.2
	10^{-4}	12.3 ± 1.1	19.8 ± 2.0
	10^{-4} Methysergide	21.2 ± 1.3	30.8 ± 2.6
	10^{-4} Cyproheptadine	20.8 ± 1.8	31.7 ± 3.0

[a] Intact epitrochlaris preparations of rat skeletal muscle were obtained and incubated for 1 h at 37° C under an atmosphere of 95% O_2 − 5% CO_2 in 0.5 ml of Krebs-Henseleit buffer (pH 7.4) containing HEPES (5 m*M*) and glucose (5 m*M*). Serotonin and antagonists were added to the final concentration indicated. Following incubation, muscles were rapidly removed, rinsed, blotted, and frozen in liquid nitrogen. Alanine and glutamine released to the medium were determined enzymatically. Values are means ± SE for at least six experiments

skeletal muscle (Table 2). These actions of serotonin on alanine and glutamine synthesis and release from skeletal muscle appear to be mediated by specific D-serotonergic receptors since the inhibition of amino acid production was blocked by cyproheptadine and methysergide, but was not influenced by morphine sulfate or low concentrations of tetrodotoxin. This pattern of inhibition defines the serotonergic receptor participating in the skeletal muscle action of serotonin as a D-serotonergic receptor similar to that found in the central nervous system. As with catecholamines, the inhibition of skeletal muscle alanine and glutamine synthesis produced by serotonin was accompanied by a concentration-dependent increase in muscle cAMP levels. This effect of serotonin on cAMP levels was not the result of an inhibition of cAMP degradation as would be expected by inhibition of cAMP phosphodiesterase, but was instead the direct effect of a stimulation of skeletal muscle adenylyl cyclase by serotonergic actions as demonstrated by in vitro preparations of sarcolemma studied with serotonin. Neither serotonin nor catecholamines, such as epinephrine, inhibit the stimulated rates of alanine and glutamine release produced by saturating quantities of exogenous amino acids, such as branched-chain amino acids, suggesting that the effects of these inhibitors do not localize to the enzymes controlling the conversion of precursor amino acids to substrates for glutamine or alanine synthesis. Instead, data obtained using skeletal muscle preparations obtained from rats prelabeled in vivo with [guanidine-^{14}C]-arginine suggest that a decreased degradation of skeletal muscle proteins is produced in vitro in the presence of serotonin or epinephrine as compared to control incubation rates of degradation.

Because these effects of epinephrine and serotonin were previously unsuspected, but were clearly associated with cAMP mediated action on skeletal muscle protein and amino acid metabolism, potential effects of other agonists linked to cGMP metabolism were also studied on skeletal muscle protein and

Fig. 3. Effect of carbamylcholine on skeletal muscle glutamine synthesis and release in vitro. Epitrochlaris skeletal muscle preparations were obtained and incubated with varying concentrations of carbamylcholine. Following 1 h of incubation, preparations were rapidly removed from the media, rinsed, blotted, and frozen in liquid nitrogen. Glutamine release to the media (●——●) was determined enzymatically as were glutamine tissue levels in perchloric acid extracts of each muscle preparation (○---○). Values shown are the means (± SE) for at least eight experiments and are expressed as nmol/min/g wet wt of muscle, or μmol/g wet wt of muscle, as indicated [17]

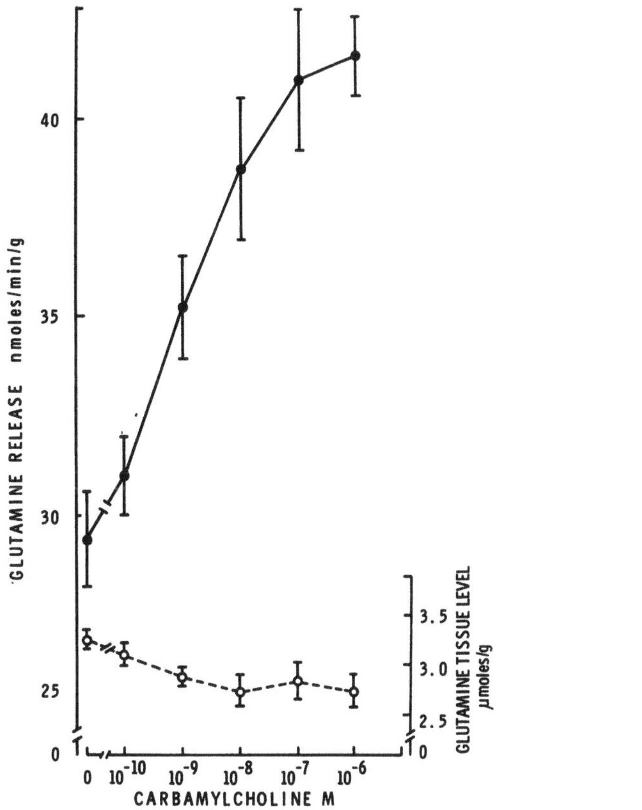

amino acid metabolism. Muscarinic cholinergic agonists have been demonstrated to produce marked elevations of cGMP levels in cardiac muscle and in gastrointestinal smooth muscle [15, 16]. In skeletal muscle, acetylcholine, in combination with neostigmine, or carbamylcholine was found to increase the synthesis and release of glutamine as well as alanine [17]. This effect was dependent upon the concentration of carbamylcholine studied. Concentrations of carbamylcholine as low as $10^{-10}\,M$ significantly altered the output of glutamine and alanine from skeletal muscle preparations as studied in vitro (Fig. 3). This increased amino acid output appeared to derive in part from a facilitated amino acid transport since glutamine levels in skeletal muscle declined in a carbamylcholine concentration-dependent fashion. Nevertheless, the quantity of increased glutamine output attributable to the increased transport and disposal of the preexisting intracellular glutamine pool accounted for only 39% of the total augmentation of glutamine release from muscle. More than 61% of the increased glutamine release produced by cholinergic agonists, such as carbamylcholine, appears to result instead from an increased net synthesis of this amino acid in skeletal muscle. This effect was not the result of concomitant skeletal muscle contraction produced by cholinergic agonists since muscle glycogen levels were unchanged in the presence of any concentration of cholinergic agonists studied in this system. The increased glutamine and alanine synthesis produced by cholinergic agonists in the presence of added amino acids was no greater than in the absence of these amino acids, suggesting that cholinergic action increasing glutamine synthesis was not the result of a direct activation of those enzymes participating in the synthesis of alanine and glutamine from precursor amino acids. Instead, it seems more likely that cholinergic agonists must stimulate skeletal muscle proteolysis, thereby increasing the availability of endogenous amino acids serving as substrates for glutamine synthesis in muscle. Support for this hypothesis is provided by the finding that cholinergic agonists also facilitate alanine synthesis, presumably by supplying an increased availability of other precursors which are different from those used for glutamine synthesis. Direct experimental testing of this hypothesis was performed using skeletal muscle preparations obtained from animals prelabeled in vivo with [guanidino-^{14}C]arginine as described earlier. When incubated in vitro with cholinergic agonists, there was a greatly increased transfer of amino acid label from the acid insoluble protein fraction of these muscle preparations to the acid soluble intra- and extracellular compartments during in vitro incubation studies. Thus, it seems evident that cholinergic agonists stimulate alanine and glutamine synthesis, and therefore, have actions which antagonize those of adrenergic and serotonergic agonists.

Although cholinergic action on cardiac and smooth muscle requires a muscarinic cholinergic receptor as noted above, cholinergic action on skeletal muscle protein and amino acid metabolism is not mediated by a muscarinic cholinergic receptor. Instead, a nicotinic cholinergic receptor appears to be the mechanism participating in the cholinergic augmentation of glutamine and alanine synthesis and release from skeletal muscle [18]. Carbamylcholine stimulation of glutamine synthesis was completely antagonized by added hexamethonium or decamethonium, but not by atropine. On the other hand, these same actions of carbamylcholine were reproduced by tetramethylammonium hydroxide, but not by pilocarpine addition. Thus, it seems clear that the cholinergic receptor participating in the increased glutamine and alanine synthesis and release in skeletal muscle is nicotinic rather than muscarinic in nature. Although this is a somewhat unusual receptor requirement for cholinergic action on cyclic nucleotide metabolism, cholinergic agonists, such as carbamylcholine, clearly produced a concentration-dependent increase in skeletal muscle cGMP levels. As noted in Fig. 4, the lowest concentration of carbamylcholine which increased cGMP levels was 10^{-9} M and the highest increments were found using 10^{-5} M carbamylcholine. These concentrations agree well with those concentrations of carbamylcholine increasing glutamine synthesis and release in skeletal muscle. The association between increased cGMP levels and increased glutamine and alanine synthesis and release in muscle was supported by studies with cGMP phosphodiesterase inhibitors which at low concentrations did not of themselves alter amino acid release or basal cGMP levels. However, the addition of submaximal concentrations of carbamylcholine in the presence of these inhibitors produced augmented cGMP responses and a concomitant augmentation of glutamine release when such inhibitors were present. It is also noteworthy that nicotinic cholinergic

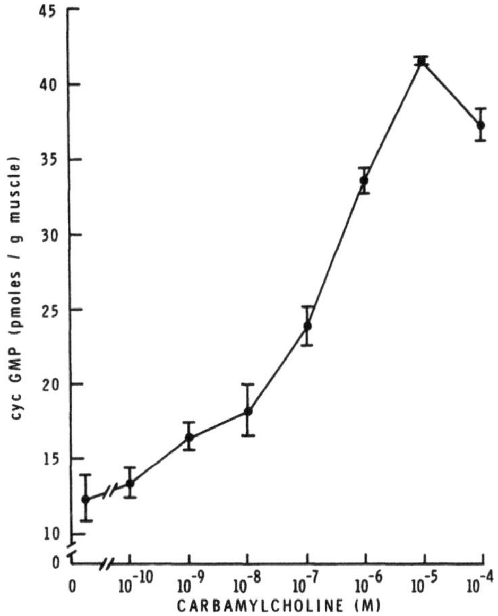

Fig. 4. Effect of carbamylcholine on intracellular cGMP levels in skeletal muscle. Epitrochlaris skeletal muscle preparations were obtained and incubated with varying concentrations of carbamylcholine in Krebs-Henseleit bicarbonate buffer (pH 7.4). Following a 10 min incubation, muscles were rapidly removed from the media, rinsed, blotted, and frozen in liquid nitrogen. cGMP lvels were determined in trichloroacetic acid extracts of each skeletal muscle preparation, using double antibody radioimmunoassay techniques. Values shown are the means (± SE) for at least 10 experiments and are expressed as pmol/g wet wt of muscle [18]

antagonists blocked both the increased cGMP levels and the increased glutamine synthesis, whereas nicotinic agonists, such as tetramethylammonium hydroxide, reproduced the increased cGMP levels and glutamine synthesis produced by carbamylcholine. On the other hand, muscarinic cholinergic agonists had no effect on cGMP levels nor did they augment glutamine synthesis in skeletal muscle. The direct addition of cGMP derivatives to skeletal muscle incubations in vitro appears to reproduce the effects of cholinergic agonists on alanine and glutamine synthesis and release from muscle and on amino acid transport.

As a result of these studies, it seems evident that there are in skeletal muscle two coordinate and apparently independent cyclic nucleotide associated mechanisms regulating skeletal muscle protein and amino acid metabolism. Agonists linked through cyclic AMP and causing a stimulation of adenylyl cyclase in muscle produce a marked inhibition of muscle proteolysis and as a result decrease the synthesis and release of glutamine as well as alanine from skeletal muscle. Counterbalancing these regulatory mechanisms, agonists generating increased cGMP levels in skeletal muscle produce a substantial stimulation of muscle proteolysis and thereby of glutamine synthesis and release. The physiologic significance of these pathways can be best assessed by studies of their participation in pathologic derangements of muscle wasting as discussed below.

Derangements of Cyclic Nucleotide Regulation of Glutamine Metabolism in Muscle Wasting States

The physiologic significance of cyclic nucleotide-associated mechanisms controlling glutamine synthesis and release in skeletal muscle can be assessed in part by studies of the regulation of muscle protein and amino acid metabolism in models of muscle wasting, such as chronic renal failure, uncontrolled diabetes mellitus, and muscular dystrophy. In all three disorders, clearly recognizable defects in cyclic nucleotide regulation of protein and amino acid metabolism appear to underlie the accelerated alanine and glutamine production in muscle of these states. Furthermore, these abnormalities of cyclic nucleotide formation and action may also account, at least in part, for the muscle wasting, muscle dysfunction, and the associated hyperglycemia noted in these disorders. For purposes of this discussion, we will focus principally on our investigations using skeletal muscle of experimentally uremic rats and also provide additional focus using in vivo human isotope kinetic studies to assess amino acid flux. In our studies using chronically uremic rats, skeletal muscle preparations were found to have an increased rate of glutamine and alanine synthesis and release. This increased release did not derive from diminished glutamine or alanine reutilization in terms of muscle protein resynthesis or oxidation to CO_2 [17]. Moreover, the latter pathways are quantitatively insignificant in muscle when compared to the net rate of glutamine or alanine synthesis and release. The increased glutamine synthesis and release did not appear to result from an increased maximal capacity of the tissue to synthesize this amino acid from added precursor amino acids, since further increases in glutamine release by exogenous, saturating concentrations of added amino acids were not produced. On the other hand, studies attempting to quantitate net muscle proteolysis in

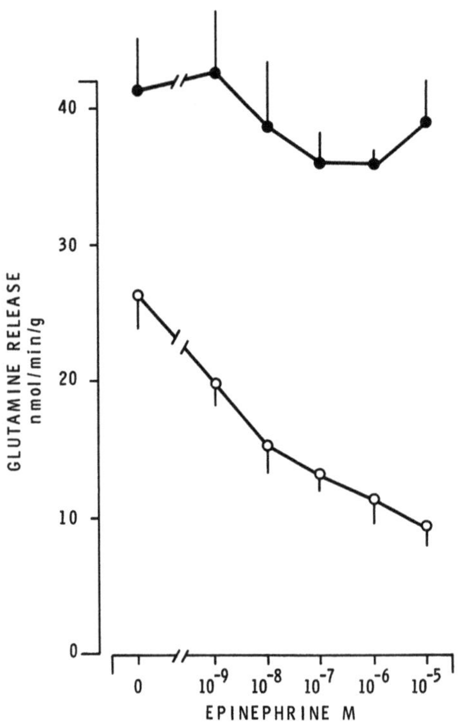

Fig. 5. Suppression of glutamine formation and release by epinephrine in skeletal muscle preparations of control and uremic rats. Epitrochlaris skeletal muscle preparations from control (○———○) and chronically uremic rats (●———●) were obtained and incubated with varying concentrations of epinephrine. Following a 1 h incubation, muscles were rapidly removed from the media, rinsed, blotted, and frozen in liquid nitrogen. Release of glutamine to the incubation media was determined enzymatically using semimicrofluorometric enzymatic techniques. Values shown are the means (± SE) for at least 10 experiments and are expressed as nmol/min/g wet wt of muscle

uremic and control rats prelabeled in vivo with [guanidino-^{14}C]arginine demonstrated accelerated net protein breakdown in uremic as compared to control muscle.

Abnormalities of insulin action did not appear to account for the increased proteolysis and amino acid output from uremic muscle. Instead, defects in cAMP-associated inhibition of proteolysis and of alanine and glutamine synthesis were clearly demonstrable. Indeed, there was marked insensitivity or resistance to catecholamine-mediated suppression of protein and amino acid metabolism in muscle of uremic as compared to control rats (Fig. 5). The dose response curve of the inhibition of glutamine synthesis by epinephrine in uremic muscle was shifted upward from the dose response curve with control muscle by nearly 50%. Moreover, this same curve was shifted to the right, to higher concentrations of epinephrine, by approx. 3 log unit intervals in the epinephrine dose response curve. A similar shift was also obtained for inhibition of alanine release by epinephrine. Furthermore, a virtually identical shift in the dose response curve of glutamine released to added serotonin was found. These findings indicate a loss of adrenergic and serotonergic action on glutamine and alanine synthesis and release in uremic muscle. This was not the result of an absent responsiveness to cAMP or its derivatives, since dibutyryl cAMP inhibited glutamine and alanine synthesis equally in both control and uremic muscle. Rather, the shift in the dose response curve of amino acid release to catecholamines appeared to reflect a primary resistance of cAMP generation in uremic as compared to control muscle. As noted in Table 3, smaller stimulations of cAMP levels were produced in uremic as compared to control muscle at all concentrations of epinephrine or serotonin studied. Indeed, two orders of magnitude higher concentrations of epinephrine or serotonin were required to produce equal elevations of cAMP in uremic as compared to control muscle. This resistance to cAMP accumulation as stimulated by adrenergic or

Table 3. Effect of epinephrine and serotonin on cAMP levels in skeletal muscle of control and uremic rats[a]

Addition	Concentration	cAMP level (pmol/g muscle)	
		Control	Uremic
Control		609 ± 39.5	543 ± 30.9
Epinephrine	10^{-8}	726 ± 32.1	595 ± 40.1
	10^{-6}	1,710 ± 71.2	794 ± 61.3
	10^{-5}	2,246 ± 95.2	1,091 ± 84.0
Serotonin	10^{-7}	995 ± 65	649 ± 40
	10^{-5}	1,381 ± 45	734 ± 18

[a] Epitrochlaris preparations from control and uremic rats were incubated with varying concentrations of epinephrine and serotonin as indicated. After a 2 min incubation, the preparations were rapidly removed from their media, rinsed, blotted, and frozen in liquid nitrogen. Levels of cAMP were determined in ether-washed trichloroacetic acid extracts of muscle using a sensitive double-antibody radioimmunoassay. Values shown are the means (± SEM) for at least 10 experiments

serotonergic agonists appeared to be the result primarily of a diminished activity of the hormone sensitive adenylyl cyclase complex in uremic muscle. Whereas significant alterations of basal or fluoride stimulable adenylyl cyclase activity were not noted in uremic as compared to control muscle, the magnitude of stimulation by catecholamines was reduced by 50% or more in uremic sarcolemma.

The finding of accelerated amino acid production from muscle of uremic animals correlates closely with in vivo isotope kinetic studies performed in normal and uremic patients. As noted in Table 4, alanine production was increased by more than twofold in uremic as compared to normal subjects. As a consequence, increased gluconeogenesis from alanine and increased total glucose production were observed. Thus, it seems reasonable to conclude that as a result of the failure of cAMP-associated regulation of skeletal muscle protein

Table 4. Glucose and alanine kinetics in control and azotemic subjects[a]

Subjects	Glucose	Alanine	Glucose produced from alanine
	Production μmol/min		
Normal	664 ± 34	163 ± 19	21.7 ± 2.24
Azotemic	1,034 ± 76	545 ± 114	69.6 ± 18.9

[a] Rates of alanine and glucose turnover and precursor-product inter-relationships were determined in patients on chronic hemodialysis and in matched controls using simultaneous primed injection-continuous infusions of [U-^{14}C]alanine and [2-^{3}H]glucose. Eight chronically dialyzed patients were studied before their first dialysis of the week

Table 5. Effect of bovine parathyroid hormone (1–84) and the synthetic (1–34) fragment on amino acid release from skeletal muscle of control and uremic rats[a]

Addition	Concentration (U/ml)	Release			
		Alanine		Glutamine	
		Control	Uremic	Control	Uremic
		nmol/min/g muscle			
Control	–	19.8 ± 2.6	29.7 ± 1.3	29.4 ± 2.5	39.9 ± 2.6
Parathyroid hormone (1–84)	50	33.1 ± 2.4	29.2 ± 2.0	46.9 ± 6.4	40.2 ± 2.1
Synthetic 1–34 peptide	57	31.1 ± 3.3	29.5 ± 1.6	48.0 ± 5.0	41.0 ± 3.0

[a] Epitrochlaris preparations from control and chronically uremic rats were obtained and incubated for 1 h at 37° C in Krebs-Henseleit bicarbonate buffer (pH 7.4) containing glucose (5 mM) and highly purified intact bovine parathyroid hormone or synthetic 1–34 bovine parathyroid hormone at the concentrations indicated. Following conclusion of the 1 h incubation, muscles were rapidly removed from the media, rinsed, blotted, and frozen in liquid nitrogen. Alanine and glutamine released to the media were determined enzymatically using microfluorometric techniques. Values shown are the means (± SEM) for at least eight experiments and are expressed as nmol of amino acids released per min/g muscle, wet wt

metabolism, overproduction of gluconeogenic and potentially gluconeogenic amino acids, such as alanine and glutamine, results. As a consequence of this overproduction, inappropriately accelerated gluconeogenesis ensues and a pseudodiabetic state thereby results.

The physiologic basis for the desensitization or loss of responsiveness of the catecholamine stimulable adenylyl cyclase in muscle of uremic animals was further studied with an initial focus on the potential role of parathyroid hormone in the pathogenesis of the abnormal cAMP metabolism of the uremic state. Both chronic renal failure as well as primary hyperparathyroidism are associated with strikingly elevated levels of parathyroid hormone and with a diffuse neuromuscular defect, prominent muscle wasting and with a tendency towards carbohydrate intolerance and even overt diabetes mellitus [19–24]. As seen in Table 5, both intact bovine parathyroid hormone [1–24] as well as the

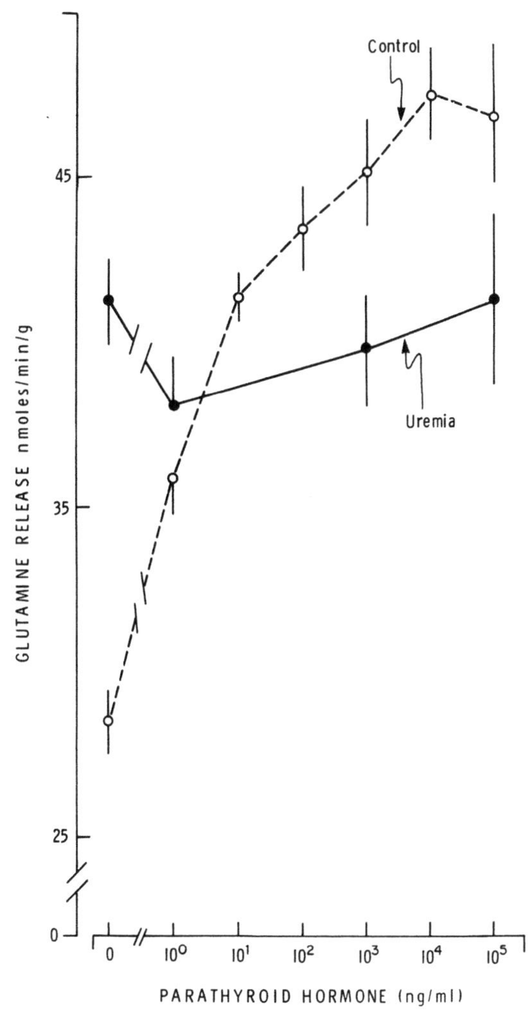

Fig. 6. Effect of parathyroid hormone on glutamine release from skeletal muscle. Epitrochlaris preparations were obtained from control (○) and chronically uremic (●) rats and incubated at 37° C for 1 h in Krebs-Henseleit bicarbonate buffer (pH 7.4) containing glucose (5 m*M*) and varying concentrations of synthetic 1–34 bovine parathyroid hormone as indicated. After incubation, the muscles were rapidly removed, rinsed, blotted, and frozen in liquid nitrogen. Glutamine released to the media was determined enzymatically using microfluorometric techniques. Values shown are the means ± SEM for at least eight experiments and are expressed as nmol alanine released per min per g muscle, wet wt

biosynthetic 1–34 fragment of parathyroid hormone stimulated the synthesis and release of both glutamine and alanine from normal, but not uremic skeletal muscle preparations [25]. This effect of parathyroid hormone was not limited solely to the single concentration studied, but indeed appeared to be dependent upon the concentration of parathyroid hormone added to incubation media (Fig. 6). This stimulation of glutamine release by added 1-34 parathyroid hormone appeared to be ultimately saturable at hormone concentrations of 10 μg/ml or higher. At no concentration of hormone studied were alterations of amino acid release in uremic muscle observed. This stimulation by parathyroid hormone of glutamine and alanine synthesis and release was not the result of a

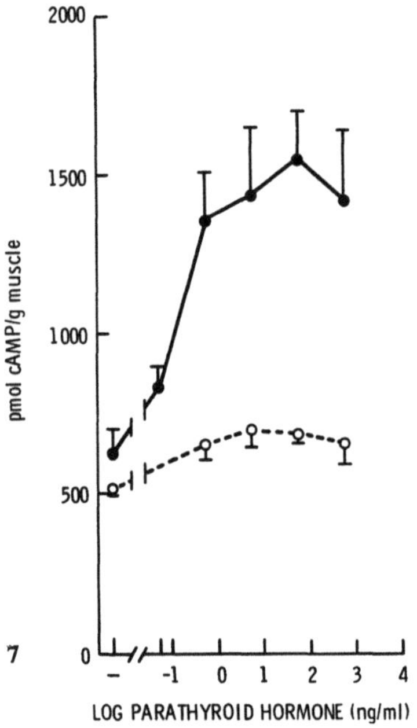

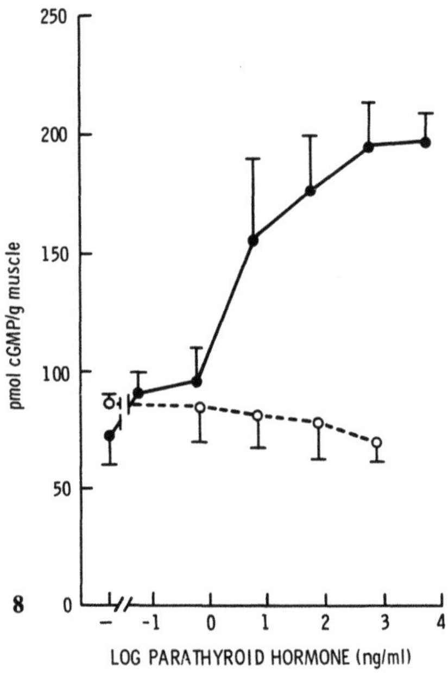

Fig. 7. Effect of parathyroid hormone on cAMP levels in skeletal muscle of control and uremic rats. Epitrochlaris preparations from control (●) and chronically uremic (○) rats were obtained and incubated with varying concentrations of parathyroid hormone in Krebs-Henseleit bicarbonate buffer (pH 7.4) containing glucose (5 m*M*). After the conclusion of the incubation, each muscle was rapidly removed, rinsed, blotted, and frozen in liquid nitrogen. Levels of cAMP in trichloroacetic exracts of each skeletal muscle preparation were then determined by double antibody radioimmunoassay. Values shown are the means (± SEM) for at least 10 experiments and are given as pmol of cAMP per g muscle, wet wt

Fig. 8. Effect of parathyroid hormone on cGMP levels and skeletal muscle of control and uremic rats. Epitrochlaris preparations from control (●) and chronically uremic (○) rats were obtained and incubated with varying concentrations of parathyroid hormone in Krebs-Henseleit bicarbonate buffer (pH 7.4) containing glucose (5 m*M*). After the conclusion of the incubation, each muscle was rapidly removed, rinsed, blotted, and frozen in liquid nitrogen. Levels of cGMP in trichloroacetic extracts of each skeletal muscle preparation were then determined by double antibody radioimmunoassay. Values shown are the means (± SEM) for at least 10 experiments and are given as pmol of cGMP per g muscle, wet wt

decreased reutilization of these amino acids by skeletal muscle preparations since parathyroid hormone did not, at these concentrations, alter amino acid oxidation to CO_2, amino acid reincorporation into protein, or net protein synthesis. Effects of parathyroid hormone on skeletal muscle cyclic nucleotide levels were also assessed. As noted in Figs. 7 and 8, parathyroid hormone increased muscle cAMP as well as cGMP levels at those concentrations of the hormone altering amino acid metabolism. These effects of parathyroid hormone on muscle cyclic nucleotide levels were observed only with control muscle preparations. No effects of parathyroid hormone on skeletal muscle preparations of uremic animals could be detected. With respect to these alterations in cyclic nucleotide levels, it is noteworthy that although parathyroid hormone increased the levels of both cAMP and cGMP, the predominant nucleotide effect with respect to amino acid metabolism appeared to be that of the increased cGMP levels. As noted earlier, increased cGMP levels produced by nicotinic cholinergic agonists are clearly associated with accelerated protein degradation and amino acid synthesis and release. A similar effect is therefore attributable to the increased cGMP levels produced by parathyroid hormone.

Regardless of the precise mechanism by which parathyroid hormone accelerates muscle proteolysis and increases the synthesis of glutamine in skeletal muscle, it is clear that parathyroid hormone addition to normal muscle reproduced the insensitivity to catecholamine suppression of glutamine synthesis and release observed in uremic muscle (Fig. 9). As shown in this figure, the addition of 1–34 bovine parathyroid hormone (1 µg/ml) increased basal glutamine release to rates observed in uremic muscle. Increasing concentrations of epinephrine did not inhibit glutamine release at concentrations less than

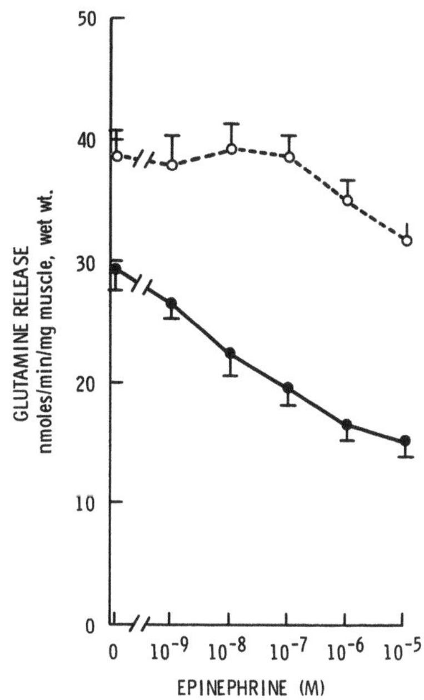

Fig. 9. Effect of parathyroid hormone on the epinephrine-induced inhibition of glutamine release from skeletal muscle. Epitrochlaris preparations were obtained from normal rats and were incubated for 1 h at 37° C under 95% O_2 – 5% CO_2 in Krebs-Henseleit bicarbonate buffer (pH 7.4) containing varying concentrations of epinephrine as indicated either in the presence (○) or the absence (●) of added 1–34 bovine parathyroid hormone (1,000 ng/ml). At the conclusion of the 1 h incubation, muscles were removed and rapidly rinsed, blotted and frozen in liquid nitrogen. Glutamine released to the media was determined enzymatically using microfluorometric techniques. Values shown are the means (± SEM) for at least six experiments and are expressed as nmol of glutamine released per min per g muscle, wet wt

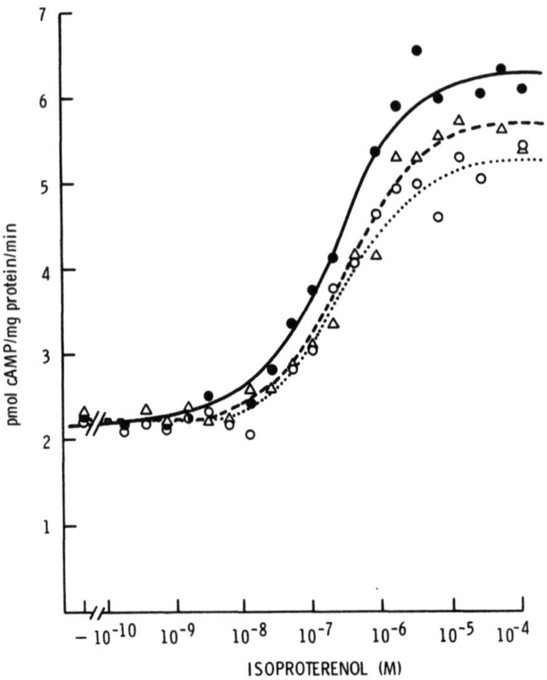

Fig. 10. Effect of preincubation with parathyroid hormone on isoproterenol stimulable adenylyl cyclase activity in rat sarcolemma. Skeletal muscle preparations were obtained from control rats and crude preparations of sarcolemma prepared by differential centrifugation. These sarcolemmal preparations were incubated for 10 min in adenylyl cyclase assay media omitting [α-^{32}P]ATP, but containing either no added 1–34 bovine parathyroid hormone (●), 1,000 ng/ml parathyroid hormone (△– – –△), or 10,000 ng/ml of 1–34 parathyroid hormone (○). After conclusion of the preincubation period, the sarcolemma was used in an adenylyl cyclase activity assay containing varying concentrations of isoproterenol as indicated. Values for adenylyl cyclase activity are the means (± SEM) for at least three experiments and are expressed as pmol of cAMP formed/min/mg protein

10^{-6} *M*. This is a shift in the dose response curve of glutamine release by three orders of magnitude. This decreased responsiveness is virtually identical to that observed in uremic muscle. The metabolic basis for the desensitization or loss of glutamine responsiveness to epinephrine suppression was further investigated. Cyclic AMP accumulation in the presence of added parathyroid hormone was less at any concentration of epinephrine or isoproterenol studied as compared to muscles incubated in the absence of added parathyroid hormone. Similarly, rat sarcolemma incubated with 1–34 parathyroid hormone showed a decreased responsiveness of the isoproterenol stimulable adenylyl cyclase activity as compared to sarcolemma preincubated in control incubation media or with isoproterenol (Fig. 10). This desensitization of skeletal muscle adenylyl cyclase activity by parathyroid hormone appeared to be dependent upon the amounts of parathyroid hormone added to the media in the initial preincubation.

Based on the results of these studies, it is possible to hypothesize that the hyperparathyroidism associated with chronic renal disease may result in heterologous desensitization of the hormone sensitive adenylyl cyclase complex in skeletal muscle of uremic animals. As a consequence, diminished catecholamine responsiveness and hence decreased adrenergic regulation of muscle proteolysis ensue. Thus, unrestrained muscle proteolysis results. The data of these studies suggests strongly that therapeutic maneuvers aimed at reducing the hyperparathyroidism of the uremic state may be efficacious in ameliorating the muscle wasting and muscle dysfunction and may also have beneficial effects on the glucose overproduction associated with uremia.

Acknowledgements. Supported in part by Grants NS 15950 and RR00350 from the National Institutes of Health and by the Fondren Foundation and the Muscular Dystrophy Association.

References

1. Garber AJ, Menzel PH, Boden G, Owen OE (1974) Hepatic ketogenesis and gluconeogenesis in humans. J Clin Invest 54: 981–989
2. Ruderman NB, Lund P (1972) Amino acid metabolism in skeletal muscle: Regulation of glutamine and alanine release in the perfused rat hindquarters. Isr J Med Sci 8: 295–302
3. Garber AJ, Karl IE, Kipnis DM (1976) Alanine and glutamine synthesis and release from skeletal muscle: I. Glycolysis and amino acid release from skeletal muscle. J Biol Chem 251: 826–835
4. Kominz DR, Hugh A, Symond P, Laki K (1954) The amino acid composition of actin, myosin, tropomyosin, and the meromyosins. Arch Biochem Biophys 50: 148–159
5. Goldstein L, Newsholme EA (1976) The formation of alanine from amino acids in diaphragm muscle of the rat. Biochem J 154: 555–560
6. Garber AJ, Karl IE, Kipnis DM (1976) Alanine and glutamine synthesis and release from skeletal muscle: II. The precursor role of amino acids in alanine and glutamine synthesis. J Biol Chem 251: 836–843
7. Snell K, Duff DA (1977) The release of alanine by rat diaphragm muscle in vitro. Biochem J 162: 399–400
8. Snell K (1979) Alanine as a gluconeogenic carrier. Trends Biochem Sci 4: 124–128
9. Owen OE, Felig P, Morgan AP, Wahren J, Cahill GF Jr (1969) Liver and kidney metabolism during prolonged starvation. J Clin Invest 48: 574–583
10. Windmueller HG, Spaeth AE (1974) Uptake and metabolism of plasma glutamine by the small intestine. J Biol Chem 249: 5070–5079
11. Felig P, Wahren J, Karl I, Cerasi E, Luft R, Kipnis DM (1973) Glutamine and glutamate metabolism in normal and diabetic subjects. Diabetes 22: 573–576
12. Karl IE, Garber AJ, Kipnis DM (1976) Alanine and glutamine formation and release in skeletal muscle: III. Dietary and hormonal regulation. J Biol Chem 251: 844–850
13. Garber AJ, Karl IE, Kipnis DM (1976) Alanine and glutamine synthesis and release from skeletal muscle. IV. β-Adrenergic inhibition of amino acid release. J Biol Chem 251: 851–857
14. Li JB, Jefferson LS (1977) Effect of isoproterenol on amino acid levels and protein turnover in skeletal muscle. Am J Physiol 256: 243–249
15. Yamashita K, Field JB (1972) Elevation of cyclic guanosine 3′,5′-monophosphate levels in dog thyroid slices caused by acetylcholine and sodium fluoride. J Biol Chem 247: 7065–7066
16. Van Sande J, De Coster C, Dumont JE (1975) Control and role of cyclic 3′,5′-guanosine monophosphate. Biochem Biophys Res Commun 62: 1168–1175
17. Garber AJ, Harari Y, Entman ML (1978) Cholinergic stimulation of alanine and glutamine formation and release from skeletal muscle. J Biol Chem 253: 7918–7923
18. Garber AJ, Entman ML, Birnbaumer L (1978) Cholinergic stimulation of skeletal muscle alanine and glutamine formation and release: Evidence for mediation by a nicotinic cholinergic receptor and guanosine 3′,5′-monophosphate. J Biol Chem 253: 7924–7930

19. Horton ES, Johnson C, Lebevitz HE (1968) Carbohydrate metabolism in uremia. Ann Intern Med 68: 63–74
20. Perkoff GT, Thomas CL, Newton JD, Sellman JC, Taylor FH (1958) Mechanisms of impaired glucose tolerance in uremia and experimental hyperazotemia. Diabetes 7: 375–383
21. Garber AJ, Bier DM, Cryer PE, Pagliara AS (1974) Hypoglycemia in compensated chronic renal insufficiency: substrate limitation of gluconeogenesis. Diabetes 23: 982–986
22. Slatopolsky E, Martin K, Hruska K (1980) Parathyroid hormone metabolism and its potential as a uremic toxin. Am J Physiol 239: F1–F12
23. Patten BM, Bilezikian JP, Mallette LE, Prince A, Engel WK, Aurbach GD (1974) Neuromuscular disease in primary hyperthyroidism. Ann Intern Med 80: 182–193
24. Choeod EG, Haust MD, Hudson AJ, Lewis FN (1970) Myopathy in primary familial hyerthyroidism. Am J Med 48: 700–707
25. Garber AJ (1983) Effects of parathyroid hormone on skeletal muscle protein and amino acid metabolism in the rat. J Clin Invest 71: 1806–1821

Cerebral Glutamine/Glutamate Interrelationships and Metabolic Compartmentation

S. Berl[1] and D. D. Clarke[2]

Introduction

The fact that in animals the concentration of glutamate and glutamine are higher in brain than in any other tissue initially attracted interest in attempting to understand the function or functions of these amino acids in the central nervous system. They were evidently required for protein synthesis and early studies established their close relationship to the Krebs tricarboxylic acid cycle and energy metabolism in brain. Glutamine formation in brain was also shown to be sensitive to elevations in plasma ammonia levels and to serve as the primary method for detoxifying this substance in brain in contrast to urea formation in liver. Also noted was the interesting observation that the increased levels of glutamine that occurred in hepatic encephalopathy was not accompanied by a concomitant decrease in glutamic acid levels. The level of these two amino acids were apparently not tightly coupled despite high levels of glutamine synthetase activity in brain and glutamic acid is the only established immediate precursor of glutamine. In recent years increasing evidence pointed to the probability that glutamic acid also functions as an excitatory neurotransmitter in the central nervous system. It appears that glutamate serving these various functions is compartmented into pools of different sizes which turn over at different rates and are not in rapid equilibrium with each other.

Studies with [^{14}C]-Labeled Amino Acids

Early studies showed that raising the circulating level of glutamate or glutamine 30–50 times did not significantly elevate the brain level of glutamic acid (Schwerin et al. 1950). This indicated a marked blood-brain barrier to the entry of glutamate from the circulation. Other organs, e.g., liver and kidney, did not show this penetrance barrier to glutamate.

Administration of [U^{14}C]-L-Glutamic Acid Intravenously

When ^{14}C-labeled glutamate, glutamine, and aspartate became available similar experiments were repeated with tracer doses of the amino acids which did not

1 Department of Neurology and Neurobiology Program Mount Sinai School of Medicine, New York, NY 10029, USA
2 Department of Chemistry, Forham University, Bronx, NY 10458, USA

Glutamine Metabolism in Mammalian Tissues
Edited by D. Häussinger and H. Sies

alter significantly their physiological levels in the plasma (Lajtha et al. 1959). The presence of a blood-brain barrier to entry of glutamate from the circulation was confirmed in these experiments (Table 1). Only a small amount of labeled amino acid was found in the brain. The liver and kidney were much more proficient in their uptake of labeled glutamate from the circulation and the glutamate was also converted to glutamine. It was noted that frequently, particularly at the earlier time points, the plasma glutamine was of higher specific activity than that of the glutamate and glutamine in these tissues. Since liver is a major source of plasma metabolites, in other experiments the liver was

Table 1. Administration of [U-^{14}C]-L-glutamic acid i.v. into mice

Time[a] (min)	Organ	Specific activity[b]	
		Glutamic acid	Glutamine
2	Plasma	29,000	3,600
	Brain	50	35
	Liver	2,000	1,300
	Red blood cells	1,900	220
	Kidney	700	390
5	Plasma	14,000	3,000
	Brain	93	57
	Liver	2,400	2,000
10	Plasma	8,000	3,000
	Brain	920	230
	Liver	9,200	11,000

[a] Each time point is the average of two experiments. Four adult mice were used for each experiment. 0.1 ml [U-^{14}C]-glutamic acid was injected. Each animal received 0.5 μCi and 7 μg glutamic acid corresponding to 30% of the plasma glutamic acid content

[b] Specific activity: cpm/μmol

Table 2. Administration of [U-^{14}C]-L-glutamic acid i.v. into rats with restricted circulation[a]

	Organ	Specific activity[b]	
		Glutamic acid	Glutamine
Unrestricted circulation	Plasma	520,000	38,000
	Red blood cells	27,000	8,200
	Lung	68,000	34,000
	Heart	7,800	5,300
	Muscle (front legs)	3,800	2,000
	Brain	300	550
Restricted circulation	Liver	1,800	790
	Kidney	17	190
	Muscle (hind legs)	100	54

[a] The inferior vena cava and the abdominal aorta were tied off in rats and 0.1 ml ^{14}C glutamic acid (5 μC; 1 μmol) was injected into the subclavian vein, and the animals sacrified after 5 min

[b] Specific activity: cpm/μmol

largely eliminated from the circulation by tying off the inferior vena cava and the abdominal aorta in rats (Lajtha et al. 1960). When labeled glutamate was administered i.v., the plasma glutamine was again higher in specific activity (counts/min/μmol; SA) than any other organ examined (Table 2). The specific activities of the lung glutamate and glutamine were higher than in other organs. Whether or not the lung glutamine equilibrates rapidly with plasma glutamine or is the source of plasma glutamine, remains unknown. The data suggested that glutamine was being formed from glutamate in an organ or organs and rapidly released to the plasma before mixing with the total glutamine of the organ. This compartmentation of glutamate to glutamine metabolism was suggested as occurring in organs other than brain even before studies established its presence in brain.

Administration of [U-^{14}C]-L-Glutamic Acid Intracisternally

Since our interests centered mainly on brain metabolism we decided to bypass the blood-brain barrier and administered the labeled glutamate into the cerebrospinal fluid via the cisterna magna (Berl et al. 1961). After short intervals of time following injection of the [U-^{14}C]-L-glutamate, blood plasma, liver and brain were analyzed for labeled glutamic acid and glutamine (Table 3). The radioactivity in these two amino acids in brain and liver were very different from that in the plasma and did not equilibrate with the plasma amino acids. The brain glutamine SA was greater than that of the plasma glutamine at all time points, even at 1 min; therefore, the labeled glutamine in the brain must have been synthesized there and not derived from the blood. The conversion of glutamate to glutamine was very rapid in the brain. By 1 min the relative specific activity (RSA) of the glutamine (glutamate = 1) was greater than 1 and reached values of 5 by 5 min. During this time period, the SA of glutamate decreased, while that of glutamine increased, although the sum of the counts in these two amino acids/g of fresh tissue remained relatively constant; the radioactivity was redistributed between the glutamate and glutamine. It is theoretically not probable for contributions of glutamate and glutamine from other sources, such as protein metabolism, carbohydrate metabolism, or citric and cyclic metabolites to explain the fivefold difference between the SA activity of glutamate and glutamine (Berl et al. 1961). The explanation offered was that the higher SA of brain glutamine relative to that of glutamate resulted from the conversion of a small pool or compartment of glutamate of high SA to glutamine. Extraction of the tissue would dilute this pool by the total amount in the tissue to a value less than that of the glutamine. It would have been anticipated that in a more homogeneous product-precursor system, the SA of the precursor glutamate would fall, while that of the product glutamine would rise and equal or slightly rise above that of glutamate, but no exceed it by fivefold (Reiner 1953). The liver showed a more conventional product-precursor relationship of glutamine to glutamate, although in this organ as well metabolic compartmentation of these metabolites is also very likely, as discussed later. Upon intracisternal injection of the ^{14}C-glutamic acid, the labeled amino acid was present throughout the brain

Table 3. Administration of [U-^{14}C]-L-glutamic acid intracisternally in the rat

Time (min)	Substrates[a]	SA			RSA Brain	Total counts/g tissue[b]	
		Plasma	Liver	Brain		Brain	Total GA + Gm
0.25	GA			7,700	0.43	77,000	92,000
	Gm			3,300		15,000	
1	GA	180,000	200	5,500	1.3	51,000	84,000
	Gm	1,200	66	7,300		33,000	
2	GA	130,000	800	3,200	3.8	32,000	91,000
	Gm	2,200	600	12,000		59,000	
5	GA	34,000	700	2,700	5.2	27,000	97,000
	Gm	6,100	760	14,000		70,000	
15	GA	8,900	190	2,200	5.0	22,000	77,000
	Gm	6,500	280	11,000		55,000	
30	GA	3,300	180	2,100	3.5	21,000	58,000
	Gm	870	240	7,300		37,000	
60	GA	1,500	61	1,900	2.6	19,000	44,000
	Gm	140	130	5,000		25,000	

[a] SA: Specific activity (cpm/μmol); GA: glutamine acid; Gm: glutamine; RSA: relative specific activity $= \frac{\text{Gm}}{\text{GA}}$

[b] Results are the average of two experiments. In each experiment, two rats were injected via the cisterna magna with [U-^{14}C]-L-glutamatic acid (0.02 ml; 40 μCi, 4 μmol/ml in normal saline)

Table 4. Administration of [U-^{14}C]-L-glutamic acid intracisternally in the rat: area metabolism

Brain area	SA/GA[a]	Gm	RSA/GSH[b]	GABA
Cerebrum	2,065	4.3	1.1	0.21
Cerebellum	7,000	5.3	1.0	0.34
Pons medulla	14,000	5.0	1.1	0.23

[a] See Table 3 for abbreviations and experimental conditions. Rats were killed after 2 min. Average of four experiments

[b] GSH: glutathione

and rapidly converted to glutamine as well as other metabolites, such as glutathione and GABA (Berl et al. 1961, Table 4). The radioactivity was highest in the cerebellum and brain stem, the areas closest to the site of injection. The RSA of glutamine (relative to that of glutamate) was approx. 4–5, that of the glutamate in glutathione approx. 1 and that of GABA approx. 0.2–0.3. An experiment of the same type in a monkey showed similar results in 14 brain areas and the spinal cord (Berl et al. 1961).

Table 5. Administration of [U-^{14}C]-L-glutamine intracisternally in the rat: area metabolism

Brain area	SA[a]		RSA	
	Gm	GA	GSH	GABA
Cerebrum	4,980	418	1.6	0.7
Cerebellum	9,900	860	1.7	0.9
Pons medulla	12,000	930	2.2	0.6

[a] See Tables 3 and 4 for experimental details and abbreviations. Animals killed 2 min after injection of U-^{14}C-L-glutamine (0.02 ml; 50 μCi, 12 μmol/ml in normal saline). Average of three experiments, two animals per experiment

Administration of [U-^{14}C]-L-Glutamine Intracisternally

Labeled glutamine was injected into the cisterna magna of rats and after 2 min the animals were killed and glutamine, glutamic acid, glutathione (glutamic acid), GABA, and aspartic acid were isolated and counted (Berl et al. 1961, Table 5). In these experiments, the SA of the glutamine was considerably higher than that of glutamate. However, the RSAs of glutathione and GABA were higher than those in the experiments in which labeled glutamate was the administered tracer (compare with Table 4). The radioactivity in the aspartate was similar to that in the GABA (data not shown). The data suggested that glutamine may serve as a source of glutamate which can form GSH or GABA better than administered glutamate. This early data may be of particular significance in the light of more recent suggestions that formation of glutamine in brain tissue may function to neutralize glutamate released as a neurotransmitter at synaptic clefts and also to conserve the carbon skeleton for use again as transmitter glutamate or GABA (Bradford et al. 1978, Hamberger et al. 1979a, b).

Administration of [U-^{14}C]-L-Aspartic Acid Intracisternally

Following the injection of labeled aspartic acid into the cisterna magna of rats, the label quickly appeared in glutamic acid and its metabolites (Berl et al. 1961, Table 6). The pattern which emerged was similar to that following administration of radioactive tracer glutamate; the RSA of glutamine was approx. 4–5.0 throughout the brain, that of GABA 0.1–0.2 and that of GSH 0.2–0.5. The brain asparagine was also significantly labeled, but its SA relative to that of aspartic acid was quite low 0.01–0.02. As anticipated (Goldstein 1959), N-acetyl aspartic acid had only trace amounts of radioactivity (data not included).

With aspartate as the labeled tracer, if the metabolic pattern was that of an homogeneous system, it would have been anticipated that as the SA of the glutamate rose that of the glutamine would also rise, but at a slower rate and eventually equal or slightly exceed that of the glutamic acid (Reiner 1953). The measured RSA value of approx. 5 was again explained by the proposal that the

Table 6. Administration of [U-^{14}C]-L-aspartic acid intracisternally in the rat: area metabolism

Brain area	SA[a]		RSA[b]			
	AA	GA	Asn	Gm	GABA	GSH
Cerebrum	37,000	1,110	0.02	4.7	0.10	0.37
Cerebellum	160,000	3,400	0.02	4.7	0.20	0.52
Pons medulla	70,000	4,500	0.01	4.0	0.07	0.20
Whole brain	55,000	1,600	0.02	5.0	0.11	0.36

[a] See Tables 3 and 4 for experimental details and abbreviations
[b] Asn: asparagine; AA: aspartic acid

$$\text{RSA of Asn} = \frac{\text{SA of Asn}}{\text{SA of AA}}$$

Animals killed 2 min after injection of [U-^{14}C]-L-aspartic acid (0.02 ml; 50 µCi, 10 µmol/ml in normal saline)

newly formed glutamate from the administered aspartate mixed with only a portion of the total tissue glutamate which was the source of labeled glutamine. When the total tissue content of the amino acids was extracted, the radioactivity in the glutamate was greatly reduced by unlabeled glutamate and its SA measured as less than that of the glutamine.

Ammonia Metabolism

The increased levels of ammonia and glutamine that occur in hepatic coma (Gilon et al. 1959) and in animals with porto-caval anastomosis or hepatectomy (Flock et al. 1953, Bollman et al. 1957) points to a close association between ammonia and glutamine/glutamate metabolism. Therefore, the administration of toxic levels of ammonia was utilized for a study of the role of these amino acids in detoxification of ammonia in vivo and for further understanding of the relationship between ammonia metabolism and glutamate-glutamine metabolic compartmentation (Takagaki et al. 1961, Berl et al. 1962a).

Administration of [^{15}N]-Ammonium Acetate in the Cat Intraarterially

In this set of experiments, [^{15}N]-ammonium acetate was administered into the carotid artery of the cat at a constant rate for periods of time up to 25 min. The results were, therefore, obtained by a constant feed of label rather than by pulse labeling. The data reflected accumulation of the labeled metabolites rather than decay as in the previous experiments.

The results of a representative experiment in which the $^{15}NH_3$ was infused for 20 min is shown in Table 7. In the blood, brain, and liver, the ammonia concentration was markedly elevated (20-fold or more), but the ^{15}N concentration in the free ammonia was less than that in the administered ammonium acetate. This indicated a rapid dilution of the administered labeled ammonia by

Table 7. Administration of ^{15}N-ammonium acetate into the cartoid artery of the cat

	Cerebral cortex		Blood		Liver	
	μmol/g	^{15}N	μmol/g	^{15}N[a]	μmol/g	^{15}N
Ammonia	6.7	65	6.2	81	13.3	81
Glutamine						
Amide	9.6	44.9	0.43	13.7	1.3	26.1
α-Amine	9.6	12.8	0.43	2.7	1.3	6.4
Glutamic acid	9.1	1.6	0.09	9.9	2.7	35.2
Aspartic acid	2.1	1.0	0.05	9.6	4.0	46.1
GABA	1.3	1.5				
Urea	5.5	1.3	5.9	11.2	6.3	17.5
Glutathione	1.8	0.43			4.8	2.3

[a] ^{15}N Ammonium acetate (99.6 atom % excess, 1 mmol/ml) was infused at a constant rate of 1 ml/min for 20 min. ^{15}N data expressed as atom % excess

the tissue ammonia, probably as a result of tissue metabolism. In the cerebral cortex, following free ammonia, the highest concentration of ^{15}N was in the amide group of glutamine followed by that in the α-amino groups of glutamine. The latter was approximately eight times greater than the ^{15}N concentration in the α-amino nitrogen of glutamic acid. Since glutamine is synthesized directly from glutamic acid, the labeled brain glutamine must have been derived from a pool or compartment of brain glutamic acid that did not exchange or equilibrate readily with the total tissue glutamic acid. The ^{15}N concentrations in the amide and α-amino moieties of brain glutamine were higher than in the glutamine isolated from blood or liver and, therefore, it was made in the brain and not derived from the blood. The α-amino nitrogen concentration of the brain glutamine was also above that present in the blood glutamic acid; the latter could not have been a source of the former. Lesser amounts of ^{15}N were also present in brain aspartic acid, GABA, glutathione (glutamic acid), and urea. The ^{15}N level in the brain urea was much below that in the blood or liver urea and could have been derived from the blood. Urea is not a major pathway for ammonia detoxification in brain.

The data suggest that in brain, glutamate dehydrogenase is associated with the formation of the small pool of glutamic acid preferentially used in the formation of glutamine, since this enzyme is responsible for the direct incorporation of ammonia into α-ketoglutarate to form glutamic acid.

The distribution of $^{15}NH_3$ in liver metabolites is very different than that in brain. In this organ the ^{15}N content in glutamic acid was higher than in the α-amino or amide moiety of glutamine. In fact, it was the aspartic acid which was most highly labeled with ^{15}N (next to free ammonia). The α-amino group is introduced into aspartic acid by transamination of oxaloacetic with glutamic acid or perhaps glutamine. Therefore, in liver, there may be pools of highly labeled glutamic acid or glutamine or both that do not exchange rapidly with the remainder of the liver glutamate or glutamine, but transaminate preferentially with oxaloacetate. These pools may be associated more directly with urea formation which is the major pathway for ammonia detoxification in the liver.

Other investigators, in a study of ammonia metabolism in brain, infused radioactive [^{13}N]-ammonia via the carotid artery of the rat (Cooper et al. 1979). In those studies, trace amounts of ammonia were administered and the physiological level was not altered. Their results essentially corroborated ours and were consistent with the interpretation that ammonia entering the brain from the blood was metabolized in a small pool of glutamate that was turning over very rapidly and distinct from a larger tissue pool of glutamate.

Carbon Dioxide Fixation in Brain

The results of the ammonia infusion experiments demonstrated, as had been reported previously (Flock et al. 1953, Bollman et al. 1957), that elevated ammonia levels lead to increased levels of glutamine without a corresponding decrease in glutamic acid in brain. If the synthesis of glutamic acid required for the synthesis of the newly formed glutamine was to occur without drastic depletion of the intermediates of the Krebs' tricarboxylic acid cycle 4-carbon units of the cycle had to be replenished. In liver, members of the Krebs cycle are formed by fixation of CO_2 to pyruvate or phosphoenolpyruvate to form malate and oxaloacetate. Therefore, studies were undertaken to evaluate the extent of CO_2 fixation in brain tissue (Berl et al. 1962b, Waelsch et al. 1964).

Administration of [^{14}C]-Sodium Bicarbonate Intravenously

^{14}C-labeled sodium bicarbonate was injected at a constant rate for time periods of up to 18 min via the inferior vena cava in the cat. A representative experiment is shown in Table 8. The aspartic acid, glutamic acid, and glutamine isolated from both liver and brain were labeled. The radioactivity in aspartic acid in both organs was higher than that in glutamic acid or glutamine. This was to be expected if in both organs CO_2 was fixed at the oxaloacetate or malate level. However, the pattern of labeling of glutamate and glutamine was different in these two organs. In brain, the glutamine had a higher specific activity than glutamic acid; the opposite occurred in the liver. This again was in accord with the conclusion that in brain a small pool of glutamic acid was preferentially directed toward the formation of glutamine, whereas in liver, glutamine was formed from a large compartment of glutamic acid. The rate of CO_2 fixation in brain was somewhat more than 10% of that in liver, the major CO_2 fixing organ.

Since α-ketoglutarate served as a precursor for the glutamic acid it would have been expected to have contained considerably higher radioactivity than glutamate or glutamine in brain and liver. In brain, the high SA of glutamine relative to that of α-ketoglutarate (Tables 8 and 9) pointed to the conclusion that

Table 8. Administration of [^{14}C]-sodium bicarbonate into the inferior vena cava of the cat[a]

	Brain		Liver	
	μmol/g	SA	μmol/g	SA
Glutamic acid	10.2	890	3.4	16,910
Glutamine	5.7	1,640	2.8	5,180
Aspartic acid	2.3	7,110	0.7	43,360
α-Ketoglutarate	–	1,070	–	6,250
Pyruvate	–	820	–	1,800
Tissue CO_2	14.2	29,000	13.4	25,000
$^{14}CO_2$ fixed	1.63		13.0	

[a] 1.7 ml of bicarbonate solution (0.06 *M*; 0.29 mCi/ml in saline) was given i.v. during the first 10 s to bring the specific activity of the blood CO_2 rapidly to a precalculated level. Infusion was continued at the rate of 0.2 ml/min for 18 min. SA: Specific activity, cpm/μmol

Table 9. Administration of [^{14}C]-sodium bicarbonate and ammonium acetate into the inferior vena cava of the cat[a]

	Brain	
	μmol/g	SA
Glutamatic acid	10.4	1,520
Glutamine	7.3	7,860
Aspartic acid	2.1	5,900
α-Ketoglutarate	–	3,190
Pyruvate	–	840
Tissue CO_2	11.4	55,000
NH_3	6.9	

[a] The ammonium acetate was administered as a 2.5 *M* solution containing $NaH^{14}CO_3$ (0.06 *M*; 0.29 mCi/ml). This solution was infused at a rate of 0.2 ml/min for 15.5 min immediately following the injection of the 1.7 ml of the bicarbonate solution (see Table 8)

the keto acid in brain was divided into at least two pools, one of which was more highly labeled and associated with rapid glutamine formation. In liver, this keto acid was of much lower SA than glutamate which suggested the presence of a small pool of labeled keto acid of high SA from which glutamate was derived. Such compartmentation of pools of α-ketoglutarate indicated the existence of at least two separate Krebs cycles in brain (Gaitonde 1965). The above data suggested that the same may be true in liver.

Administration of [^{14}C]-Sodium Bicarbonate and Ammonium Acetate Intravenously

In cats in which ^{14}C-sodium bicarbonate and ammonium acetate were given together via the inferior vena cava, the labeling pattern of the amino acids was altered in brain, predominantly in favor of the incorporation of label into newly formed glutamine. The glutamine achieved the highest SA, approximately five times greater than that of glutamic acid and also higher than that of aspartic acid. The data suggested that in the presence of elevated levels of tissue ammonia, oxaloacetate formed by CO_2 fixation was preferentially utilized for the formation of α-ketoglutarate and in turn of glutamate to be converted to glutamine. Calculation of the amount of CO_2 incorporated into the amino acids in brain in the presence or absence of increased ammonia levels indicated that in the former condition the rate of CO_2 fixation was increased, almost all in glutamine, as compared to the latter condition (Waelsch et al. 1964, Berl 1971).

In Vitro Studies

In the study of whole animals, the question was often raised as to the contribution of other organs to the pattern of metabolic compartmentation seen in brain. This led to the study of metabolic compartmentation in brain slice preparations (Berl and Clarke 1969). Using isotopically labeled glutamate, aspartate, bicarbonate, acetate, leucine, and other precursors, the metabolic picture described above in living animals was confirmed. Hence, there could be no doubt that this compartmentation of glutamate-glutamine metabolism seen in brain was characteristic of this tissue.

Location of the Compartments

This chapter presents some of the data on which the conclusion is based that metabolites are separated into pools or compartments within the tissue. In brain, one is faced with a very complex mixed cell population of several different types of glia and neurons and their axonal and dendritic extensions as well as possibly different populations of mitochondria. Compartmentation of the glutamate-glutamine systems has never been demonstrated in brain tissue homogenates. Therefore, such phenomena are very likely to be associated with anatomical cellular differentiation.

Most of the data can be explained by the very oversimplified assumption of two or three cellular compartments. Based on a large variety of data, the small pool of glutamate active in glutamine formation has been placed in the astrocytic glia, the large pool of glutamate in the neurons, and perhaps a subpool of the larger pool in nerve endings (Balazs et al. 1973). Developmental studies support such a localization since the evidence for metabolic compartmentation begins to be manifested during the development of the neuronal cell bodies, their axons, and dendrites (Berl 1965, Patel and Balazs 1970), marking the development of the large compartment. However, glial development is also present during this period.

Based on the metabolic studies, it would be anticipated that the small pool would be associated with glutamine synthetase, glutamate dehydrogenase, and with acetyl-CoA synthetase, the enzyme that converts acetate to acetyl-CoA (acetate is also preferentially metabolized to glutamine in the small compartment) (Berl and Frigyesi 1969). In animals in which ammonia was injected or hepato-portal shunt performed (Cavanagh 1974) as well as in humans with hepatic encephalopathy (Cavanagh 1974), it was the astrocytes rather than the neurons which were most severely affected. In similar animal studies, histochemical evaluations indicated increased glutamate dehydrogenase activity localized to the astrocytes (Norenberg 1976). Also, glutamate dehydrogenase was found to be low in synaptosomes (Neidle et al. 1969, Wilson and Barch 1971) and nerve cell bodies (Kuhlman and Lowry 1956). Other studies found brain glutamate dehydrogenase and acetyl-CoA synthetase to be enriched in a fraction of mitochondria that sedimented in a sucrose gradient more rapidly than the average of all the mitochondria (Reijnierse et al. 1975). Although there is no direct evidence of its source, this fraction of mitochondria may be predominantly from glial cells.

The placement of the small compartment in the glia was greatly enhanced by the immunohistochemical study in which glutamine synthetase was found to be localized in the glia and not evident in neuronal cell bodies (Martinez-Hernandez et al. 1977). Similar immunohistochemical techniques also showed that rat retina glutamine synthetase was concentrated in the Muller cells, the glial cells of the retina (Riepe and Norenberg 1977). G. Gambos (personal communication 1983), using similar methods, found the glutamine synthetase principally in the Bergmann glial cells of the rat cerebellum.

These many metabolic and anatomic studies provide some of the basis for the idea that glia function in the uptake and metabolism of exogenous or extracellular substances, e.g., transmitter amino acids and ammonia. Thus, they provide protective mechanisms as well as methods for the conservation and recycling of carbon skeleton and amino nitrogen.

References

Balazs R, Patel AJ, Richter D (1973) Metabolic compartments in brain: Their properties and relation to morphological structures. In: Balazs R, Cremer JE (eds) Metabolic compartmentation in the brain. Macmillan, London, pp 167–184

Berl S (1965) Compartmentation of glutamic acid metabolism in developing cerebral cortex. J Biol Chem 240: 2047–2054

Berl S (1971) Cerebral amino acid metabolism in hepatic coma. In: Polli E (ed) Neurochemistry of hepatic coma. Experimental biology and medicine, vol IV. Karger, Basel, pp 71–84

Berl S, Clarke DD (1969) Compartmentation of amino acid metabolism. In: Lajtha (ed) Handbook of neurochemistry, vol II. Plenum Press, New York, pp 447–472

Berl S, Frigyesi TL (1969) The turnover of glutamate, glutamine, aspartate and GABA labeled with [^{14}C]-acetate in caudate nucleus, thalamus and motor cortex (cat). Brain Res 12: 444–455

Berl S, Lajtha A, Waelsch H (1961) Amino acid and protein metabolism – VI. Cerebral compartments of glutamic acid metabolism. J Neurochem 7: 186–197

Berl S, Takagaki G, Clarke DD, Waelsch H (1962a) Metabolic compartments in vivo. Ammonia and glutamic acid metabolism in brain and liver. J Biol Chem 237: 2562–2569

Berl S, Takagaki G, Clarke DD, Waelsch H (1962b) Carbon dioxide fixation in the brain. J Biol Chem 237: 2570–2573

Bollman JL, Flock EV, Grindlay JH, Bickford RG, Lichtenfield FR (1957) Coma with increased amino acids of brain and cerebrospinal fluid in dogs with Eck's fistula. Arch Surg 75: 405–412

Bradford HF, deBelleroche JS, Ward HK (1978) On the metabolic and intrasynaptic origin of amino acid transmitters. In: Fonnum F (ed) Amino acids as chemical transmitters. Plenum Press, New York, pp 367–377

Cavanagh JB (1974) Liver bypass and the glia. In: Plum F (ed) Brain dysfunction in metabolic disorders. Raven Press, New York, pp 13–38

Cooper AJL, McDonald JM, Gelbard AS, Gledhill RF, Duffy TE (1979) The metabolic fate of ^{13}N-labeled ammonia in rat brain. J Biol Chem 254: 4982–4992

Flock EV, Block MA, Grindlay JH, Mann FC, Bollman JL (1953) Changes in free amino acids of brain and muscle after total hepatectomy. J Biol Chem 200: 529–536

Gaitonde MK (1965) Rate of utilization of glucose and compartmentation of α-oxoglutarate and glutamate in rat brain. Biochem J 95: 803–810

Gilon E, Szeinzberg A, Tauman G, Bodonyi E (1959) Glutamine estimation in cerebrospinal fluid in cases of liver cirrhosis and hepatic coma. J Lab Clin Med 53: 714–719

Goldstein FB (1959) Biosynthesis of N-acetyl-L-aspartic acid. J Biol Chem 234: 2701–2706

Hamberger A, Chiang GH, Nylen ES, Scheff SW, Cotman CW (1979a) Glutamate as a CNS transmitter. I. Evaluation of glucose and glutamine as precursors for the synthesis of preferentially released glutamate. Brain Res 168: 513–530

Hamberger A, Chiang GH, Sandoval E, Cotman CW (1979b) Glutamate as a CNS transmitter. II. Regulation of synthesis in the releasable pool. Brain Res 168: 531–541

Kuhlman RE, Lowry OH (1956) Quantitative histochemical changes during the development of the rat cerebral cortex. J Neurochem 1: 173–180

Lajtha A, Berl S, Waelsch H (1959) Amino acid and protein metabolism of the brain – IV. The metabolism of glutamic acid. J Neurochem 3: 322–332

Lajtha A, Berl S, Waelsch H (1960) Compartmentation of glutamic acid metabolism in the central nervous system. In: Roberts E, Baxter CF, Van Harreveld A, Wiersma CAG, Adey WR, Killian KF (eds) Inhibition in the nervous system and gamma-aminobutyric acid. Pergamon Press, New York, pp 460–467

Martinez-Hernandez A, Bell KP, Norenberg MD (1977) Glutamine synthetase: Glial localization in brain. Science 195: 1356–1358

Neidle A, Van den Berg CJ, Grynbaum A (1969) Heterogeneity of rat brain mitochondria isolated in continuous sucrose gradients. J Neurochem 16: 225–234

Norenberg MD (1976) Histochemical studies in experimental porto-systemic encephalopathy. Arch Neurol 33: 265–269

Patel AJ, Balazs R (1970) Manifestation of metabolic compartmentation during the maturation of the rat brain. J Neurochem 17: 955–971

Reijnierse GLA, Veldstra H, Van den Berg CJ (1975) Short-chain fatty acid synthesis in brain. Biochem J 152: 477–484

Reiner JM (1953) The study of turnover rates by means of isotopic tracers. Arch Biochem Biophys 46: 53–99

Riepe RE, Norenberg MD (1977) Muller cell localization of glutamine synthetase in rat retina. Nature (London) 268: 654–655

Schwerin P, Bessman SP, Waelsch H (1950) The uptake of glutamic acid and glutamine by brain and other tissue of the rat and mouse. J Biol Chem 184: 37–44

Takagaki G, Berl S, Clarke DD, Waelsch H (1961) Glutamic acid metabolism in brain and liver during infusion with ammonia labeled with nitrogen^{-15}. Nature (London) 189: 326

Waelsch H, Berl S, Rossi CA, Clarke DD, Purpura DD (1964) Quantitative aspects of CO_2 fixation in mammalian brain in vivo. J Neurochem 11: 717–728

Wilson JE, Barch D (1971) Heterogeneity of rat brain mitochondria. Fed Proc (Abstr) 30: 1139

Glutamine Metabolism in Lymphoid Tissues

M. S. M. ARDAWI and E. A. NEWSHOLME[1]

Introduction

When lymphocytes are suitably stimulated either specifically (by an antigen) or non-specifically (by mitogens), they are transformed into a state of high biochemical activity which initiates production of various mediators of immunity including antibodies if they are B-lymphocytes and the mediators of cellular immunity (e.g., lymphotoxin, chemotactic factors, mitogenic factors) if they are T-lymphocytes. To perform this activity lymphocytes require an increased rate of ATP generation and it has generally been assumed that glucose was the only quantitatively important fuel (see Hume and Weidemann 1980). Recently Ardawi and Newsholme (1982) have provided evidence to suggest that other fuels can be utilised by lymphocytes. This suggestion is based on the maximal activities of key enzymes of the energy-producing pathways in lymphocytes, the ability of these cells to utilise glutamine, ketone bodies (3-hydroxybutyrate and acetoacetate) and long-chain fatty acids (oleate and palmitate) (Table 1) and the effects of these fuels on oxygen consumption by lymphocytes. Of these fuels glutamine may be quantitatively the most important.

Glutamine Utilisation by Lymphocytes

There is considerable evidence that glutamine utilisation is important, if not essential, for rapidly dividing cells (for review see Krebs 1980); this is not surprising since it provides nitrogen for a number of important precursors for the synthesis of macromolecules, including purines, pyrimidines, amino sugars and some amino acids (Tate and Meister 1978). The evidence that glutamine metabolism is important for lymphocytes and hence for the immune response of the body is as follows:

1. Lymphocytes and other lymphoid tissues possess high activities of phosphate-dependent glutaminase (hereafter referred to as glutaminase) (Table 2).
2. Glutaminase activity was increased during an in vivo immunological challenge (Ardawi and Newsholme 1982).

1 Department of Biochemistry, University of Oxford, South Parks Road, Oxford OX1 3QU, England

Glutamine Metabolism in Mammalian Tissues
Edited by D. Häussinger and H. Sies

Table 1. Rates of utilisation of various substrates by various in vitro tissue preparations of the rat in comparison to lymphocytes

Tissue preparation	Conditions	Substrate			
		Glucose	Glutamine	Aceto-acetate	Oleate
		Rates of utilisation (μmol/min/g dry wt)			
Lymphocytes[a]	Isolated cells 100% O_2	0.62	2.70	0.42	0.58
Thymocytes[b]	Isolated cells 100% O_2	0.37	–	0.57	–
Enterocytes[c]	Isolated cells O_2/CO_2 (19 : 1)	12.00	11.00	0.73[d]	–
Colonocytes[e]	Isolated cells O_2/CO_2 (19 : 1)	5.60	3.70	2.04	–
Intact heart[f]	Perfused isolated working heart O_2/CO_2 (19 : 1)	7.60	–	10.80	2.90

Data are obtained from:
[a] Ardawi 1983
[b] Hume et al. 1978
[c] Watford et al. 1979a
[d] Hanson and Parsons 1978
[e] Roediger 1982
[f] Taegtmeyer et al. 1980

Table 2. Activities of glutaminase in lymphoid tissue of different animals[a]

Animal	Tissue	Glutaminase activity (μmol/min/g fresh wt at 37° C)
Gerbil	Mesenteric lymph node	5.6
Guinea pig	Mesenteric lymph node	6.0
Hamster	Mesenteric lymph node	6.6
Mouse	Mesenteric lymph node	9.2
Rabbit	Mesenteric lymph node	4.6
Rat	Mesenteric lymph node	8.7
	Thymus	6.0
	Spleen	6.1
	Peyer's patches	10.6
	Bone marrow	3.7

[a] The results are means from at least six separate pools of tissue obtained from at least six animals (for details, see Ardawi and Newsholme 1984a)

3. Lymphocytes isolated from rat mesenteric lymph nodes and incubated under aerobic conditions utilised glutamine at a rate of 2.7 μmol/min per g dry wt and this was increased by 51% after mitogenic stimulation by Concanavalin-A (Con A) (Ardawi and Newsholme 1983).

Table 3. Rates of glutamine utilisation and product formation in incubated rat lymphocytes[a]

Rates of utilisation or production (μmol/h/g dry wt)							
Glutamine	Glutamate	Alanine	Aspartate	Asparagine	Ammonia	Pyruvate	Lactate
– 160[b]	99.0	1.9	43.2	4.03	99.7	8.1	6.4

[a] Lymphocytes (5×10^7 cell/ml) were incubated under 100% O_2 for 1 h in the presence of 2 m*M* glutamine in phosphate-buffered saline (pH 7.2) plus 10% (v/v) foetal bovine serum. Rates are given as means of at least six separate experiments (for details see Ardawi and Newsholme 1983)

[b] A negative sign indicates utilisation

4. In the absence of glutamine, [^{3}H]-thymidine incorporation into lymphocytes cultured for 48 h was very low and was similar to that observed in the absence of Con A; maximal rates of incorporation were observed when glutamine was present in the culture medium at a concentration of 0.3 m*M* glutamine (Ardawi and Newsholme 1983).

The Pathway of Glutamine Metabolism in Lymphocytes

The major end products of glutamine metabolism in lymphocytes are glutamate, aspartate and ammonia (Table 3). This is in contrast to the cells of the small intestine which produce alanine rather than aspartate from glutamine (see, e.g., Hanson and Parsons 1977, Watford et al. 1979a). Of the glutamine utilised by lymphocytes, ammonia production accounted for about 30% of the glutamine-nitrogen which, since the glutaminase reaction produces ammonia, suggests that glutamate is metabolised via a transaminase reaction rather than by glutamate dehydrogenase. This is expected from the low activity of the latter enzyme in lymphocytes (Ardawi and Newsholme 1982). For the glutamine metabolised beyond the stage of glutamate, about 70% was accounted for as aspartate and only 3% as alanine. This finding is expected from the maximum activities of aspartate aminotransferase and alanine aminotransferase in rat lymphocytes (see Table 4). In some experiments, in which a complete profile of amino acids during glutamine metabolism in lymphocytes was obtained (by amino acid autoanalysis), similar changes in amino acids to those described above were observed (Ardawi 1983).

On the basis of end products of metabolism and maximal activities of some enzymes (Table 4) (Ardawi 1983), it is considered that glutamine is utilised by lymphoid tissues and lymphocytes via a pathway which is similar to that proposed for the kidney (e.g., Goldstein 1976) or the small intestine (Hanson and Parsons 1980) but with some important differences (see Fig. 1).

1. The low activity of glutamine transaminase in lymphocytes (Ardawi 1983) suggests that most of the glutamine is converted into glutamate by the action of glutaminase. In the kidney, glutamine transaminase may be important at low rates of glutamine utilisation (Goldstein 1975).

Table 4. Maximal activities of glutaminase and other enzymes involved in glutamine metabolism in lymphocytes

Enzyme	Maximal activities (μmol/min/g fresh wt at 25° C)
Glutaminase[a]	8.7
Glutamate dehydrogenase	3.3
Aspartate aminotransferase	2.6
Alanine aminotransferase	0.17
Oxoglutarate dehydrogenase	0.53
Citrate synthase	10.3
Isocitrate dehydrogenase (NAD^+-plus $NADP^+$-linked)	1.6
Malate dehydrogenase ($NADP^+$-linked)	0.14
Phosphoenolpyruvate carboxykinase[a]	0.68
Pyruvate kinase	22.7
Pyruvate carboxylase	1.9

[a] Activity refers to 37° C (for further details see Ardawi and Newsholme 1982)

2. In lymphocytes, glutamate is metabolised via transamination rather than oxidative deamination via glutamate dehydrogenase. This is suggested from the stoichiometry between glutamine utilisation and ammonia production (Table 3) and from the low maximal activity of glutamate dehydrogenase (Table 4).
3. In lymphocytes, aspartate rather than alanine is produced which is expected from the maximal activities of aspartate aminotransferase and alanine aminotransferase (Table 4).

On the basis of maximal enzyme activities, it has been suggested that the conversion of oxaloacetate to phosphoenolpyruvate catalysed by phosphoenolpyruvate carboxykinase (PEPCK) is part of the pathway for glutamine oxidation (Ardawi and Newsholme 1982). By using 3-mercaptopicolinate, which is an inhibitor of PEPCK, the participation of PEPCK in the oxidation of glutamine was confirmed (Ardawi and Newsholme 1983). This is similar to the pathway of glutamine oxidation in the kidney (Watford et al. 1979b, 1980), but unlike that postulated for the small intestine in which 3-mercaptopicolinate had no effect on the metabolism of glutamine (Hanson and Parsons 1977). However, this experiment does not exclude the participation of other reactions [e.g., $NAD(P^+)$-linked malate dehydrogenase, oxaloacetate decarboxylase] in the conversion of malate or oxaloacetate to pyruvate.

The above evidence suggests the existence of a pathway for the conversion of glutamine to pyruvate in lymphocytes but the precise amount of pyruvate produced from glutamine and the fate of the pyruvate are questions that at present cannot be completely answered. In particular, it is unclear whether the carbon skeleton of glutamine is converted into acetyl-CoA via pyruvate, that is, is the carbon of glutamine oxidised by the normal TCA-cycle?

A summary of the possible fates of the carbon of glutamine is as follows:

1. For complete oxidation of glutamine to carbon dioxide, 2-oxoglutarate derived from glutamine would be metabolised via a pathway which involves five

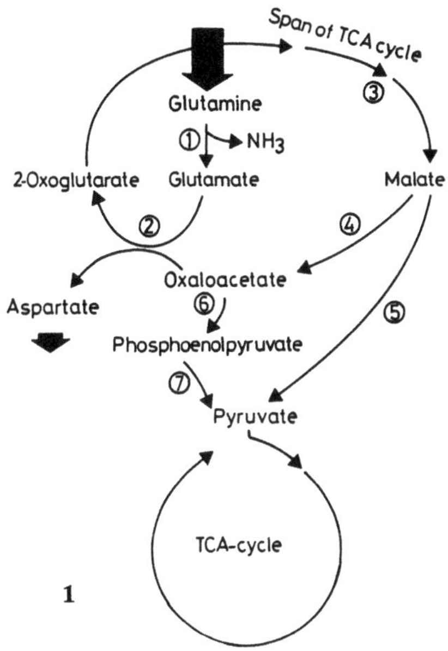

1

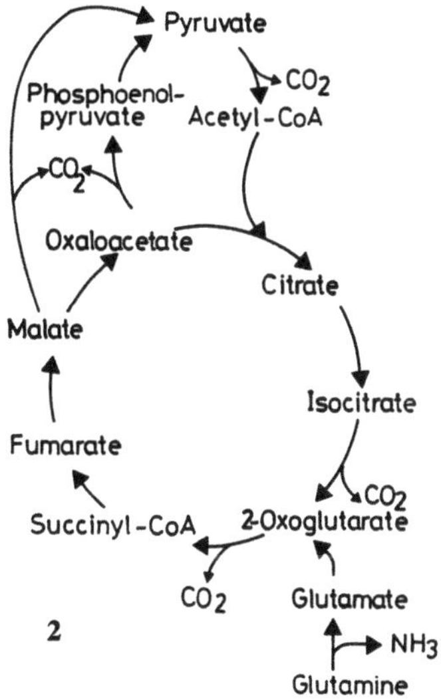

2

Fig. 1. The proposed pathway for glutamine metabolism in lymphocytes. *1* Glutaminase; *2* aspartate aminotransferase; *3* enzymes of the TCA-cycle converting 2-oxoglutarate into malate; *4* malate dehydrogenase (NAD^+-linked); *5* malic enzyme [$NAD(P)^+$-linked]; *6* phosphoenolpyruvate carboxykinase; *7* pyruvate kinase. The oxidation via the TCA-cycle requires convertion of pyruvate to acetyl-CoA via pyruvate dehydrogenase

Fig. 2. Possible pathway(s) for the partial or complete oxidation of glutamine in lymphocytes

decarboxylation reactions; three within the TCA-cycle [oxoglutarate dehydrogenase (twice) plus isocitrate dehydrogenase] and PEPCK plus pyruvate dehydrogenase (Fig. 2).

2. For partial oxidation of glutamine, 2-oxoglutarate derived from glutamine would be metabolised to oxaloacetate and then to aspartate (see Fig. 2) and only 20% of the carbon of glutamine would be recovered as CO_2. Alternatively, or in addition, the oxaloacetate could be metabolised to pyruvate via PEPCK [or an $NAD(P)^+$-linked malate dehydrogenase] and the pyruvate converted to alanine or lactate; in this case only 40% of the carbon of glutamine would be recovered as CO_2.

Unfortunately, it is difficult to decide if the major fate of glutamine is partial rather than complete oxidation, although the authors consider that partial oxidation is most likely. Although all the major end products can be measured in an incubation system and a balance sheet constructed, these end products are produced in the absence of added glutamine and it is unclear if this "endogenous" metabolism is suppressed by the addition of exogenous glutamine. Further, of the glutamine utilised by lymphocytes 89% can be accounted for as glutamate, aspartate, asparagine, alanine and lactate. This suggests that 11% is oxidised and this difference may be due to experimental error in the measurement of glutamate, aspartate etc. rather than oxidation via acetyl-CoA.

If the carbon skeleton of glutamine is not oxidised by the "normal" TCA-cycle, it raises important questions concerning the regulation of pyruvate availability for pyruvate dehydrogenase and the regulation of the activity of this latter enzyme.

The Physiological Significance of Glutamine for Lymphocytes

Glutamine can be considered to be a unique amino acid since its oxidation provides energy and its carbon and nitrogen provide precursors for a number of biosynthetic processes (see Kovacevic and McGivan 1983). Glutamine is essential for proliferation of lymphocytes in culture; it could not be replaced by glucose, 3-hydroxybutyrate or any other amino acid (Ardawi and Newsholme 1983).

An important observation with lymphocytes was that glutamine is utilised at a high rate (about fourfold higher than that of glucose) (Ardawi and Newsholme 1983) and the question arises as to which aspect of glutamine metabolism is important for lymphocytes: provision of energy, provision of carbon and nitrogen precursors for biosynthetic processes or both?

The rate of glutamine utilisation at the physiological blood concentration (0.5 m*M*) is found to be 55 μmol/h per 10^{10} cells. Knowledge of the amount of DNA and RNA in lymphocytes enables the amount of glutamine required for the biosynthesis of purines and pyrimidines needed for doubling the number and size of cells to be calculated. For the total in vivo population of lymphocytes of a 200 g rat (about 4×10^9 lymphocytes including those found in lymph nodes, spleen, thymus, bone marrow and peripheral circulation) the rate of utilisation for this purpose is about 4% of the in vitro rate of glutamine utilisation by incubated lymphocytes.

Calculations based on the results presented in Table 3 and oxygen uptake data (see Ardawi and Newsholme 1983) indicate that glutamine can contribute 53% to the fuel of respiration assuming complete oxidation of glutamine or 11%–22% assuming only partial oxidation. Calculations based on the formation of $^{14}CO_2$ indicate that glutamine can contribute between 10%–30%, and calculations based on oxygen uptake data indicate that the contribution of glutamine to the ATP-generating capacity of lymphocytes, when compared to that of glucose and oleate, is about 37% (Ardawi 1983).

Consequently, although glutamine must provide precursors for the various biosynthetic processes, during proliferation, the high rate of utilisation of this amino acid appears to be necessary since it functions as a fuel for lymphocytes: indeed, this may be the case for all rapidly dividing cells.

Despite the obvious difference in quantitative importance, it is possible that the two roles of glutamine, the provision of energy via its oxidation and the provision of carbon and nitrogen precursors for biosynthetic processes are intimately metabolically linked. The use of glutamine as a fuel necessitates a high rate of glutamine utilisation so that this amino acid and the intermediates it produces are always available to satisfy the requirements for precursors for DNA and RNA synthesis during proliferation. Could the demand for glutamine during cell division be easily accomodated and regulated if this amino acid was not being used for energy formation? Only further work on glutamine metabolism in rapidly dividing cells will answer this question.

The Control of the Pathway of Glutamine Utilisation in Lymphocytes

Since glutamine metabolism is important in lymphocytes (resting or stimulated), it is therefore important to understand how the flux through the pathway of glutamine utilisation is controlled.

Although a complete answer is not yet possible, the question is tackled here in relation to the principles developed for metabolic control in general (see Newsholme and Crabtree 1979, Nesholme 1980).

Equilibrium Nature of Reactions

In order to acquire information on the control of the flux through any metabolic pathway, it is esential to define the structure of the pathway; that is, to identify the near- and non-equilibrium reactions and the flux-generating step (Newsholme and Crabtree 1979).

One means of indicating the equilibrium of reactions involves measurement of the concentrations of substrates and products of the individual reactions. Mass action ratio (Γ) for a particular reaction in a pathway is compared with the apparent equilibrium constant (K'_{eq}) of the reaction. The experimental methods for this have been described in detail by Newsholme and Start (1973).

Since information is not available for concentration of most of the intermediates of the pathway in lymphocytes, a comparison of the values of the

Table 5. Apparent equilibrium constants and mass action ratios of some reactions involved in the pathway of glutamine utilisation in the kidney[a]

Reaction catalysed by	Apparent equilibrium constant (K_{eq})	Calculated mass action ratio (Γ)	Suggested equilibrium nature
Glutaminase	320	4.98	Non-equilibrium
Glutamate dehydrogenase	3.87×10^{-3}	0.027	Near-equilibrium
Aspartate aminotransferase	0.17	1.16	Near-equilibrium
Oxoglutarate dehydrogenase	2.04×10^{5}	1.3×10^{-3}	Non-equilibrium
Fumarase	0.23	0.17	Near-equilibrium
Malate dehydrogenase (NADP-linked)	5.1×10^{-5}	0.095	Near-equilibrium
Phosphoenolpyruvate carboxykinase	0.37	0.0014	Non-equilibrium

[a] It is assumed that 1 g of tissue contains 0.75 ml of water (Hems and Brosnan 1970). The intramitochondrial concentrations of metabolites were assumed to equal the intracellular concentrations. For further details see Ardawi (1983)

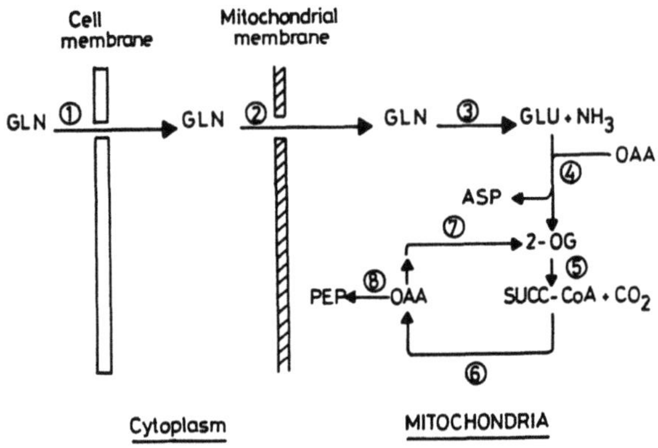

Fig. 3. Key reactions that may be involved in the regulation of glutamine utilisation in lymphocytes. *1* Glutamine transport across the cellular membrane; *2* glutamine transport across the mitochondrial membrane; *3* glutaminase; *4* aspartate aminotransferase; *5* oxoglutarate dehydrogenase; *6* enzymes of the TCA-cycle converting succinyl-CoA to oxaloacetate; *7* enzymes converting oxaloacetate to 2-oxoglutarate; *8* phosphoenolpyruvate carboxykinase. *GLN* glutamine; *GLU* glutamate; *OAA* oxaloacetate; *ASP* aspartate; *2-OG* 2-oxoglutarate; *SUCC-CoA* succinyl-CoA; *PEP* phosphoenolpyruvate

apparent equilibrium constants and mass-action ratios of some of the reactions involved in the similar if not identical pathway of glutamine utilisation in the kidney are presented in Table 5. The authors consider that the equilibrium of the reactions involved in the glutamine pathway in lymphocytes is similar: that is, the transport of glutamine across the cellular and the mitochondrial membranes plus the reactions catalysed by glutaminase (see Ardawi and Newsholme 1984a), oxoglutarate dehydrogenase and PEPCK are non-equilibrium (Fig. 3).

Flux-Generating Step

The flux-generating step is the first non-equilibrium reaction in the pathway that is saturated with pathway substrate. Two lines of evidence suggest that none of the reactions in the glutamine pathway is saturated with substrate – that is, the pathway in the lymphocytes does not possess a flux-generating step.

1. The rate of glutamine metabolism by lymphocytes is increased by raising the concentration of glutamine in the incubation medium above the normal plasma value (see Ardawi and Newsholme 1983).
2. The K_m values of the enzymes that catalyse non-equilibrium reactions in the glutamine pathway are similar or greater than the concentration of their pathway substrates (Ardawi 1983).

Apart from the glutamine that is produced in the lumen of the small intestine, all of which is probably utilised by the enterocytes, most of the glutamine present in the blood is synthesized in and released by skeletal muscle (Felig et al. 1973, Garber 1980, Schröck et al. 1980, Snell 1980, Brosnan et al. 1983). Hence, it is

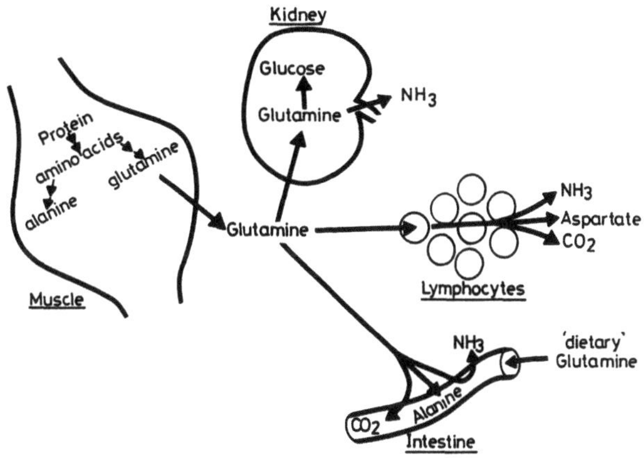

Fig. 4. Relationship between glutamine production in muscle and glutamine utilisation in various tissues

likely that most or all the glutamine utilised by the lymphocytes originates in the muscle and, therefore, the flux-generating step for the glutamine-utilisation pathway must reside in skeletal muscle (Fig. 4).

External Regulation

Reactions of a pathway that are subject to external regulation, are indicated by the "crossover" approach: external regulation is indicated if the concentrations of the intermediates of such reactions are either unchanged or decreased when the flux is increased or vice versa. It was demonstrated that the rate of glutamine utilisation by lymphocytes can be increased during incubation by addition of glucose or Con A. Hence, there is external regulation of the glutamine pathway in lymphocytes (see Ardawi and Newsholme 1983). Under these conditions, the concentrations of glutamine and 2-oxoglutarate in the lymphocyte decreased. This suggests that the non-equilibrium reactions of the glutamine pathway are subject to external regulation. Similar conclusions have also been made on the evidence that is available for the glutamine pathway in the kidney (Newsholme and Leech 1983).

Candidates for the external regulators of the processes are obtained from a study of the properties of the non-equilibrium processes in vitro. For lymphocytes, only the process of glutamine transport across the cell membrane and glutaminase have been studied in vitro. The properties of other non-equilibrium processes (mitochondrial transport of glutamine and oxoglutarate dehydrogenase) have been studied for the kidney and these will be used in this paper to construct a theory for the control of the pathway of glutamine utilisation in lymphocytes.

The properties that are relevant to metabolic control are given in Table 6.

Table 6. Some properties of "non-equilibrium" enzymes and transport system of the glutamine pathway[a]

Non-equilibrium enzymes or transport systems	Activators	Inhibitors
Cellular transport of glutamine[b]	Not known	Leucine
Mitochondrial transport of glutamine[c]	Not known	2-Oxoglutarate
Glutaminase[b]	Phosphate, ATP, GTP, Phosphoenolpyruvate, Leucine	Glutamate, 2-Oxoglutarate
Oxoglutarate dehydrogenase[c]	Ca^{++}, H^{+}	Not known

[a] For further details see Ardawi (1983); Ardawi and Newsholme (1984a)
[b] Lymphocytes
[c] Kidney

Glutamine Metabolism in Lymphocytes and Its Relationship to that in Muscle and Other Tissues

Using the observed rates of glutamine utilisation by incubated lymphocytes (at 0.5 m*M*), the rate of glutamine utilisation by the total lymphocyte population in vivo can be calculated. Assuming a 200 g rat possesses approx. 4×10^9 lymphocytes, the rate of glutamine utilisation is about 24 µmol/h. It is known that muscle is capable of releasing approx. 42 µmol/h (see Schröck et al. 1980). Hence, the magnitude of glutamine required by lymphocytes population can be satisfied by that released by muscle. However, if the demand for glutamine by the lymphocytes was to be dramatically increased could the muscle respond? Alternatively, if the rate of glutamine release by muscle was to be dramatically reduced could the immune system suffer? Thus, whether the synthesis of glutamine by muscle can limit the response of the immune system to an infection or attack of the host is a question that may be of considerable importance in the maintenance of the health of man.

The turnover of glutamine is extremely high when compared with the glutamine requirements of lymphocytes and the cells of the small intestine. In a rat weighing 200 g, the turnover of glutamine was estimated to be approx. 550 µmol/h (see Lund 1980). Since the glutamine requirements of both the lymphocyte population and the small intestine can be calculated to be approx. 100 µmol/h (based on the results of Ardawi and Newsholme 1983, and Windmueller and Spaeth 1974, respectively), this represents only about 20% of the estimated glutamine turnover. It is unlikely that substrate cycling between glutamine and glutamate can account for the rest of glutamine turnover, because it would involve no loss of carbon. Protein turnover, which amounts to about 8.8 g/day in a 200 g rat (Garlick et al. 1975), provides the most likely explanation. Thus, the continual turnover of proteins most of which probably occurs in muscle, serves to maintain a pool of available amino acids such as those which can be converted into glutamine (branched-chain amino acids, glutamate) if necessary. This may be a means of ensuring that the supply of glutamine from the glutamine-producing tissues is normally sufficient for the important glutamine-utilising tissues.

Acknowledgements. We thank Professor R. R. Porter, F.R.S., for interest and encouragement, and the financial support of the Ministry of High Education (Kingdom of Saudi Arabia).

References

Ardawi MSM (1983) Some aspects of energy metabolism in lymphocytes. D Phil Thesis, Univ Oxford

Ardawi MSM, Newsholme EA (1982) Maximum activities of some enzymes of glycolysis, the tricarboxylic acid cycle and ketone-body and glutamine utilisation pathways in lymphocytes of the rat. Biochem J 208: 743–748

Ardawi MSM, Newsholme EA (1983) Glutamine metabolism in lymphocytes of the rat. Biochem J 212: 835–842

Ardawi MSM, Newsholme EA (1984a) Intracellular localization and properties of phosphate-dependent glutaminase of rat mesenteric lymph nodes. Biochem J 217: 289–296

Ardawi MSM, Newsholme EA (1984b) Glutamine transport into lymphocytes. Biochem J (submitted)

Brosnan JT, Man KC, Hall DE, Colbourne SA, Brosnan ME (1983) Interorgan metabolism of amino acids in streptozotocin-diabetic ketoacidotic rat. Am J Physiol 244 (endocrinol Metab 7): E151–E158

Felig P, Wahren J, Räf L (1973) Evidence of Interorgan amino acid transport by blood cells in humans. Proc Natl Acad Sci USA 70: 1775–1779

Garber AJ (1980) Glutamine metabolism in skeletal muscle. In: Mora J, Palacios R (eds) Glutamine metabolism, enzymology and regulation. Academic Press, London New York, pp 259–284

Garlick PJ, Millward DJ, James WPT, Waterlow JC (1975) The effect of protein deprivation and starvation on the rate of protein synthesis in tissues of the rat. Biochim Biophys Acta 414: 71–84

Goldstein L (1975) Regulation of renal glutamine deamination. Med Clin North Am 59: 667–680

Goldstein L (1976) Ammonia production and excretion in the mammalian kidney. In: Thurau K (ed) Int Rev Physiol (Kidney and urinary tract physiology II), vol XI. Univ Park Press, Baltimore, pp 283–316

Hanson PJ, Parsons DS (1977) Metabolism and transport of glutamine and glucose in vascularly perfused small intestine of the rat. Biochem J 166: 509–519

Hanson PJ, Parsons DS (1978) Factors affecting the utilisation of ketone bodies and other substrates by rat jejunum: effects of fasting and of diabetes. J Physiol (London) 278: 55–67

Hanson PJ, Parsons DS (1980) The interrelationship between glutamine and alanine in the intestine. Biochem Soc Trans 8: 506–509

Hems DA, Brosnan JT (1970) Effects of Ischaemia on content of metabolites in rat liver and kidney in vivo. Biochem J 120: 105–111

Hume DA, Weidemann MJ (1980) Mitogenic lymphocyte transformation. Elsevier/North Holland, Amsterdam

Hume DA, Radik JL, Ferber E, Weidemann MJ (1978) Aerobic glycolysis and lymphocyte transformation. Biochem J 174: 703–709

Kovacevic Z, McGivan JD (1983) Mitochondrial metabolism of glutamine and glutamate and its physiological significance. Physiol Rev 63: 547–605

Krebs HA (1980) Glutamine metabolism in the animal body. In: Mora J, Palacios R (eds) Glutamine metabolism, enzymology and regulation. Academic Press, London New York, pp 319–329

Lund P (1980) Glutamine metabolism in the rat. FEBS Lett 117: K86–K92

Newsholme EA (1980) Reflections on the mechanism of action of hormones. FEBS Lett 117: K121–K134

Newsholme EA, Crabtree B (1979) Theoretical principles in the approaches to control of metabolic pathways and their application to glycolysis in muscle. J Mol Cell Cardiol 11: 839–856

Newsholme EA, Leech AR (1983) Biochemistry for the medical sciences. Wiley, London New York

Newsholme EA, Start C (1973) Regulation on metabolism. Wiley, London New York

Roediger WEW (1982) Utilisation of nutrients by isolated epithelial cells of the rat colon. Gastroenterology 83: 424–429
Schröck H, Cha C-JM, Goldstein L (1980) Glutamine release from hindlimb and uptake by kidney in the acutely acidotic rat. Biochem J 188: 557–560
Snell K (1980) Muscle alanine synthesis and hepatic gluconeogenesis. Biochem Soc Trans 8: 205–213
Taegtmeyer H, Hems R, Krebs HA (1980) Utilisation of energy-providing substrates in the isolated working rat heart. Biochem J 186: 701–711
Tate SS, Meister A (1978) Glutamine synthetases of mammalian liver and brain. In: Prusiner S, Stadtman ER (eds) The enzymes of glutamine metabolism. Academic Press, London New York, pp 77–125
Watford M, Lund P, Krebs HA (1979a) Isolation and metabolic characteristics of rat and chicken enterocytes. Biochem J 178: 589–596
Watford M, Vinay P, Lemieux G, Gougoux A (1979b) The formation of pyruvate from citric-acid-cycle intermediates in kidney cortex. Biochem Soc Trans 7: 753–755
Watford M, Vinay P, Lemieux G, Gougoux A (1980) The regulation of glucose and pyruvate formation from glutamine and citric-acid-cycle intermediates in the kidney cortex of rats, dogs, rabbits and guinea pigs. Biochem J 188: 741–748
Windmueller HG, Spaeth AE (1974) Uptake and metabolism of plasma glutamine by the small intestine. J Biol Chem 249: 5070–5079

Glutamine Metabolism by Cultured Mammalian Cells

H. R. Zielke, C. M. Sumbilla, C. L. Zielke, J. T. Tildon, and P. T. Ozand[1]

Introduction

Cultured mammalian cells are grown in medium containing carbohydrates, amino acids, salts, vitamins, and growth factors. In the 1950's Eagle [1] identified the basic composition of tissue culture medium, except for the growth factors which were provided by added serum. Based on this work most culture media contain the essential amino acids required by humans [2] plus glutamine, tyrosine, and cystine. Depletion of any one of these amino acids results in cell death [1]. Eagle et al. [3] observed that the glutamine requirement of cells was 10- to 100-fold greater than for any other amino acid. The increased requirement for glutamine was related only marginally to the chemical instability of this amino acid since the daily spontaneous decomposition at 37° C amounted to only 2%–3% of the glutamine in the medium [4]. Although a portion of the glutamine is used for transamination reactions and direct incorporation into protein, we [4, 5] and other investigators [6–8] have proposed that glutamine is oxidized to CO_2 to provide a significant portion of the energy requirement of cultured mammalian cells.

Glutamine as a Respiratory Substrate

Anabolic and Catabolic Functions of Glucose

Cells grown in tissue culture medium have two energy sources available: glucose and glutamine. Glucose is rapidly metabolized to lactate with very little oxidation to CO_2 via the tricarboxylic acid cycle [9]. However, the rate of glucose metabolism is not directly related to cell growth [9–13]. Identical cell growth was observed when human fibroblasts were subcultured in medium with 5 m*M* glucose or in medium with 80 μ*M* glucose to which small aliquots of glucose were added daily (Fig. 1). Under the low glucose conditions the cells utilized only 7% as much glucose as is used by the cells cultured in 5 m*M* glucose medium [5]. Cellular glycogen can not compensate for the decreased level of glucose in the medium since cellular glycogen is depleted within hours under these conditions [14].

1 Division of Pediatric Research, University of Maryland, 655 W. Baltimore Street, Baltimore, MD 21201, USA

Glutamine Metabolism in Mammalian Tissues
Edited by D. Häussinger and H. Sies

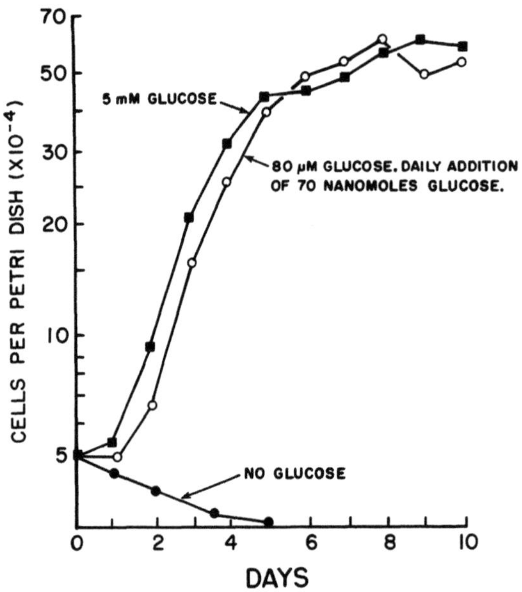

Fig. 1. Effect on cell growth of daily addition of glucose to medium containing limiting glucose. Human diploid fibroblasts were subcultured in 60 mm diameter plastic petri dishes with 5.5 ml of MEM medium supplemented with 10% dialyzed fetal bovine serum and the glucose concentrations indicated in the figure. To one set of dishes 20 μl of sterile 3.46 m*M* glucose solution (70 nmol) was added daily. Adapted from [5]

The results raised important questions about the essential role of glucose in tissue culture medium and about the possibility that other components of tissue culture medium serve as energy sources for the cells. Earlier studies had demonstrated that at 55 μ*M* glucose the primary products of glucose metabolism were the ribose moiety of nucleoside triphosphates and nucleic acids rather than lactate [15]. Consequently, Zielke et al. [4] proposed that cell death in the absence of glucose resulted from its anabolic rather than its catabolic role. This hypothesis was verified by the growth of cells in glucose-free medium in which some of the anabolic functions of glucose were replaced by low levels of nucleosides. Under these conditions cells utilized approximately twice as much glutamine per cell division as was used in medium with 5.5 m*M* glutamine. These data clearly support a respiratory role for glutamine in human diploid fibroblasts. Earlier Kovačević and Morris [6] had shown that glutamine was a potentially important substrate for energy metabolism for Ehrlich ascites tumor cells. Donnely and Scheffler [7], using a mutant Chinese hamster cell line with a respiratory deficiency, observed increased glucose utilization and greatly reduced glutamine oxidation. They concluded that normal Chinese hamster cells derived approx. 40% of their energy from glutamine oxidation. Recently this genetic approach was taken further by Franchi et al. [16] who investigated two mutants of Chinese hamster lung fibroblasts: the DS7 mutant which is deficient in aerobic glycolysis and glucose transport due to a lack of phosphoglucose isomerase activity and the GSK3 mutant which has a defect in oxidative metabolism. The DS7 cells relied solely on glutamine oxidation for their energy requirements whereas the GSK3 cells derived all energy from glycolysis.

Table 1. Reciprocal inhibition of glucose and glutamine oxidation[a]

C-Substrate	Addition	$^{14}CO_2$ released (nmol/mg protein/h)
L [1-^{14}C] Glutamine	–	310
L [1-^{14}C] Glutamine	Glucose	54
L [1-^{14}C] Glutamine	HTU	276
L [5-^{14}C] Glutamine	–	80
L [5-^{14}C] Glutamine	Glucose	13
L [5-^{14}C] Glutamine	HTU	66
D [1-^{14}C] Glucose	–	30
D [1-^{14}C] Glucose	Glutamine	53
D [6-^{14}C] Glucose	–	7.5
D [6-^{14}C] Glucose	Glutamine	0.9

[a] The concentration of radioactive or nonradioactive glucose and glutamine, when present in these experiments, was 5.5 or 2.0 m*M*, respectively. HTU: Eagle's minimum essential medium minus glucose, plus 10% dialyzed bovine calf serum, 100 μ*M* hypoxanthine, 40 μ*M* thymidine, and 100 μ*M* uridine [5]

Regulation of Glutamine Oxidation by Hexoses

The rate of glutamine oxidation by cultured cells is regulated by the presence of glucose [5, 17, 18]. The addition of 5.5 m*M* glucose inhibited the rate of L[1-^{14}C]glutamine and L[5-^{14}C]glutamine oxidation by 80% (Table 1). Reciprocally, D[6-^{14}C]glucose oxidation was inhibited 90% by the addition of 2 m*M* glutamine. In contrast, 2 m*M* glutamine stimulated D[1-^{14}C]glucose oxidation by 77% which is consistent with a role of both glucose and glutamine in nucleic acid biosynthesis. It was calculated that under normal conditions with both glucose and glutamine present, human diploid fibroblasts derived two-thirds of their energy requirement from glycolysis and one-third from glutamine oxidation. Since the K_i apparent for glucose inhibition of glutamine oxidation was estimated to be 95 μ*M* [17], the percent contribution to the energy requirements of human dipoid fibroblasts by glutamine probably does not increase significantly unless glucose is almost depleted. Interestingly, up to 1 μ*M* pyruvate had no effect on the rate of glutamine oxidation [19]. This suggests that the regulation of glutamine oxidation is directly interrelated to anaerobic glycolysis (the Crabtree effect).

Kuchka et al. [20] studied the influence of hexose conditions on glutamine oxidation by SV40 transformed and diploid human fibroblasts. The rate of glutamine oxidation was increased 2.8-fold in the SV40 transformed cells, but the percent inhibiton of glutamine oxidation by glucose was similar: 80% inhibition for the normal cells and about 70% inhibition for the transformed cells. Substitution of galactose for glucose allowed the cells to grow with an uninhibited rate of glutamine oxidation. Likewise, it has been estimated that HeLa cells obtain more than 98% of their energy from glutamine when galactose or fructose is the available hexose [8]. In the presence of glucose, HeLa cells obtain at least 65% of their energy from glutamine oxidation. The conclusions

Table 2. Oxidative metabolism of substrates by human diploid fibroblasts

Labelled substrate	Concentration of substrate (m*M*)	$^{14}CO_2$ released[a] (nmol/mg protein/h)
L [1-^{14}C] Glutamine	0.4	67.8 ± 1.1
L [1-^{14}C] Glutamine	2.0	98.0 ± 5.4
D [6-^{14}C] Glucose	0.3	0.27 ± 0.02
D [6-^{14}C] Glucose	1.5	0.11 ± 0.02
[3-^{14}C] Acetoacetate	0.05	0.24 ± 0.07
[3-^{14}C] Acetoacetate	0.50	1.78 ± 0.31
D-3 [3-^{14}C] Hydroxybutyrate	0.15	0.12 ± 0.01
D-3 [3-^{14}C] Hydroxybutyrate	1.50	1.44 ± 0.14
[1-^{14}C] Octanoate	0.04	0.03 ± 0.04
[1-^{14}C] Octanoate	0.20	0.30 ± 0.57
[1-^{14}C] Palmitate	0.04	0.22 ± 0.01
[1-^{14}C] Palmitate	0.20	0.33 ± 0.03

[a] The oxidative rates are expressed as mean ± SD for four samples using logarithmic phase cells in 2 ml tissue culture media supplemented with 10% dialyzed newborn calf serum, pyrimidine, and purine nucleosides, but lacking glucose and glutamine, unless otherwise indicated. The radioactive substrates used were as indicated. From [17], with permission

Table 3. Effect of other substrates on L [1-^{14}C] glutamine and L [U-^{14}C] glutamine oxidation[a]

Addition to the standard incubation medium	$^{14}CO_2$ released (nmol/mg protein/h)	
	L [1-^{14}C] Glutamine	L [U-^{14}C] Glutamine
None	68.2 ± 1.0	31.9 ± 0.4
Glucose (1.5 m*M*)	10.2 ± 1.4	4.5 ± 0.5
Acetoacetate (0.5 m*M*)	74.7 ± 1.6	29.9 ± 0.7
D-3-Hydroxybutyrate (1.5 m*M*)	70.8 ± 3.1	30.3 ± 3.8
Octanoate (0.2 m*M*)	63.7 ± 2.2	35.0 ± 1.9
Palmitate (0.2 m*M*)	66.0 ± 6.1	33.6 ± 3.3

[a] The oxidation was performed in media with isotopically labelled 0.4 m*M* glutamine under the conditions described in Table 2. The final concentrations of other additions were as indicated [17]

from this series of studies are that cultured mammalian cells require a hexose source for anabolic functions, that glucose serves both anabolic and catabolic functions, that glutamine provides between 30% and 65% of their energy in the presence of glucose and nearly 100% of their energy in the presence of either galactose, fructose, or nucleosides.

Cultured cells are exposed to several classes of potential respiratory substrates besides glutamine, namely, ketone bodies and free fatty acids. The rate of oxidation of these compounds is shown in Table 2. The low rate of oxidation of these compounds is reflected in their lack of inhibition of glutamine oxidation (Table 3). It can be concluded that glutamine is the primary oxidative substrate used by cultured cells.

Glutamine Metabolism

Glutaminase

The enzymatic reactions in the utilization of glutamine are glutaminase, glutamate dehydrogenase, and/or glutamate oxalacetate transaminase to form α-ketoglutarate. Subsequent tricarboxylic acid cycle enzymes result in its total oxidation to CO_2 and its equilibration with other TCA cycle intermediates or glycolytic intermediates as discussed subsequently. The specific activity of glutaminase increased about threefold during the early logarithmic phase of human fibroblast cell growth and then declined [21]. Glutamate dehydrogenase activity remained constant under these conditions. The increase in specific activity of glutaminase is consistent with the observation of Knox et al. [22] that there is a proportionality of glutaminase content to the growth rate of neoplasms. To date there is no evidence that the regulation of glutamine oxidation is controlled through regulation of glutaminase activity. However, the recent development of a glutaminase assay applicable to living cells may provide the means for studying glutaminase activity in its cellular environment rather than in an artifical test tube situation. This assay developed by Zielke et al. [23] measures the release of ^3H_2O from L[2-^{3}H]glutamine by the following reaction sequence:

$$\text{L}\,[2\text{-}^3\text{H}]\ \text{glutamine} \xrightarrow{\text{glutaminase}} \text{L}\,[2\text{-}^3\text{H}]\ \text{glutamate}$$

$$\text{L}\,[2\text{-}^3\text{H}]\ \text{glutamate} \xrightarrow[\text{transaminase}]{\text{glutamate-oxalacetate}} \alpha\text{-ketoglutarate} + {}^3\text{H}_2\text{O}$$

No [2-^{3}H]glutamate could be demonstrated in the medium indicating that the second reaction was not rate limiting. Furthermore, if L[2-^{3}H]glutamate were the substrate for glutamate dehydrogenase, the tritium would be transfered to NADH and likewise released as ^3H_2O. Therefore, the release of ^3H_2O should be a direct measurement of intracellular glutaminase activity. The enzymatic activity measured for purified glutaminase with the ^3H_2O assay agreed with the results obtained with a spectrophotometric assay [23]. However, it was observed that the activity measured with the ^3H_2O assay in living cells was only about 3% of that previously measured in cell extracts using a spectrophotometric assay [21]. This discrepency in glutaminase activity has now been confirmed. The ^3H_2O "flow through" assay was used to measure glutaminase activity both in growing human fibroblasts and in cell extracts of comparable cell density. The specific activity in whole cells growing 5 days in Eagle's minimum essential medium was 45 ± 5 nmol/mg protein/h while it was 166-fold higher in cell extracts, 7,450 ± 525 nmol/mg protein/h. There are several possible explanations for this difference: (1) the ^{3}H-glutamine does not equilibrate with the internal nonradioactive glutamine during the 6 h experiment; (2) the ^3H_2O produced in the cells does not equilibrate with the external water; (3) the internal concentration of glutamine is below the K_m of glutaminase; or (4) the intracellular glutaminase is highly regulated and not fully expressed. Further studies are required to differentiate between these alternative explanations. However, it should be noted that under identical conditions, the rate of L[1-^{14}C]glutamine oxidation by whole cells was 10.2 nmol/mg protein/h (Table 3) indicating that the intracellular glutaminase activity measured by ^3H_2O release is sufficient to account for the rate of glutaminase oxidation.

Products of Glutamine Metabolism

The primary metabolic products of glutamine are CO_2, glutamate, and lactate (Table 4). Both CO_2 and glutamate are consistent with the role of glutamine as a major energy source for cultured cells. Suprisingly about 13% of the metabolized glutamine was also converted to lactate by both human diploid fibroblasts [24] and HeLa cells [8]. It has been proposed that lactate is synthesized via malic enzyme [8, 24], an enzyme found both in the particulate and cytosolic fraction of human fibroblasts [25]. The conversion of glutamate to the lactate precursor pyruvate explains the high rate of glutamine incorporation into lipids which was one-half of the rate observed for acetoacetate [26]. This glutamine derived pyruvate also provides a source for acetyl CoA which is required for oxidation of the complete carbon skeleton of glutamine in the absence of glucose (Tables 1, 3).

The route of glutamine metabolism was a function of the glucose concentration in the medium. In medium with 55 μ*M* glucose, the amount of aspartate synthesized was seven-fold greater than that in presence of 5.5 m*M* glucose [24, 27]. Although in low glucose medium, aspartate was synthesized from glutamine, in the presence of 5.5 m*M* glucose, aspartate was synthesized from glucose carbons. Kovačević [28] utilizing Ehrlich ascites tumor cells

Table 4. Products of glutamine metabolism by cultured mammalian cells

Product	Percent of metabolized glutamine	References
CO_2	35–56	[8, 36]
Glutamate	17	[24]
Lactate	13	[8, 24]
Pyruvate	2	[8]
Citrate	3	[8]
Malate	1	[8]
Aspartate	< 1	[8]
Cellular material	2	[8, 24]

similarly observed an inhibitory effect of glucose on aspartate formation from glutamine. The explanation for this observation is probably related to the decreased level of NADH observed in the absence of glucose [29] which results in an elevated NAD/NADH ratio. A decreased availability of NADH may alter the malate-aspartate shuttle function by preventing the cytosolic conversion of aspartate to oxalacetate and malate and the subsequent re-entry of malate into the mitochondrion. The results of such events is an accumulation of aspartate in the medium.

The available data suggest that glutamine may also be an important respiratory substrate in other tissues of the body. Glutamine is one of the most abundant amino acids in blood [30] and brain [31]. It is released along with alanine from muscle during fasting [32] and it is the major respiratory substrate of the small intestine [33], reticulocytes [34], and oocytes [35].

Acknowledgement. Supported in part by NIH research grant HD16596.

References

1. Eagle H (1959) Amino acid metabolism in mammalian cell cultures. Science 130: 432–437
2. Rose WC (1938) The nutritive significance of the amino acids. Physiol Rev 18: 109–136
3. Eagle H, Oyama VI, Levy M, Horton CL, Fleischman R (1956) The growth response of mammalian cells in tissue culture to L-glutamine and L-glutamic acid. J Biol Chem 218: 607–616
4. Zielke HR, Ozand PT, Tildon JT, Sevdalian DA, Cornblath M (1976) Growth of human fibroblasts in the absence of glucose utilization. Proc Natl Acad Sci USA 73: 4110–4114
5. Zielke HR, Ozand PT, Tildon JT, Sevdalian DA, Cornblath M (1978) Reciprocal regulation of glucose and glutamine utilization by cultured human diploid fibroblasts. J Cell Physiol 95: 41–48
6. Kovačević Z, Morris HP (1972) The role of glutamine in the oxidative metabolism of malignant cells. Cancer Res 32: 326–333
7. Donnelly M, Scheffler IE (1976) Energy metabolism in respiration-deficient and wild type Chinese hamster fibroblasts in culture. J Cell Physiol 89: 39–52
8. Reitzer LJ, Wice BM, Kennell D (1979) Evidence that glutamine, not sugar, is the major energy source for cultured HeLa cells. J Biol Chem 254: 2667–2676
9. Morell B, Froesch ER (1973) Fibroblasts as an experimental tool in metabolic and hormone studies. Eur J Clin Invest 3: 112–118
10. Bissell MJ, Hatie C, Rubin H (1972) Patterns of glucose metabolism in normal and virus-transformed chick cells in tissue culture. J Natl Cancer Inst 49: 555–565
11. Bryant JC, Schilling EL, Earle WR (1958) Massive fluid suspension cultures of certain mammalian tissue cells. Glucose utilization and cell proliferation. J Natl Cancer Inst 21: 349–364
12. Graff S, Moser H, Kastner O, Graff AM, Tannenbaum M (1965) The significance of glycolysis. J Natl Cancer Inst 34: 511–519
13. Rheinwald JG, Green H (1974) Growth of cultured mammalian cells on secondary glucose sources. Cell 2: 287–293
14. Sevdalian DA, Zielke HR (1978) Depletion of cellular glycogen during the early logarithmic growth phase of human fibroblasts. Experientia 34: 384–385
15. Renner ED, Plagemann PGW, Bernlohr RW (1972) Permeation of glucose by simple and facilitated diffusion by Novikoff rat hepatoma cells in suspension culture and its relationship to glucose metabolism. J Biol Chem 247: 5765–5776
16. Franchi A, Silvestre P, Pouysségur J (1981) A genetic approach to the role of energy metabolism in the growth of tumor cells: tumorigenicity of fibroblast mutants deficient either in glycolysis or in respiration. Int J Cancer 27: 819–827

17. Sumbilla CM, Zielke CL, Reed WD, Ozand PT, Zielke HR (1981) Comparison of the oxidation of glutamine, glucose, ketone bodies and fatty acids by human diploid fibroblasts. Biochim Biophys Acta 657: 301–304
18. Kvamme E, Svenneby G (1961) The effect of glucose on glutamine utilization by Ehrlich ascites tumor cells. Cancer Res 21: 92–98
19. Zielke HR, Zielke CL, Ozand PT (1984) Glutamine: a major energy source for cultured mammalian cells. Fed Proc 43: 121–125
20. Kuchka M, Markus HB, Mellman WJ (1981) Influence of hexose conditions on glutamine oxidation of SV-40-transformed and diploid fibroblast human cell lines. Biochem Med 26: 356–364
21. Sevdalian DA, Ozand PT, Zielke HR (1980) Increase in glutaminase activity during the growth cycle of cultured human diploid fibroblasts. Enzyme 25: 142–144
22. Knox WE, Horowitz ML, Friedell GH (1969) The proportionality of glutaminase content to growth rate and morphology of rat neoplasms. Cancer Res 29: 669–680
23. Zielke CL, Zielke HR, Ozand PT (1982) A radioisotope assay for L-glutaminase: the measurement of tritiated water formation from the reaction of L-2-^{3}H glutamine with the L-glutaminase and glutamate oxalacetate transaminase couple. Anal Biochem 127: 134–142
24. Zielke HR, Sumbilla CM, Sevdalian DA, Hawkins RL, Ozand PT (1980) Lactate: a major product of glutamine metabolism by human diploid fibroblasts. J Cell Physiol 104: 433–441
25. Sumbilla CM, Ozand PT, Zielke HR (1981) Activities of enzymes required for the conversion of 4-carbon TCA cycle compounds to 3-carbon glycolytic compounds in human diploid fibroblasts. Enzyme 26: 201–205
26. Reed WR, Zielke HR, Baab PJ, Ozand PT (1981) Ketone bodies, glucose, and glutamine as lipogenic precursors in human diploid fibroblasts. Lipids 9: 677–684
27. Zielke HR, Sumbilla CM, Ozand PT (1981) Effect of glucose on aspartate and glutamate synthesis by human diploid fibroblasts. J Cell Physiol 107: 251–254
28. Kovačević Z (1971) The pathway of glutamine and glutamate oxidation in isolated mitochondria from mammalian cells. Biochem J 125: 757–763
29. Schwartz JP, Johnson GS (1976) Metabolic effects of glucose deprivation and of various sugars in normal and transformed fibroblast cell lines. Arch Biochem Biophys 173: 237–245
30. Meister A (1956) Metabolism of glutamine. Physiol Rev 36: 103–127
31. Perry TL, Berry K, Diamond S, Mak C (1971) Regional distribution of amino acids in human brain obtained at autopsy. J Neurochem 18: 513–519
32. Felig P (1975) Amino acid metabolism in man. Annu Rev Biochem 44: 933–955
33. Windmueller HG, Spaeth AE (1974) Uptake and metabolism of plasma glutamine by the small intestine. J Biol Chem 249: 5070–5079
34. Rapoport S, Rost J, Schultze M (1971) Glutamine and glutamate as respiratory substrates of rabbit reticulocytes. Eur J Biochem 23: 166–170
35. Bae I-H, Foote RH (1975) Carbohydrate and amino acid requirements and ammonia production of rabbit follicular oocytes matured in vitro. Exp Cell Res 91: 113–118
36. Stoner GD, Merchant DJ (1972) Amino acid utilization by L-M strain mouse cells in a chemically defined medium. In Vitro 7: 330–343

Clinical Aspects

Ammonia Detoxication and Glutamine Metabolism in Severe Liver Disease and its Role in the Pathogenesis of Hepatic Encephalopathy*

W. GEROK and D. HÄUSSINGER[1]

Introduction

The term ammonia is used for the sum of the nonionized NH_3 and the ionized NH_4^+. Ammonia is mainly produced by degradation of amino acids in the human organism. About 1,100 mmol ammonia arise in the catabolism of 100 g protein having an average amino acid composition. Other sources of ammonia are nucleosides and nucleotides by hydrolytic deamination of their purines and pyrimidines, further amines by oxidative deamination. In addition, enteral microorganisms produce ammonia, which can be reabsorbed from the gut.

In spite of this high and continuous ammonia production, the concentration of ammonia in blood and tissue is very low (about 10 μM in blood plasma). At the pH of arterial blood 1%–3% of ammonia is in the nonionized and approx. 98% in the ionized form. The low concentration of ammonia in blood and inter- and intracellular fluid is essential because already with a five- to tenfold increase of ammonia concentration toxic effects arise, especially as functional disturbances of the central nervous system. Effective mechanisms for the detoxication of ammonia are, therefore, indispensable, i.e., metabolic reactions by which ammonia is transformed to a nontoxic substance, which can be rapidly excreted.

The biosynthesis of urea and of glutamine are the two main ways for the detoxication of ammonia. Urea synthesis by the Krebs-Henseleit cycle [30] is localized only in the liver. The detoxication of ammonia through this pathway is definitive. In contrast, formation of glutamine is not a liver-specific process and the detoxication of ammonia is only preliminary. Glutamine serves to transport ammonia as a nontoxic metabolite from peripheral organs, especially muscle, kidney, gut, and brain to the liver, where the definitive detoxication through urea synthesis takes place. In the kidney ammonia produced by degradation of glutamine can be directly excreted via urine.

In the following we describe

1. Alterations of urea and glutamine synthesis in liver injury and severe liver disease.
2. Consequences of these alterations for the glutamine metabolism in muscle.

* Our research was supported by Deutsche Forschungsgemeinschaft and by SANDOZ-Stiftung für therapeutische Forschung

1 Medizinische Universitätsklinik Freiburg, Hugstetterstr. 55, D-7800 Freiburg, FRG

Glutamine Metabolism in Mammalian Tissues
Edited by D. Häussinger and H. Sies

3. Significance of these metabolic abnormalities for pathogenesis of hepatic encephalopathy.

Alterations of Urea and Glutamine Synthesis in Liver Injury and Severe Liver Disease

Hepatocyte Heterogeneity in Regard to Urea Synthesis and Glutamine Metabolism

Besides the urea cycle a second cycling process in the liver is involved in the detoxication of ammonia: the glutamine cycle (Fig. 1). Glutamine is synthesized by amidation of glutamate with glutamine-synthetase [29]; this is associated with simultaneous glutamine degradation by glutaminase [18]. Both cycles are partly localized in the cytosol, partly in the mitochondria; both cycles are energy dependent. But in the urea cycle ammonia is irreversibly transformed to urea at the expense of energy, whereas in the glutamine cycle there is no net flux of metabolites in spite of expenditure of energy. Therefore, the glutamine cycle fulfills the criteria of a "futile cycle", if only net reaction is considered.

Häussinger [16] demonstrated heterogeneity of rat hepatocytes regarding urea synthesis and glutamine metabolism through measurement of absolute flux rates dependent on the direction of liver perfusion. In contrast to other methods for demonstration of metabolic hepatocyte heterogeneity – histochemistry, localized intoxication, microdissection – this procedure can be performed with the structurally and metabolically intact liver. These studies showed the periportal localization of urea synthesis and glutaminase activity, in contrast to the perivenous localization of glutamine synthetase activity (Fig. 2). So the

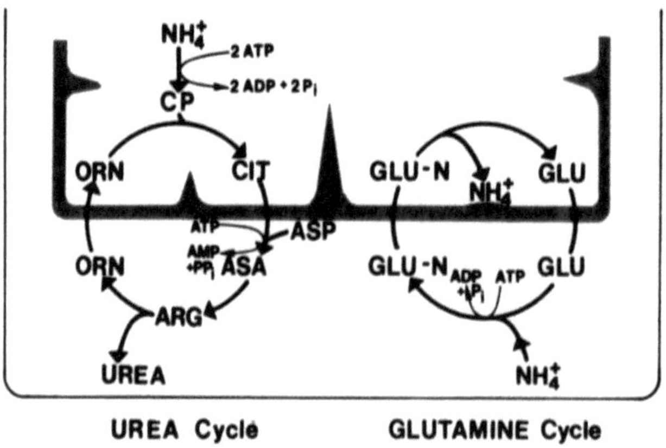

Fig. 1. Urea cycle and glutamine cycle. *Hatched area* marks schematically the mitochondrial membrane. Both cycles are energy-dependent and partly localized in the cytosol, partly in the mitochondria. In contrast to the urea cycle, the glutamine cycle is a futile cycle, if only the net reaction is considered. *CP* carbamoylphosphate; *CIT* citrulline; *ASA* argininosuccinate; *ARG* arginine; *ORN* ornithine; *GLU* glutamate; *GLU-N* glutamine

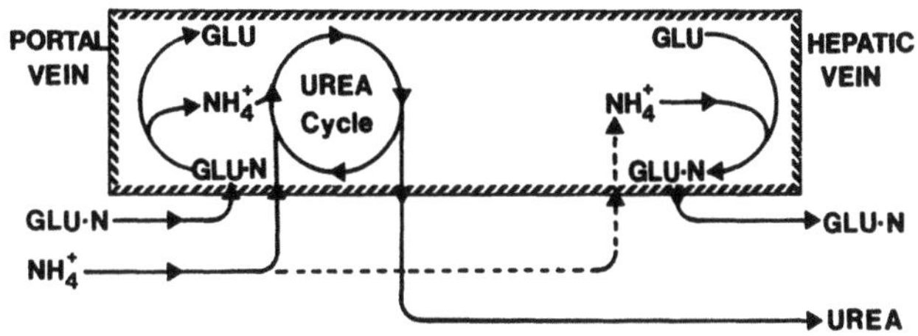

Fig. 2. Hepatocyte heterogeneity in regard to urea cycle and glutamine cycle. *Hatched area* marks schematically the hepatocytes from the portal vein *(left)* to the central (hepatic) vein zone *(right)*. Urea synthesis and glutaminase reaction are localized in the periportal zone, glutamine synthesis in the perivenous zone. Instead of an intracellular "futile" glutamine cycle an intercellular useful cycle exists. (See Sies and Häussinger, this Vol., p. 78 f.)

simultaneous activity of glutaminase and glutamine synthetase provides an intercellular glutamine cycling.

This hepatocyte heterogeneity with regard to urea synthesis and glutamine metabolism provides two advantages for physiological liver function:

1. Ammonia Detoxication is Improved. Urea synthesis proceeds only with ammonia supply in high concentration due to the high K_m value of carbamylphosphate synthetase (1–2 m*M*), the rate limiting enzyme for ammonia entry into the urea cycle [35]. In contrast, glutamine synthetase with its lower K_m value (0,3 m*M*) binds ammonia also in very low concentration [10]. The decreasing ammonia concentration in the blood from periportal to perivenous area corresponds to the K_m values of the enzymes for ammonia detoxication. Small amounts of ammonia escaping the detoxication by urea cycle in the periportal area can be removed by glutamine synthesis in the perivenous zone.

Furthermore, this hepatocyte heterogeneity explains the well-known phenomenon that glutamine stimulates urea synthesis. Because of the periportal, intramitochondrial localization of glutaminase, not only ammonia from portal blood, but also ammonia derived from glutamine will be the substrate for urea synthesis, so increasing the flux rate through the urea cycle. Resynthesis of glutamine in the perivenous zone ensures that the glutamine concentration in the effluent and influent is nearly identical, so preventing a loss of glutamine under conditons of physiological pH.

Therefore, hepatocyte heterogeneity provides a means for effective and economic ammonia detoxication.

2. Hepatic Role in pH Regulation. As pointed out by Bourke et al. [42] and by Atkinson and Camien [2], a major function of hepatic urea synthesis is not only the removal of toxic ammonia, but also the disposal of bicarbonate. Glutamine synthesized in the liver is the source for ammonium ions produced by renal glutaminase and excreted via urine. In contrast to urea synthesis ammonia removal by glutamine synthesis is not associated with disposal of bicarbonate.

With regard to this switching from ammonia and bicarbonate disposal by urea synthesis to the ammonia removal by glutamine synthesis, the two differently localized reactions of the glutamine cycle offer two regulatory sites. Experimental results [18] agree with this concept. The hepatocyte heterogeneity in regard to urea synthesis and glutamine metabolism gives the liver an important role in the stabilization of blood pH (see also Sies and Häussinger, this Vol., p. 78 f.).

Urea Synthesis and Glutamine Metabolism in CCl_4-Induced Liver Injury

Intoxication with CCl_4 results in a pericentral (perivenous) liver cell necrosis. Ammonia uptake by perfused rat liver is significantly decreased after pretreatment of the rat with CCl_4 (Fig. 3). The decreased ammonia uptake by the liver after CCl_4 treatment is due to an almost complete inhibition of glutamine synthesis (Fig. 4). Urea production from added ammonium ions in the concentration range from 0–0.5 m*M* was not influenced by CCl_4-induced liver cell necrosis (Fig. 5). The result is explained by an impairment of perivenously localized glutamine synthetase activity as a consequence of the perivenous liver cell necrosis induced by CCl_4. In contrast, urea production localized in the periportal zone is uninfluenced by CCl_4 pretreatment [17].

Urea Synthesis and Glutamine Metabolism in Severe Chronic Liver Disease, Especially Liver Cirrhosis

In severe chronic liver diseases – chronic active hepatitis, liver cirrhosis – the lesions of hepatocytes and cell necroses are predominantly localized in the periportal area.

Urea synthesis in patients with liver cirrhosis was measured with radioactive labelled urea. Jones et al. [26] using this method found a reduction of urea synthesis rate in cirrhotics to about a half of the normal value. Another procedure for determination of the rate of urea synthesis is its calculation from urea excretion taking into consideration the urea distribution space and intestinal urea breakdown. With this method and following an increasing oral protein load Rudman et al. [46] found a maximal rate of urea synthesis at plasma α-amino concentrations above 11 mmol/l in healthy subjects of 65 mg urea-N/h/kg body wt and a decrease of this value in cirrhotics to 27 mg urea-N/h/kg body wt. Vilstrup [53], increasing α-amino concentration by infusion of amino acids, observed a linear relationship between urea synthesis rate and α-amino concentration. The slope of this relationship as an expression of the hepatic conversion of α-amino nitrogen to urea nitrogen – the so-called functional hepatic nitrogen clearance – was decreased in cirrhotics (13.7 l/h) compared to control subjects (22.4 l/h) and was correlated with quantitative measures of liver function (galactose elimination capacity, antipyrin clearance). But in contrast to the study of Rudman et al. [46] saturation of urea synthesis could not be demonstrated by Vilstrup [53]. He suggests that hepatic urea

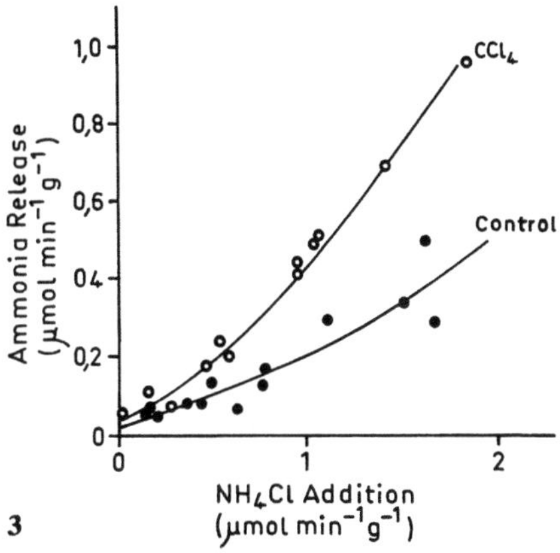

3

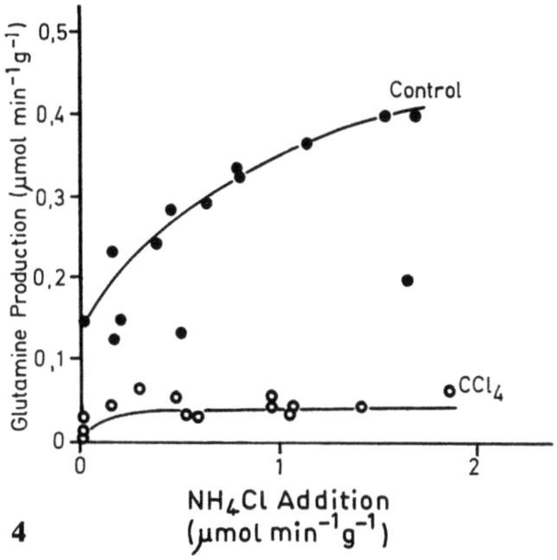

4

Fig. 3. Ammonia release by perfused liver of normal (●) and CCl_4 pretreated (○) rats in relation to the supply of ammonia to the liver. The uptake (detoxication) of ammonia by the liver is diminished in CCl_4-pretreated rats. An ammonia addition of 1 μmol · min^{-1} · g^{-1} corresponds to an influent concentration of about 0.25 mM

Fig. 4. Glutamine production by perfused liver of normal (●) and CCl_4-pretreated (○) rats in relation to ammonia supply. Perivenous liver cell necrosis induced by CCl_4 results in decreased perivenously localized glutamine synthesis

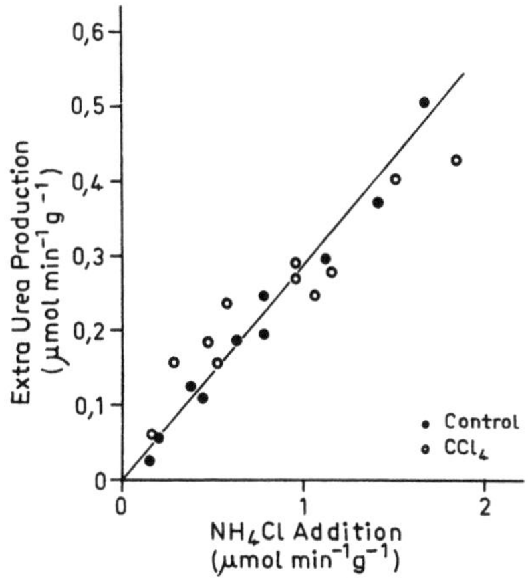

Fig. 5. Extra urea production by perfused liver of normal (●) and CCl_4-pretreated (○) rats in relation to ammonia supply. By perivenous liver cell necrosis the periportally localized urea synthesis is uninfluenced

synthesis is a saturable process, but apparently maximum rates require α-amino nitrogen concentrations in blood that are unfeasibly high in clinical studies because of the nausea and vomiting that they cause. Recently, members of Rudman's group using a kinetic urea tracer method could not demonstrate a maximal rate of urea synthesis according to saturation of the metabolic process [47].

However, these measurements of urea synthesis in the intact organism allow no decision as to whether the produced rate of urea synthesis is the consequence of porto-systemic shunts through which the blood bypasses the liver or of a decreased flux through the pathway of urea synthesis in the liver itself.

6 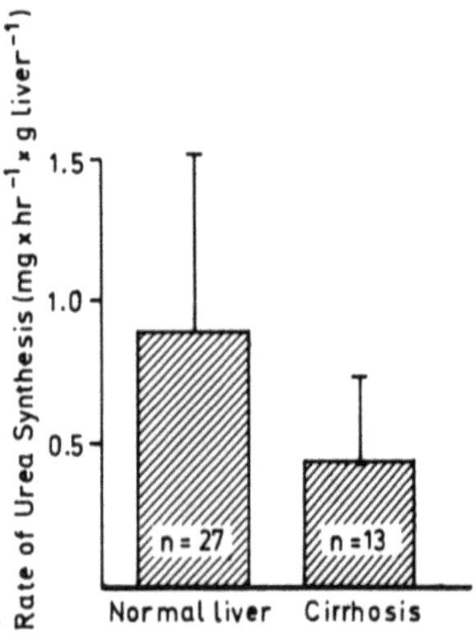

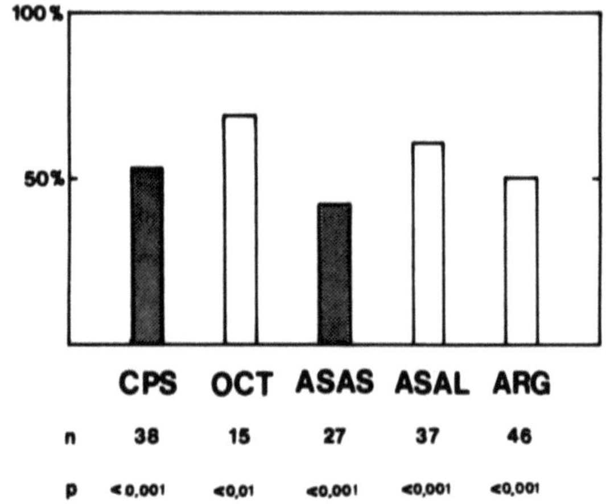 7

Fig. 6. Rate of urea synthesis (mg $\times$ h^{-1} $\times$ g liver^{-1}), measured in biopsy specimens of normal and cirrhotic liver. In liver cirrhosis the rate of urea synthesis is diminished under conditions of optimal substrate supply and with independence on hemodynamic influences ($p < 0.025$)

Fig. 7. Activity of urea cycle enzymes measured in biopsy specimens of cirrhotic liver as percentage of the activity in normal liver (= 100%). Note the significant decrease of activity in liver cirrhosis, especially for the two rate-limiting enzymes CPS and ASAS. *CPS* carbamoylphosphate synthetase (EC.2.7.2.5); *OCT* ornithine carbamoyltransferase (EC.2.1.3.3); *ASAS* argininosuccinate synthetase (EC.6.3.4.5); *ASAL* argininosuccinate lyase (EC.4.3.2.1); *ARG* arginase (EC.3.5.3.1)

Furthermore, these methods of determining the rate of urea synthesis require a high protein intake or amino acid infusion in high concentration, both combined with a great strain for the cirrhotic patient. In addition, in patients with ascites or disturbances of renal function, these measurements can not be performed.

Urea synthesis measured in slices of liver biopsies bypasses these limitations and objections [36]. In the tissue slices of cirrhotic liver, urea synthesis is lowered by more than half in comparison with normal liver tissue (Fig. 6). It should be noted that this reduction of urea synthesis was measured under optimal substrate supply and was independent of hemodynamic influences. DNA-content and the activity of nucleotide pyrophosphatase – two markers of the number of cells – did not differ significantly in biopsy specimens of normal and cirrhotic liver. Hydroxyproline content and the collagen moiety are significantly higher in cirrhosis compared to normal liver, as is to be expected.

Also the activities of urea cycle enzymes are significantly lower in biopsies taken from patients with liver cirrhosis compared to a control group with normal liver (Fig. 7). The decrease in activity is significant regardless whether the reference is liver wet weight or DNA content. In view of the decreased urea formation rate, it is of particular importance that the activities of the rate limiting enzymes of the urea cycle, carbamoylphosphate synthetase and argininosuccinate synthetâse, are greatly reduced in the cirrhotic liver. It is noteworthy that already in a prestate of cirrhosis, the chronic active hepatitis, this decrease of the activities of urea cycle enzymes can be demonstrated [36].

As yet not results on glutamine synthesis in biopsy specimens of severe chronic liver diseases are available. But Ansley and co-workers [1] compared the plasma concentration of glutamine with the rate of urea synthesis in patients with liver cirrhosis and control subjects. They found a significant inverse correlation between urea synthesis and glutamine concentration in plasma. This demonstrates that in patients with severe chronic liver disease, in whom the urea synthesis is decreased, hepatic or extrahepatic glutamine synthesis presumably expands.

The impaired urea synthesis in liver cirrhosis explains not only the arise of hyperammonemia, but also the pathogenesis of alkalosis of the cirrhotic patient. Alkalosis is the consequence of diminished disposal of bicarbonate via urea synthesis.

Concluding Remarks

The metabolic alterations in toxic liver injury and severe chronic liver disease, especially liver cirrhosis, agree with the concept of hepatocyte heterogeneity. Liver injury with perivenous necrosis of hepatocytes induced by CCl_4 intoxication results in decreased formation of glutamine and unaltered urea synthesis. Liver cirrhosis with predominantly periportal lesions of the hepatocytes is followed by a reduced urea synthesis presumably with increase of glutamine synthesis.

Therefore, lesions of hepatocytes limited to special zones of the liver lobule can produce distinct alterations of urea synthesis and glutamine metabolism. This explains not only hyperammonemia, but also disturbances of pH-regulation in liver injury and liver cirrhosis.

Glutamine and Ammonia Metabolism of Skeletal Muscle in Severe Chronic Liver Disease

Metabolism under Normal Conditions

Studies on the dynamics of ammonia metabolism in man with ^{13}N labeled ammonia showed that about 50% of the arterial ammonia was metabolized by skeletal muscle [32]. This underlines the importance of skeletal muscle in the intermediary metabolism of ammonia.

In healthy persons the arteriovenous concentration difference for ammonia across normal resting muscle is close to zero indicating a very small ammonia net uptake [15, 25]. Because there is a rapid metabolic trapping of ammonia by muscle, a continual consumption and production of ammonia by the muscle must take place. During exercise the net release of ammonia from muscle may increase markedly [33]. Little is known on the factors regulating the uptake and release of ammonia by the muscle.

Skeletal muscle contains glutamine synthetase [27]. Normal subjects show a small, but significant arterio-femoral-venous difference according to a small net

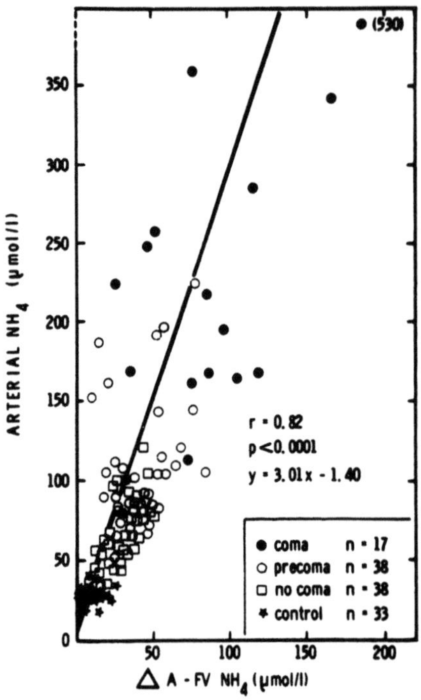

Fig. 8. Relationship between arterial ammonia concentration and arterio-femoral-venous ammonia differences (ΔA-FVNH$_4$) [25]

release of glutamine by the muscle [25]. Alanine is the amino acid with the highest arterio-femoral-venous difference in man under these conditions.

Metabolism in Liver Cirrhosis

In patients with cirrhosis and hyperammonemia, the net uptake of ammonia by the skeletal muscle increases (Fig. 9). This increase of arterio-femoral-venous difference grows progressively from the state of cirrhosis without hepatic encephalopathy to the precoma and coma state [25]. There is a linear correlation between the arterial ammonia concentration and the arterial-femoral-venous ammonia difference (Fig. 8). Muscle wasting and cachexia, frequent symptoms in severe cirrhosis cause a reduction in ammonia uptake by skeletal muscle [15, 32].

For glutamine Ganda and Ruderman [15] found an arteriovenous difference across the forearm in cirrhotics comparable to that reported in normal subjects. In contrast, in an extensive study of Imler et al. [25] the arterio-femoral-venous difference of glutamine concentrations were significantly higher in the cirrhotics compared to healthy controls. This augmented release of glutamine was independent of the precoma or coma state of the cirrhotic patient (Fig. 9).

Blood flow through the muscle was not measured in both studies. Since the peripheral blood flow is approximately doubled in patients with cirrhosis [28], it seems likely that the absolute rate of glutamine release by skeletal muscle in cirrhotics is much greater than in normal subjects.

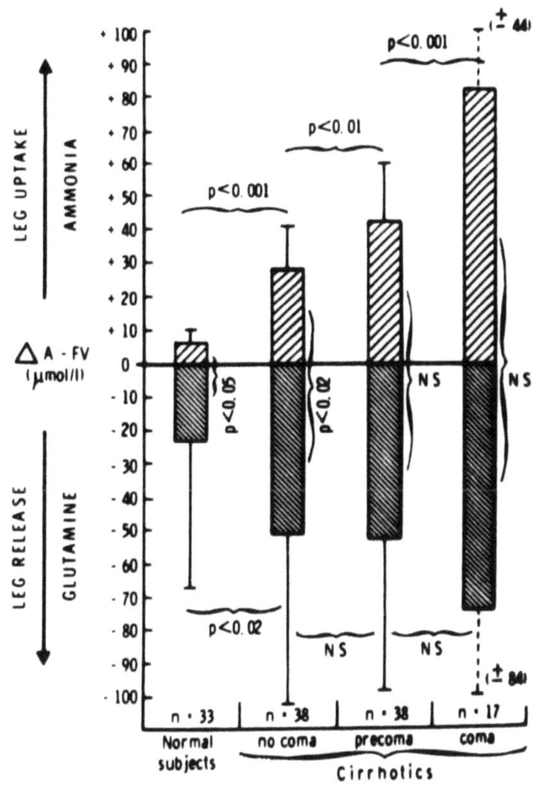

Fig. 9. Comparison of the positive arterio-femoral-venous difference (uptake) for ammonia to the negative arterio-femoral-venous difference (release) for glutamine in normal subjects and in three groups of cirrhotics [25]

The comparison (Fig. 9) between the positive arterio-femoral-venous ammonia difference, which represents ammonia uptake by the muscle, and the negative arteriovenous glutamine difference, which means glutamine release from the muscle, shows that in normal subjects and cirrhotics glutamine release is greater than ammonia uptake. However, in the precomatose and comatose patients there is a balance between ammonia and glutamine release.

Whereas in healthy subjects, the skeletal muscle releases preferentially alanine and then glutamine, in patients with cirrhosis the arteriovenous difference for glutamine across the muscle is threefold greater than that of alanine. For all amino acids except glutamine and alanine the muscle arteriovenous differences in cirrhotics with hyperammonemia are very low [15].

Concluding Remarks

Ammonia is taken up by skeletal muscle and utilized for the synthesis of glutamine. The product of ammonia detoxication, glutamine, may be carried to the liver and after deamidation serve there as a precursor and stimulator of urea synthesis.

In liver cirrhosis with hyperammonemia uptake of ammonia as well as release of glutamine are increased. Muscle can act as a "trap" to limit the rise of blood ammonia. Thus, in patients with cirrhosis and hyperammonemia skeletal muscle becomes an important organ in ammonia homeostasis. However, if muscle wasting or cachexia take place in the cirrhotic patient, this function of skeletal muscle in trapping ammonia is strongly impaired.

Ammonia and Glutamine Metabolism in Relation to the Pathogenesis of Hepatic Encephalopathy

Characterization of Hepatic Encephalopathy and Hypotheses on its Pathogenesis

Hepatic encephalopathy, a disturbance of brain function, is a frequent feature of severe chronic liver diseases, especially of cirrhosis with portosystemic anastomoses. It results from metabolic abnormalities and is potentially entirely reversible with appropriate treatment.

With regard to the symptoms it is a neuropsychiatric syndrome, i.e., abnormal mental reactions are combined with neurological disturbances. In the mildest form only psychometric tests show an abnormality, more severe forms are characterized by inappropriate behavior, intellectual deterioration, personality changes, inarticulate speech, and neuromotoric disturbances. In the severest form there exists a deep coma without response to painful stimuli. Variability in the pattern of symptoms is a main feature, and the clinical course is fluctuant as well.

There are currently three hypotheses on the pathogenesis of hepatic encephalopathy:

1. The first hypothesis recognizes that substances acting as neurotoxins are not removed or detoxified by the liver and accumulate in the brain. Such substances are suggested to be ammonia, mercaptans, and short- or medium-chain fatty acids. There is strong evidence that these substances can interact synergistically to produce alterations which are out of proportion to the effect of the individual substance [56]. Accumulation of ammonia and its effects on the nervous system have a central role in this synergism.
2. According to the second hypothesis alterations in amino acid metabolism result in alterations of neurotransmitters in the central nervous system. Two possibilities of these alterations are in discussion: (a) the increased production of a normal transmitter (i.e., serotonin) in brain and (b) the formation of false neurotransmitters (i.e., octopamine and phenylethanolamine) instead of normal neurotransmitters (i.e., dopamine, norepinephrine).
3. The third hypothesis postulates that the inhibitory neurotransmitter γ-aminobutyric acid produced by enteral microorganisms escapes the degradation in the liver with cirrhosis (shunts) and enters the brain through a disturbed blood-brain barrier.

A recently postulated "unified hypothesis" suggests that hepatic encephalopathy is the complex result of hyperammonemia, altered transport of amino acids and amines through the blood-brain barrier, altered amino acid metabolism with altered plasma pattern, and an increased exchange of amino acids for brain glutamine. So no single factor, but a network of conditions would cause hepatic encephalopathy.

Ammonia and Other Neurotoxins

Ammonia

Evidence for the important role of ammonia in the pathogenesis of hepatic encephalopathy stems primarily from clinical observations. Ingesting ammonia generating substances – proteins, peptides, amino acids, urea, ammonium salts – the cirrhotic patient frequently shows the symptoms of hepatic encephalopathy. The removal of these substances or of enteral bacteria which produce ammonia can reverse the encephalopathy. Also the inherited disorders of the urea cycle, the therapy with glutaminase or asparaginase, and Reye's syndrome are associated with hyperammonemia and neuropsychiatric disturbances similar to hepatic encephalopathy. Furthermore, experimentally induced hyperammonemia in animals leads to morphological alterations in the brain, the Alzheimer type II of astrocytes, which characterize hepatic encephalopathy. Finally, there is a significant correlation between the degree of encephalopathy and the concentrations of glutamine and α-ketoglutaramate in cerebrospinal fluid; these concentrations reflect ammonia in the brain [56].

Two major objections against the ammonia hypothesis are in discussion. First, there is only a poor correlation between the ammonia concentration in blood and the degree of hepatic encephalopathy. But only a small amount of ammonia passes the blood-brain barrier and can reflect the ammonia concentration within the brain. The second objection is based on the difference between the hyperkinetic preconvulsive or convulsive state induced by acute ammonia intoxication in experimental animals and the hypokinetic syndrome in hepatic encephalopathy. But there is strong evidence that acute and chronic ammonia intoxication are not comparable. Further, superimposition of ammonia load on a chronic low-grade ammonia intoxication enhances the toxic effect of the additional ammonia.

In summary, there is strong evidence for ammonia as a pathogenetic factor in hepatic encephalopathy.

The pathway of ammonia metabolism in the brain is summarized in Fig. 10. Ammonia is taken up by the brain from the blood; 11% of the ammonia present in arterial blood is extracted [32]. The two main reactions, by which ammonia is metabolized, are (a) reductive amination of α-ketoglutarate to glutamate and (b) ATP-dependent amidation of glutamate to form glutamine.

The flux through the conversion of glutamine to α-ketoglutaramate and hydrolysis of this metabolite to α-ketoglutarate is low in normal individuals, but increased concentrations of glutamine and α-keto acids in brain of patients with hepatic encephalopathy may promote α-ketoglutaramate formation.

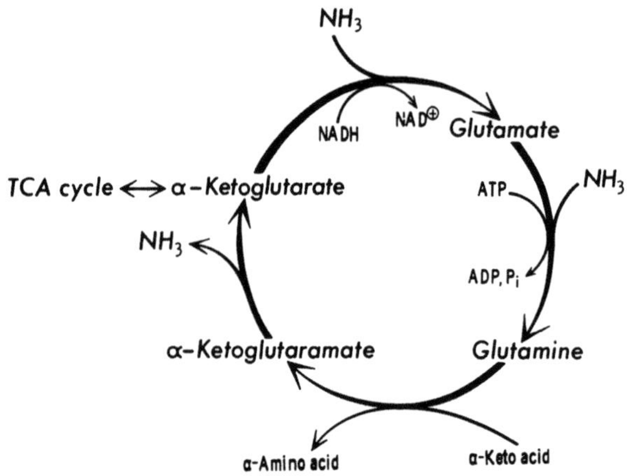

Fig. 10. Pathway of ammonia metabolism in brain

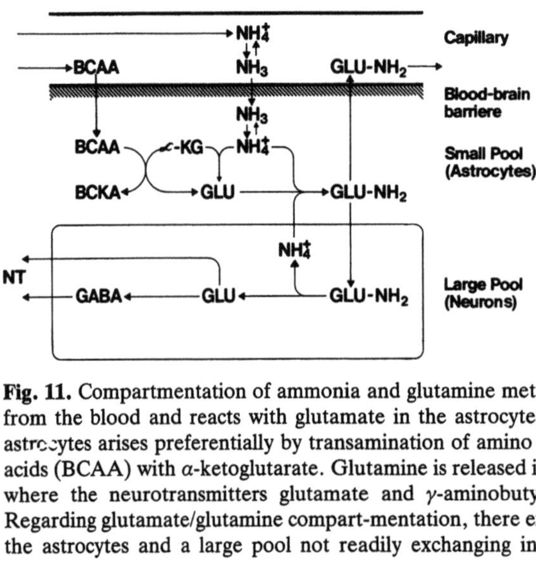

Fig. 11. Compartmentation of ammonia and glutamine metabolism in brain. Ammonia is taken up from the blood and reacts with glutamate in the astrocytes to form glutamine. Glutamate in the astrocytes arises preferentially by transamination of amino acids, especially branched-chain amino acids (BCAA) with α-ketoglutarate. Glutamine is released in the blood or diffuses into the neurons where the neurotransmitters glutamate and γ-aminobutyric acid are formed from glutamine. Regarding glutamate/glutamine compart-mentation, there exists a small metabolically active pool in the astrocytes and a large pool not readily exchanging in the neurons

Two compartments are the site of ammonia-glutamate-glutamine metabolism in brain as Berl and Clarke [3] have shown (Fig. 11): astrocytes and neurons. By the astrocytes, which have a tight connection to the blood-brain barrier, ammonia is taken up from the blood and incorporated rapidly into glutamine. Astrocytes contain most, if not all, of the brain glutamine synthetase activity [38, 41]. Glutamate is formed within the astrocytes by transamination of α-ketoglutarate with amino acids that are taken up from blood [5].

The neurons contain the large pool of glutamate that did not exchange rapidly with the glutamate in astrocytes. Glutamate in the neurons is formed from glutamine with a brain-specific glutaminase and is released as neurotransmitter or serves as precursor for formation of γ-aminobutyric acid.

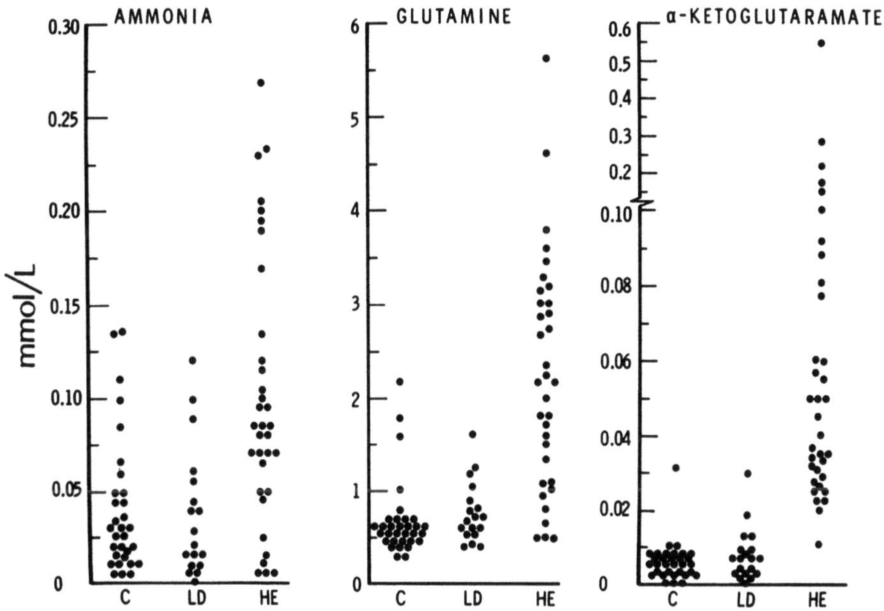

Fig. 12. Concentrations of ammonia, glutamine, and α-ketoglutaramate in human CSF. *C* control with normal liver function; *LD* liver disease without encephalopathy; *HE* liver disease with hepatic encephalopathy [11]

In patients with hepatic encephalopathy, the concentrations of ammonia, glutamine, and α-ketoglutaramate in CSF are elevated [11]. Glutamine and α-ketoglutaramate in CSF are presumably the spillover of these metabolites from the brain. Figure 12 shows that the glutamine concentration in CSF is about 30 to 50 times that of α-ketoglutaramate.

The mechanism by which ammonia causes hepatic encephalopathy is unknown. Three hypotheses are in discussion.

An early hypothesis suggested that amination of α-ketoglutarate by ammonia depletes this intermediate of the tricarboxylic acid cycle and, thus, inhibits oxidative energy production [4]. However, in animals with acute ammonia intoxication the concentration of α-ketoglutarate in brain is not decreased [11].

A second hypothesis deals with α-ketoglutaramate. Its concentration in CSF may be increased 10- to 50-fold in patients with hepatic encephalopathy [12]. Infusion of this metabolite into the cerebral ventricles induces a disturbance of cerebral function and behaviour of rats, but the concentrations usually found in CSF of patients with hepatic encephalopathy are exceedingly smaller. Therefore, it seems unlikely that the compound is a significant factor in the pathogenesis of hepatic encephalopathy [12].

According to a third hypothesis [21], ammonia may interfere with the malate-aspartate shuttle by promoting the synthesis of glutamine from cytoplasmic glutamate, thereby decreasing the pool of glutamate available for

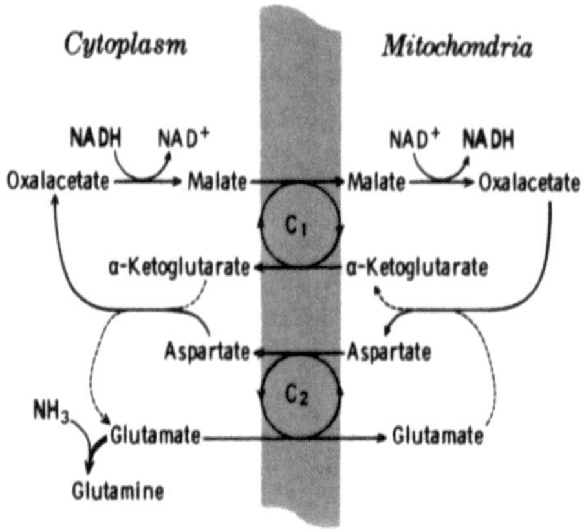

Fig. 13. Interference by ammonia in the transport of reducing equivalents from cytoplasm to mitochondria via the malate-aspartate shuttle. C_1 and C_2 are membrane carriers for malate-α-ketoglutarate and glutamate-aspartate transport, respectively. Ammonia may interfere with the shuttle by promoting the synthesis of glutamine from cytoplasmic glutamate, thereby decreasing the pool of glutamate available for exchange with aspartate [21]

exchange with aspartate (Fig. 13). In agreement with this hypothesis is the fact that in animals with portocaval shunt and with ammonia challenge the lactate/pyruvate ratio shifts towards a more reduced state and the concentrations of ATP and phosphocreatine decrease [21]. Furthermore, if the synthesis of glutamine is blocked by methionine sulfoximine, animals survive a lethal dose of ammonium salts [20, 55].

Mercaptans

Mercaptans (methyl mercaptan, ethyl mercaptan, dimethyl sulfide) are neurotoxins. Methyl mercaptan, the most toxic, arises in the metabolism of methionine, especially through methionine degradation by enteral bacteria, and is normally detectable in the breath in minimal amounts. Ingestion of excess methionine by the cirrhotic produces hepatic encephalopathy. Furthermore, patients with cirrhosis and hepatic encephalopathy excrete increased quantities of the three mercaptans in the breath and urine. The plasma concentration of mercaptans is increased. It is generally assumed that fetor hepaticus is due to a mixture of the three mercaptans in the breath. Finally, in rats with fulminant hepatic necrosis a fivefold increase of methyl mercaptan in the brain was demonstrated with simultaneously augmented plasma concentration, which is comparable to those seen in hepatic encephalopathy in man [39].

Objections against a pathogenetic role of mercaptans are based on the hyperkinetic convulsive state, induced in experimental mercaptan intoxication, in contrast to the hypokinetic state in hepatic encephalopathy. Furthermore, there is no correlation between blood levels of mercaptans and the degree of encephalopathy [45].

The mechanism of toxicity is still unknown. Methyl mercaptan inhibits the Na^+, K^+- activated ATPase in brain [43], but only in much higher concentrations than those normally seen in severe liver cirrhosis with hepatic encephalopathy.

Short/Medium Chain Fatty Acids

These substances show increased plasma concentrations in acute liver failure and cirrhosis with hepatic encephalopathy. They are formed by enteral microorganisms from intestinal lipids and by an incomplete degradation of long-chain fatty acids in severe liver disease [44]. Coma and alterations of EEG resembling those of hepatic encephalopathy in man are induced by the infusion of short-chain fatty acids in experimental animals [52].

Objections against the significance of short/medium-chain fatty acids are the missing correlation between their increase in blood plasma and the degree of encephalopathy [40]. The plasma concentrations of short-chain fatty acids inducing coma in animals are much higher than those in the plasma of patients with hepatic encephalopathy [54].

The mechanisms by which the short-chain fatty acids exert their neurotoxic effect is poorly understood. They inhibit the Na-K-activated ATPase in biomembranes [8], thereby disturbing the nerve impulse propagation. Further, fatty acids can displace tryptophan from albumin, so mediating the cerebral disturbance through increased free tryptophan and tryptophan-derived serotonin. Finally, short-chain fatty acids can inhibit urea production and thereby potentiate the ammonia effects on brain metabolism [9].

Synergism

With combined administration of ammonia, fatty acids and mercaptans, a synergism in the induction of hepatic encephalopathy has been demonstrated [56]. Thus, the coma-producing dose of octanoic acid was decreased when subcoma doses of ammonia or methyl mercaptan were given simultaneously. In a similar way, the coma dose of ammonia was reduced by subcoma doses of fatty acids and mercaptans. In general, through synergistic action smaller amounts of each of these three neurotoxins might be necessary to produce hepatic encephalopathy than if given alone. Clinical observations are in line with this concept of synergism.

False Neurotransmitters

This hypothesis was firstly conceived by Fischer and Baldessarini in 1971 [14]. In severe chronic liver disease, an abnormal pattern of plasma amino acids is a nearly constant finding. This imbalance of amino acids is characterized by an increase of the aromatic amino acids – phenylalanine, tyrosine, and tryptophan – and a decrease of branched-chain amino acids – valine, leucine, and isoleucine.

Since these two groups of amino acids compete for entry across the blood-brain barrier using a common transport system, the concentration of aromatic amino acids in brain increases. The elevated concentration of tryptophan in brain induces serotonin synthesis. The concentrations of this

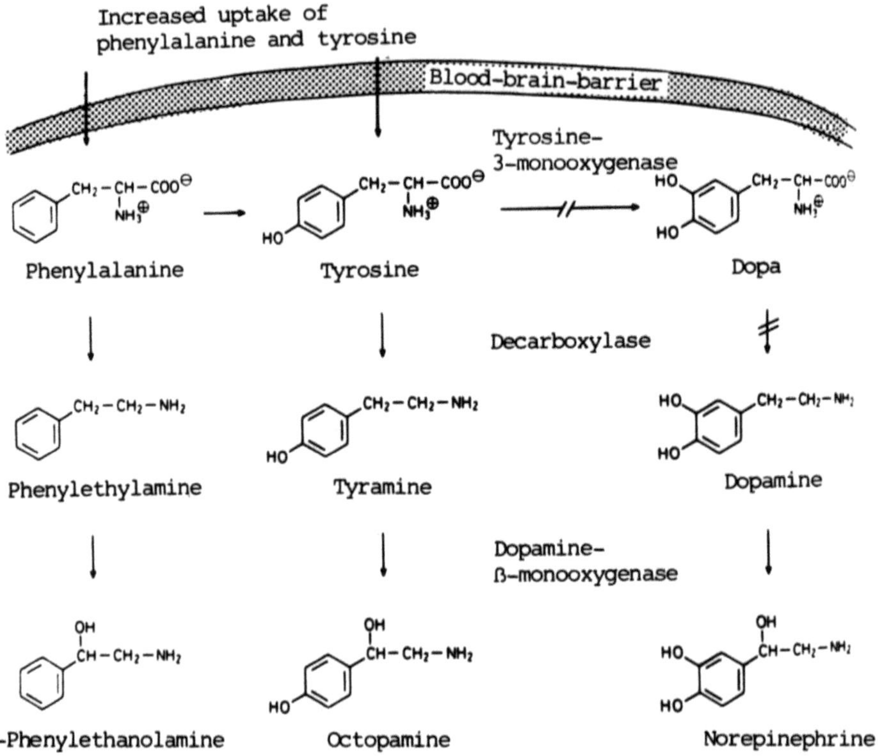

Fig. 14. Formation of false neurotransmitters (octopamine, β-phenylethanolamine) in the brain. According to the altered amino acid pattern in blood plasma, aromatic amino acids are preferentially transported through the blood-brain barrier by the common transport system for neutral amino acids. Phenylalanine and tryptophan in higher concentration inhibit the tyrosine-3-monooxygenase and thereby the synthesis of the normal neurotransmitters (dopamine and norepinephrine). The excess of phenylalanine and tyrosine is transformed to the false neurotransmitters (β-phenylethanolamine, or octopamine)

physiological neurotransmitter and its major metabolite 5-hydroxyindole acetic acid are increased in the brain of rats with portocaval shunts [7], as well as in the brain of patients with hepatic encephalopathy [24]. Phenylalanine with the highest concentration in brain inhibits the tyrosine-3-monooxygenase, and therefore, the hydroxylation of tyrosine to dopa. The biosynthesis of the normal neurotransmitters dopamine and norepinephrine is inhibited and the excess of phenylalanine and tyrosine is transformed to the false neurotransmitters β-phenylethanolamine and octopamine (Fig. 14). The sympathomimetic action of octopamine is less than 2% that of norepinephrine.

In agreement with this hypothesis, animals with experimental acute hepatic necrosis or with portocaval shunts show an increased uptake of aromatic amino acids by the brain. The concentrations of octopamine and β-phenylethanolamine are increased in brain and cerebrospinal fluid in such animals with liver failure [22]. Also patients with hepatic coma excrete octopamine in increased amounts

[37, 51]. There is a fairly good correlation between the serum octopamine concentration and the degree of neurological disturbance in such individuals [31]. Furthermore, it was demonstrated that the administration of L-dopa, which can enter the brain, may produce interruption or decline of the consciousness in patients with hepatic failure [34].

The most severe objection against this hypothesis is the result that the changes in neurotransmitter status could not cause hepatic encephalopathy in experimental animals. Decreasing dopamine and norepinephrine by more than 90% has no effect on brain function in normal rats. Also the intraventricular instillation of octopamine, raising the brain concentration of this false neurotransmitter more than 20,000-fold and decreasing norepinephrine levels by about 90% is without effect [57]. In postmortem tissues from patients with and without encephalopathy and control subjects without liver disease, the dopamine as well norepinephrine concentrations were not significantly different [6]. Finally, the beneficial therapeutic effect of L-dopa is probably unrelated to the cerebral concentrations of neurotransmitters. The protective action of L-dopa is caused by an increased renal excretion of ammonia preventing hyperammonemia. So the brain function under L-dopa therapy correlates with the brain ammonia, but not with dopamine concentration in experimental animals [57].

γ-Aminobutyric Acid

γ-Aminobutyric acid (GABA) is a potent cerebral inhibitory neurotransmitter. About 25%–45% of the nerve endings in human brain are sensitive to GABA. GABA is formed in the presynaptic neuron by decarboxylation of glutamate and released from storage sites. Its binding to receptors results in hyperpolarization and inhibition of the postsynaptic neuron. An efficient reuptake mechanism eliminates GABA quickly from the synaptic region; GABA is reincorporated into storage vesicles or catabolized, preferentially by GABA transaminase.

GABA is also a product of enteric microorganisms [49]. Normally, the resorbed GABA is inactivated by transamination in the liver. However, in liver cirrhosis the rate of GABA clearance from plasma is strikingly decreased with the consequence of increased plasma levels of GABA [13]. The augmented GABA may pass through a damaged blood-brain barrier and then exert inhibitory effects in the brain (Fig. 15). In addition, the number of low and high affinity receptors for GABA in the brain of animals with experimental liver injury is increased [48].

It is in line with this hypothesis that in acute experimental liver failure, caused by galactosamine, as well as in fulminant hepatitis and in chronic liver disease with hepatic encephalopathy, the plasma concentrations of GABA as measured with a radioreceptor assay are increased [49]. There is strong experimental evidence that this GABA is produced by enteral bacteria in the gut; the mean concentration of GABA in portal plasma was twice that in aortic plasma [49]. In normal liver GABA is degraded by GABA transaminase; the activity of this enzyme is high in the liver. In experimental hepatic failure a

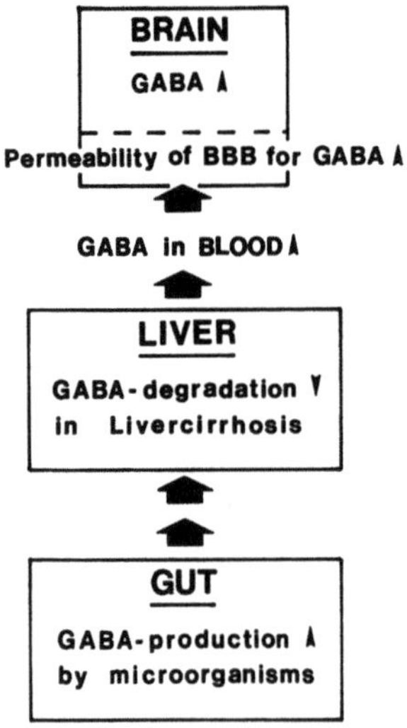

Fig. 15. Scheme of the GABA hypothesis

strong decrease of GABA clearance from blood plasma was demonstrated [13].

But a necessary prerequisite for this hypothesis is the permeability of the blood-brain barrier for GABA, which cannot enter the brain under normal conditions. With computerized imaging techniques and quantitative autoradiography, an increased permeability of the blood-brain barrier for α-aminoisobutyric acid in acute hepatic failure caused by galactosamine was demonstrated [48]. Schafer and Jones [48] assume that an analogous permeability for GABA, which does not normally cross the blood-brain barrier, may exist. But this conclusion is based on analogy, not on direct measurement. Further, most of the experiments supporting the GABA hypothesis are performed in experimental acute liver injury. Whether the alterations of GABA metabolism in chronic liver diseases with hepatic encephalopathy are comparable to those in acute experimental liver injury must be tested in further investigations. As yet an increased uptake of GABA by the brain by direct measurements in chronic liver disease with hepatic encephalopathy has not been conclusively documented.

Unified Ammonia-Neurotransmitter Hypothesis

In 1979 James et al. [23] suggested a hypothesis for the pathogenesis of hepatic encephalopathy which combines some facts of the ammonia and neurotransmitter hypothesis. In rats with portocaval shunt and hyperammonemia, a change in the blood-brain barrier was demonstrated, consisting of a selective increased uptake of neutral amino acids and decreased uptake of dibasic amino acids into the brain. Hyperammonemia increases the influx of ammonia into the brain where it is detoxified in astrocytes by formation of glutamine. Glutamine readily exchanges with the neutral amino acids phenylalanine, tyrosine, and methionine in the plasma, enhancing the concentration of these amino acids in the brain. Thus, hyperammonemia contributes indirectly to the accumulation of aromatic amino acids in the brain. The altered plasma amino acid pattern in chronic liver diseases with an increase of the aromatic amino acids and decrease of branched-chain amino acids may favor the augmented uptake of the aromatic amino acids into the brain through the changed blood-brain barrier. As a consequence, the synthesis of false neurotransmitters can occur. Also amines, such as GABA, may enter the brain through the disturbed blood-brain barrier.

According to this hypothesis, hepatic encephalopathy is not the result of a single metabolite, but arises through the combination of hyperammonemia with a disturbance of the blood-brain barrier, altered amino acid pattern in plasma, detoxication of ammonia in the brain by formation of glutamine, exchange of neutral amino acids for brain glutamine, and increased intracerebral formation of false neurotransmitters or uptake of neurotransmitter substances (GABA) from the blood through the disturbed blood-brain barrier.

Concluding Remarks

Glutamine metabolism in the brain may have significance in the pathogenesis of hepatic encephalopathy. Glutamine binds ammonia for detoxication. But this effect may be combined with depletion of glutamic acid as a neurotransmitter, as an important component of the malate-aspartate shuttle, and as a precursor for GABA. Glutamine efflux from the brain increases brain uptake of aromatic amino acids and contributes to excess production of false neurotransmitters.

References

1. Ansley JD, Isaaks JW, Rikkers LF, Kutner MH, Nordlinger BM, Rudman D (1978) Quantitative tests of nitrogen metabolism in cirrhosis: relation to other manifestation of liver disease. Gastroenterology 75:570–579
2. Atkinson DE, Camien MN (1982) The role of urea synthesis in the removal of metabolic bicarbonate and the regulation of blood pH. Curr Top Cell Regul 21:261–302
3. Berl S, Clarke DD (1969) Compartmentation of amino acid metabolism. In: Lajtha A (ed) Handbook of neurochemistry, vol II. Plenum Press, New York, pp 447–472
4. Bessman SP, Bessman AN (1955) The cerebral and peripheral uptake of ammonia in liver disease with an hypothesis for the mechanism of hepatic coma. J Clin Invest 34:622–628
5. Cooper AJL, McDonald JM, Gelbard AS, Gledhill RF, Duffy TE (1979) The metabolic fate of ^{13}N-labeled ammonia in rat brain. J Biol Chem 254:4982–4992
6. Cuilleret G, Pomier-Layrargues G, Pons F, Cadilhac J, Michel H (1980) Changes in brain catecholamine levels in human cirrhotic hepatic encephalopathy. Gut 21:565–569
7. Cummings MG, Soeters PB, James JH, Keane JM, Fischer JE (1976) Regional brain indolamine metabolism following chronic portocaval anastomosis in the rat. J Neurochem 27:501–509
8. Dahl DR (1968) Short chain fatty acid inhibition of rat brain Na-K-adenosine triphosphatase. J Neurochem 15:815–820
9. Derr RF, Zieve L (1976) Effect of fatty acids on the disposition of ammonia. J Pharmacol Exp Ther 197:675–680

10. Deuel TF, Louie M, Lerner A (1978) Glutamine synthetase from rat liver. Purification, properties and preparation of specific antisera. J Biol Chem 253: 6111–6118
11. Duffy TE, Plum F (1982) Hepatic encephalopathy. In: Arias I, Popper H, Schachter D, Shafritz DA (eds) The liver: Biology and pathobiology. Raven Press, New York, pp 693–715
12. Duffy TE, Vergara F, Plum F (1974) α-Ketoglutaramate in hepatic encephalopathy. Res Publ Assoc Nerv Ment Dis 53: 39–51
13. Ferenci P, Schafer DF, Shrager R, Jones EA (1981) Metabolism of the inhibitory neurotransmitter γ-amino-butyric acid in a rabbit model of acute hepatic failure. Hepatology 1: 509
14. Fischer JE, Baldessarini RJ (1971) False neurotransmitters and hepatic failure. Lancet II: 75–80
15. Ganda OP, Ruderman NB (1976) Muscle nitrogen metabolism in chronic hepatic insufficiency. Metabolism 25: 427–435
16. Häussinger D (1983) Hepatocyte heterogeneity in glutamine and ammonia metabolism and the role of an intercellular glutamine cycle during ureogenesis in perfused rat liver. Eur J Biochem 133: 269–275
17. Häussinger D, Gerok W (1984) Hepatocyte heterogeneity in ammonia metabolism: Impairment of glutamine synthesis in CCl_4 induced liver cell necrosis with no effect on urea synthesis. Chem Biol Interact 48: 191–194
18. Häussinger D, Gerok W, Sies H (1983) Regulation of flux through glutaminase and glutamine synthetase in isolated perfused rat liver. Biochim Biophys Acta 775: 272–278
19. Häussinger D, Gerok W, Sies H (1984) Hepatic role in pH regulation. Role of the intercellular glutamine cycle. TIBS (in press)
20. Hindtfelt B, Plum F (1975) L-Methionine-DL-sulphoximine and acute ammonia toxicity. J Pharm Pharmacol 27: 456–458
21. Hindtfelt B, Plum F, Duffy TE (1977) Effect of acute ammonia intoxication on cerebral metabolism in rats with portocaval shunts. J Clin Invest 59: 386–396
22. James JH, Hodgman JM, Funovics JM, Fischer JE (1976) Alterations in brain octopamine and brain tyrosine following portocaval anastomosis in rats. J Neurochem 27: 223–227
23. James JH, Ziparo V, Jeppson B, Fischer JE (1979) Hyperammonaemia, plasma amino acid imbalance, and blood-brain amino acid transport: A unified theory of portal systemic encephalopathy. Lancet II: 772–775
24. Jellinger K, Riederer P (1977) Brain monoamines in metabolic endotoxic coma: A preliminary biochemical study in human postmortem material. J Neural Transm 41: 275–286
25. Imler M, Schlienger JL, Frick A (1978) Comparative study of ammonia and glutamine levels in blood and in cerebrospinal fluid in hepatic encephalopathy. In: Wewalka F, Dragosics B (eds) Aminosäuren, Ammoniak und hepatische Encephalopathie. G Fischer, Stuttgart, pp 105–117
26. Jones EA, Cain GD, Dickinson G (1972) Corticosteroid-induced changes in urea metabolism in patients with hepatocellular disease. Gastroenterology 62: 612–617
27. Iqbai K, Ottoway JH (1970) Glutamine synthetase in muscle and kidney. Biochem J 119: 145–156
28. Klassen GA, Aronoff A, Karpati G (1969) Forearm metabolism in patients with chronic liver disease. Clin Sci 37: 455–470
29. Krebs HA (1935) Metabolism of amino acids IV. The synthesis of glutamine from glutamic acid and ammonia, and the enzymatic hydrolysis of glutamine in animal tissues. Biochem J 29: 1951–1969
30. Krebs HA, Henseleit K (1932) Untersuchungen über die Harnstoffbildung im Tierkörper. Hoppe-Seyler's Z Physiol Chem 210: 33–71
31. Lam KC, Tall AR, Goldstein GB, Mistilis SP (1973) Role of false neurotransmitter, octopamine, in the pathogenesis of hepatic and renal encephalopathy. Scand J Gastroenterol 8: 465–472
32. Lockwood AH, McDonald JM, Reiman RE, Gelbard AS, Laughlin JS, Duffy TE, Plum F (1979) The dynamics of ammonia metabolism in man. Effects of liver disease and hyperammonemia. J Clin Invest 63: 449–460
33. Lowenstein JM (1972) Ammonia production in muscle and other tissues: The purine nucleotide cycle. Physiol Rev 52: 382–414

34. Lunzer M, James IM, Weinman J, Sherlock S (1974) Treatment of chronic hepatic encephalopathy with levodopa. Gut 15: 555–561
35. Lusty CJ (1978) Carbamylphosphate synthetase I of rat-liver mitochondria. Purification, properties and polypeptidemolecular weight. Eur J Biochem 85: 373–383
36. Maier KP, Talke H, Gerok W (1978) Harnstoffzyklusenzyme und Harnstoffsynthese bei chronischen Lebererkrankungen. In: Wewalka F, Dragosics B (eds) Aminosäuren, Ammoniak und hepatische Encephalopathie. G. Fischer, Stuttgart, pp 33–38
37. Manghani KK, Lunzer MR, Billing BH, Sherlock S (1975) Urinary and serum octopamine in patients with portalsystemic encephalopathy. Lancet II: 943–946
38. Martinez-Hernandez A, Bell KP, Norenberg MD (1976) Glutamine synthetase: Glial localization in brain. Science 195: 1356–1358
39. McClain CJ, Zieve L, Doizaki W, Gilberstadt S, Onstad G (1978) Mercaptans in portal systemic encephalopathy due to alcoholic liver disease. Gastroenterology 74: 1065
40. Moryan MH, Bolton CH, Morris JS, Read AE (1974) Medium chain triglycerides and hepatic encephalopathy. Gut 15: 180–184
41. Norenberg MD, Martinez-Hernandez A (1979) Fine structural localization of glutamine synthetase in astrocytes of rat brain. Brain Res 161: 303–310
42. Oliver J, Koelz AM, Costello J, Bourke E (1977) Acid-base induced alterations in glutamine metabolism and ureogenesis in perfused muscle and liver of the rat. Eur J Clin Invest 7: 445–449
43. Quarfoth G, Ahmed K, Foster D, Zieve L (1976) Action of methanethiol on membrane Na^+K^+-ATPase of rat brain. Biochem Pharmacol 25: 1039–1044
44. Rabinowitz JL, Staeffen J, Blanquet P, Vincent JD, Terme R, Series C, Meyerson RM (1978) Sources of serum ^{14}C-octanoate in cirrhosis of the liver and hepatic encephalopathy. J Lab Clin Med 91: 223–227
45. Record CO, Al Mardini H, Bartlett K (1982) Blood and brain mercaptan concentrations in hepatic encephalopathy. Hepatology 2: 144
46. Rudman D, Difulco JTh, Galambos JT, Smith R, Salam A, Warren WD (1973) Maximal rates of excretion and synthesis of urea in normal and cirrhotic subjects. J Clin Invest 52: 2241–2249
47. Rypins EB, Henderson JM (1980) A tracer method for measuring rate of urea synthesis in normal and cirrhotic subjects. Gastroenterology 78: 1419–1424
48. Schafer DF, Jones EA (1982) Potential neural mechanisms in the pathogenesis of hepatic encephalopathy. In: Popper H, Schaffner F (eds) Progress in liver diseases, vol VII. Grune & Stratton, New York, pp 615–627
49. Schafer DF, Fowler JM, Jones EA (1981) Colonic bacteria: A source of γ-aminobutyric acid in blood. Proc Soc Exp Biol Med 167: 301–303
50. Schafer DF, Ferenci P, Kleinberger G, Hoofnagle JH, Jones EA (1981) Elevated serum concentrations of the inhibitory neurotransmitter γ-aminobutyric acid in patients with hepatocellular disease. Hepatology 1: 543
51. Smith AR, Rossi-Fanelli F, Ziparo V, James JH, Perelle BA, Fischer JE (1978) Alterations in plasma and CSF amino acids, amines and metabolites in hepatic coma. An Surg 187: 343–350
52. Teychenne PF, Walters I, Claveria LE, Calne DB, Price J, MacGillivary BB, Gompertz D (1976) The encephalopathic action of five-carbon-atom fatty acids in the rabbit. Clin Sci Mol Med 50: 463–472
53. Vilstrup H (1980) Synthesis of urea after stimulation with amino acids: relation to liver function. Gut 21: 990–995
54. Walker CO, McCandless DW, McGarry JD, Schenker S (1970) Cerebral energy metabolism in short-chain fatty acid induced coma. J Lab Clin Med 76: 569–583
55. Warren KS, Schenker S (1964) Effect of an inhibitor of glutamine synthesis (methionine sulfoximine) on ammonia toxicity and metabolism. J Lab Clin Med 64: 442–449
56. Zieve L, Nicoloff DM (1975) Pathogenesis of hepatic coma. Ann Rev Med 26: 143–157
57. Zieve L, Olsen RL (1977) Can hepatic coma be caused by a reduction of brain noradrenaline or dopamine? Gut 18: 688–691
58. Zieve L, Doizaki WM, Derr RF (1979) Reversal of ammonia coma in rats by L-dopa: A peripheral effect. Gut 20: 28–32

Molecular Targets of Anti-Glutamine Therapy with Acivicin in Cancer Cells

G. WEBER, M. S. LUI, J. SEBOLDT, and M. A. FADERAN[1]

Systematic studies in this laboratory showed that there was an increase in the activities of enzymes of glutamine utilization of pyrimidine and purine biosynthesis in rat hepatomas of different growth rates and in other animal and human neoplasms. We also observed that there were increased concentrations of strategic nucleotides and deoxynucleotides in hepatomas and these elevations were linked with neoplastic transformation and progression (for overview see Weber [13]). Because of this increase in enzymic activities and nucleotide pools relating to glutamine utilization, an investigation was carried out to understand the impact of the anti-glutamine agent, acivicin, on the molecular targets, namely, key enzymic activities and concentrations of strategic nucleotides. These studies revealed a profound influence of the anti-glutamine agent on the enzymology and metabolism and survival of cancer cells [1, 3, 15].

Materials and Methods

The properties of the biological model systems, including the animal and human tumors and the hepatoma 3924A tissue culture cell line, were described [12, 13]. The methods for determination of the survival of tumor cells in tissue culture (clonogenic assays) and the transplanted hepatoma (tumor volumes) were outlined [13, 15]. The methods for determination of enzymic activities, pools of nucleotides, and concentration of L-glutamine were reported elsewhere [13, 15]. The expression of the biochemical results and the anti-tumor action and statistical evaluation were as outlined [13, 15].

Results and Discussion

Our investigations revealed the presence of a profound imbalance in glutamine utilization in hepatomas of different growth rates and in the concentrations of strategic nucleotides. Evidence will now be presented documenting the presence of this imbalance, since the activities of the key enzymes and the pools of nucleotides provide molecular targets for the design of anti-cancer chemotherapy and they also yield indicators for the quantitative measure of the success of chemotherapy at the biochemical and biological level.

1 Laboratory for Experimental Oncology, Indiana University School of Medicine, Riley Cancer Wing 337, 702 Barnhill Drive, Indianapolis, IN 46223, USA

Glutamine Metabolism in Mammalian Tissues
Edited by D. Häussinger and H. Sies

Table 1. Increased specific activities of glutamine-utilizing enzymes in tumors

Tumors	Increased activities (% of control normal tissue values)[a]			
	Amido-trans-ferase	GMP synthetase	Carbamoyl-phosphate synthetase II	CTP synthetase
Rat hepatoma 3683F	212	548	946	1,061
Rat renal cell carcinoma MK-3	220	297	495	430
Rat sarcoma	1,345	218	1,805	7,768
Human renal cell carcinoma	158	167	180	181
Human colon tumor		183	329	314

[a] All activities were significantly increased as compared to those in the relevant normal control tissues ($p < 0.05$)

Evidence for the Role of Glutamine-Utilizing Enzymes in Neoplastic Transformation and Progression

Our studies showed that the specific activities of amidophosphoribosyltransferase (EC 2.4.2.14), GMP synthetase (EC 6.3.4.1), carbamoyl-phosphate synthetase II (EC 2.7.2.9), and CTP synthetase (EC 6.3.4.2) were increased in various neoplasms (Table 1). Investigations in the spectrum of hepatomas of different growth rates indicated that the behavior of amidophosphoribosyltransferase activity was transformation-linked (activity increased in all the hepatomas), whereas that of the other enzymes was transformation- and progression-linked (activities increased in all hepatomas and correlated positively with tumor growth rate) [13].

Evidence for the Role of Nucleotides in Neoplastic Transformation and Progression

Our investigations showed that in the rapidly growing rat hepatoma 3924A the concentrations of CTP, the product of CTP synthetase, and GMP, a product of GMP synthetase, were elevated. The pools of dGTP, dCTP, dATP, and dTTP, products of ribonucleotide reductase, were markedly increased. The expansion in these nucleotide pools was linked with transformation and progression (Table 2) [5, 13].

Evidence for the Role of L-Glutamine in Neoplastic Transformation and Progression

Measurement of the L-glutamine content in hepatomas of different growth rates showed that the concentration of this amino acid in the slowly growing tumors

Table 2. Concentrations of nucleotides in hepatoma 3924A

Nucleotides	Hepatoma 3924A % of liver[a]
CTP	406
GMP	325
dCTP	793
dGTP	482
dTTP	1,214
dATP	1,748

[a] All values were significantly different from normal liver (≡ 100%) ($p < 0.05$)

Table 3. Glutamine content in normal liver and hepatomas of different growth rates[a]

Tissues	Growth rate	Weeks	Glutamine	
			μmol/g	% of liver
Normal liver				
ACI/N strain			5.7 ± 0.2	100
Buffalo strain			6.1 ± 0.5	100
Hepatomas				
20	Slow	66	7.7 ± 0.7	126
9618A	Slow	52	8.4 ± 0.4	138[b]
16	Slow	41	4.3 ± 0.6	70[b]
44	Medium	19	3.5 ± 0.6	58[b]
7800	Medium	6.4	1.9 ± 0.1	31[b]
5123t.c.	Medium	5.6	2.4 ± 0.1	39[b]
3924A	Rapid	4.3	0.7 ± 0.1	12[b]
7777	Rapid	3.9	0.7 ± 0.1	11[b]

[a] Mean ± SE of three or more rats. Buffalo livers are controls for all hepatomas except 3924A, which is carried in ACI/N rats

[b] Significantly different from control liver ($p < 0.05$)

was in the range of the normal liver, it decreased in tumors of medium growth rate, and in the rapidly growing hepatoma 3924A it was 12% of the normal liver (Table 3). Thus, the glutamine concentration negatively correlated with the tumor growth rate. In the rapidly growing hepatomas the low glutamine concentration was attributed, in part at least, to the markedly increased activities of the glutamine-utilizing enzymes of purine and pyrimidine biosynthesis and to the decrease in the activity of glutamine synthetase and the increase in the activity of glutaminase (Fig. 1) [15].

The Molecular Correlation Concept and the Design of Enzyme-Pattern-Targeted Chemotherapy

The increased activities of enzymes of glutamine utilization and the increased concentrations of strategic nucleotides should confer selective advantages to

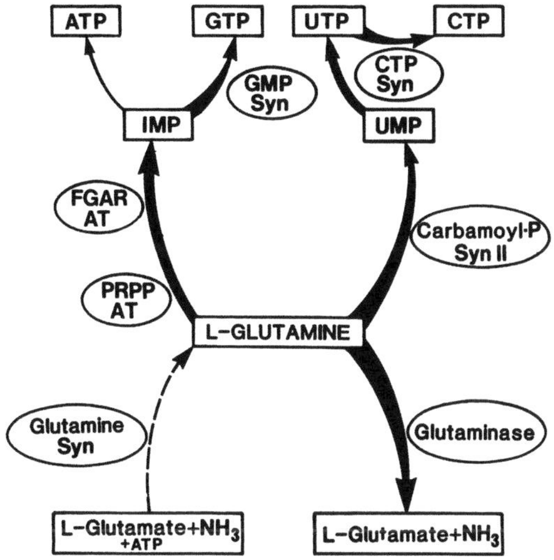

Fig. 1. Enzymic imbalance in glutamine metabolism in rapidly growing hepatoma 3924A. *GMP syn* GMP synthetase; *CTP syn* CTP synthetase; *carbamoyl-P Syn II* carbamoyl-phosphate synthetase II; *FGAR AT* phosphoribosylformylglycinamidine synthetase; *PRPP AT* amidophosphoribosyltransferase (amidotransferase); *glutamine syn* glutamine synthetase

cancer cells. The molecular correlation concept suggested that identification of the selective advantages conferred by enzymic and metabolic imbalance to cancer cells should pinpoint potentially sensitive targets for the design of chemotherapy [12, 13]. There should be particular interest in those enzymic activities and nucleotide concentrations which increased in parallel with tumor growth rate because they should be intimately linked with the biochemical expression of the neoplastic program for replication. On the basis of this reasoning, the primary molecular targets for anti-glutamine action should be the activities of GMP and CTP synthetases because of their immediate involvement in production of the elevated pools of GMP and CTP. There is also interest in the activities of other enzymes of glutamine utilization in purine and pyrimidine biosynthesis. In measuring the impact of in vivo inhibition of the key enzymes of glutamine utilization, targets and indicators are the pools of rNTP and dNTP, with primary emphasis on guanylates and cytidylates.

In the selection of the biological models, priority was accorded to the solid and tissue culture lines of rapidly growing hepatoma 3924A because the low L-glutamine content in these cells should make them more sensitive to the action of anti-glutamine drugs, whereas a control normal organ such as liver should be selectively less sensitive because of its nearly tenfold higher glutamine concentration. The survival of tissue culture cells and the change in tumor volumes in animals inoculated with solid hepatoma 3924A should indicate the extent of anti-cancer response of a drug and should provide a comparison with the biochemical impact of the anti-cancer agent.

The Anti-Glutamine Drug, Acivicin: Biochemical and Cytogenetic Action and Anti-Cancer Spectrum

For these studies, we selected the anti-glutamine agent, acivicin (NSC-163501, AT-125), L-(alphaS,5S)-alpha-amino-3-chloro-4,5-dihydro-5-isoxazoleacetic

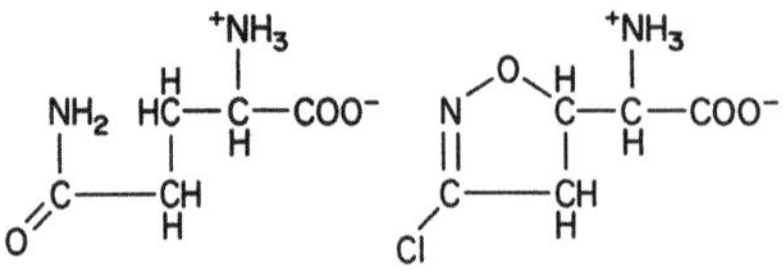

Fig. 2. Comparison of L-glutamine and acivicin

acid (Fig. 2). This antibiotic was effective against L1210 and P388 leukemias in mouse and in human lung and breast carcinoma xenografts in athymic mice (for review of literature, see Neil et al. [8]). In cultured L1210 cells Neil et al. showed that acivicin arrested cell cycle progression in the early S-phase and that L-glutamine protected these cells [6, 8]. Cooney et al. [2] and Jayaram et al. [6] demonstrated that in extracts from bacteria and fetal rat liver, acivicin competitively inhibited the activities of various enzymes that catalyzed a transfer of the amido group of L-glutamine and noted that the inhibition was not reversed by extensive dialysis [2]. Acivicin was also a competitive inhibitor of purified rat liver CTP synthetase [9].

The first systematic study of the in vivo and in vitro biochemical pharmacology of acivicin in hepatoma 3924A and in normal liver was carried out in this laboratory with emphasis on determining the impact on activities of key enzymes of pyrimidine and purine biosynthesis and on the pools of rNTP and dNTP [1, 3, 7, 13, 15, 17]. These investigations were extended to the in vivo action of acivicin in rat sarcoma and rat brain and to in vitro studies in human primary colon carcinoma.

Evidence for the Cytotoxic Action of Acivicin in Hepatoma 3924A Cells in Tissue Culture

Dose-response studies were carried out in logarithmically growing hepatoma 3924A cells and colonies were stained and counted after 7 days. The LC_{50} for acivicin was 1.4 μM and survival dropped sharply at higher concentrations of the drug. Addition of pyrimidine and purine nucleosides, cytidine, deoxycytidine, and guanosine together, completely protected the cells against the action of acivicin, but singly they provided no protection [7, 17]. These results suggested that acivicin interfered with both cytidylate and guanylate synthesis; this was confirmed by showing that the activities of the glutamine-utilizing enzymes, GMP and CTP synthetases, were inhibited and both CTP and GTP pools were depressed [13]. The demonstration of the anti-cancer action of acivicin against hepatoma cells is significant because of the notorious insensitivity of hepatocellular carcinomas to chemotherapy. These results indicated that an in-depth examination of acivicin action should provide useful information for the rational design of selective chemotherapy.

Acivicin Action in Vitro: Competitive Inhibition and Inactivation of Glutamine-Utilizing Enzymes

In vitro addition of acivicin to crude or purified extracts of liver and hepatoma 3924A resulted in inhibition of amidophosphoribosyltransferase, GMP synthe-

Table 4. Inhibition by acivicin of activities of glutamine-utilizing enzymes in hepatoma 3924A

Enzymes	Extract	Glutamine K_m (mM)	Glutamine concentration in standard assay (mM)
Amidotransferase	Crude supernatant	2.3	20
GMP synthetase	Crude supernatant	0.2	2
CTP synthetase	Purified preparation	0.5	10
Carbamoyl-phosphate syn. II	Purified preparation	0.017	1

Enzymes	Acivicin K_i (mM)	Type of inhibition	K_{inact} (μM)	T^a (min)
Amidotransferase	5.000	Competitive	–	–
GMP synthetase	0.005	Mixed	241	0.5
CTP synthetase	0.001	Competitive	250	2.5
Carbamoyl-phosphate syn. II	0.007	Competitive	90	0.7

[a] Minimum inactivation half time

tase, carbamoyl-phosphate synthetase II, and CTP synthetase activities with a K_i = 5,000, 5, 7, and 1μM drug concentration (Table 4). The inhibition mechanism was in line with the postulated role of acivicin as a competitive antagonist for the glutamine-utilizing enzymes in rat hepatoma. However, our studies showed that there was another mechanism which was more important for the biological action and this was the rapid inactivation of the enzymes. These investigations were the first to show in mammalian cells the time- and dose-dependent inactivation of purified carbamoyl-phosphate synthetase II prepared from hepatoma 3924A [1]. The time course of inactivation and the plot to determine the minimum inactivation time are shown in Fig. 3. Only glutamine and glycine provided protection from inactivation of the enzyme by acivicin. The activities of CTP synthetase, GMP synthetase, and amidophosphoribosyltransferase tested in extracts from hepatoma 3924A, normal rat liver and brain, and from human colon carcinoma were also inactivated in a dose-, time-, and temperature-dependent fashion.

Acivicin Action in Vivo: Impact on Activities of Glutamine-Utilizing Enzymes and on rNTP and dNTP Concentrations in Hepatoma 3924A

Half-Life of Acivicin in Rat Blood. Rats were injected with acivicin (25 mg/kg, i.p.) and were killed at intervals after treatment. Acivicin blood concentration reached a peak at 10, 30, and 60 min after injection and then rapidly declined with a half-life of 1.5 to 2 h. No acivicin was detectable in the blood 12 h after injection [15].

Effect of Acivicin Injection on Hepatoma Enzymes. Acivicin was injected into rats carrying hepatoma 3924A on both flanks in subcutaneous position. At the

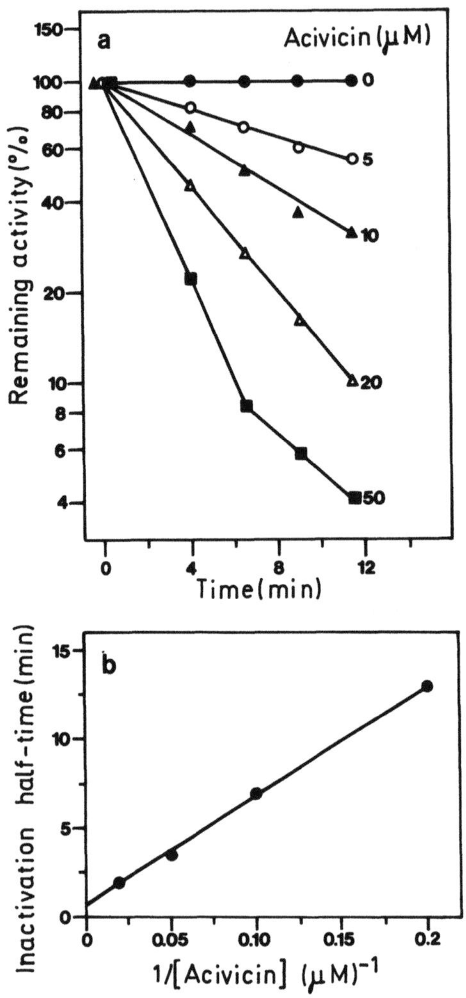

Fig. 3. Effect of acivicin concentration on inactivation of carbamoyl-phosphate synthetase II. **a** shows the time course of inactivation by different concentrations of acivicin. **b** is the plot of inactivation half-time vs 1/[acivicin]

beginning of the experiment, tumors were 14 days old and approx. 1 to $1^1/_2$ cm in diameter. Groups of rats (8 in each) were killed after injection at 10 and 30 min and at 1, 2, 6, 12, 24, and 72 h. Half of the animals were anesthetized and tumors and host liver were freeze-clamped for the nucleotide assays. The other rats were killed by decapitation and the tumors and livers were used to prepare extracts for enzyme assays. Control animals received physiological saline and were killed at the same time periods as the acivicin-treated ones. Subsequent dose-response studies showed that 5 mg per kg acivicin was as effective as 25 mg. Studies on the effect of GMP synthetase activity were carried out with the 5 mg/kg dose.

Acivicin treatment caused a strikingly rapid decline in activities of the glutamine-utilizing enzymes with significant decreases detected at 10 min after injection. By 6 h, amidotransferase, CTP synthetase, carbamoyl-phosphate synthetase II, and GMP synthetase activities in the hepatomas decreased to 34%, 4%, 6%, and 9% of the controls. Enzymic activities bottomed out at these

Table 5. In vivo effect of acivicin on enzyme activities in hepatoma 3924A[a]

Time after acivicin injected	Amidophos-phoribosyl-transferase	GMP synthetase	CTP synthetase	Carbamoyl-phosphate synthetase II
Control	142 ± 4.4[b]	55.9 ± 1.6	56.5 ± 2.6	82.0 ± 8.6
	100	100	100	100
10 min	87[c]	75[c]	20[c]	72
0.5 h	56[c]	31[c]	7[c]	50[c]
1 h	32[c]	–	3[c]	13[c]
2 h	21[c]	–	3[c]	6[c]
6 h	34[c]	9[c]	4[c]	6[c]
12 h	35[c]	13[c]	8[c]	17[c]
24 h	46[c]	35[c]	23[c]	35[c]
48 h	74[c]	86	37[c]	66
72 h	99	87	60[c]	79

[a] Specific activities (nmol/h/mg protein) are expressed as percentage of 0.9% NaCl solution-injected controls. Acivicin dose, 25 mg/kg i.p.; dose for GMP synthetase studies, 5 mg/kg
[b] Mean ± SE of four or more rats in each group
[c] Significantly different from controls ($p < 0.05$)

low levels for 12 h after injection and then slowly started to return toward normal range by 72 h (Table 5). Activities in host liver also decreased, but in a less pronounced fashion than those in hepatoma and the pattern of return was similar to that in hepatoma [3, 13, 15].

In the enzyme extracts, the decreased enzymic activity was not reversible under in vitro conditions by various procedures, indicating that the mechanism of acivicin action was primarily an inactivation of these enzymes. On the basis of a report in bacterial systems on other glutamine-utilizing enzymes [11], we postulated that the rapid action of acivicin in vivo was due to alkylation of the active center with acivicin acting as a site-directed analog of L-glutamine [1, 13]. Therefore, the L-glutamine concentration in the tissues might be a major determining factor in acivicin toxicity; the high levels of glutamine in liver might explain the less pronounced action of the drug on the glutamine-utilizing enzymes in this tissue. The clinical relevance of this observation is that the glutamine concentration of the tumors might determine sensitivity or resistance to acivicin.

The in vivo selectivity of acivicin was shown by the fact that the ammonia-utilizing and the aspartate carbamoyltransferase activities of the enzyme complex were not inhibited by acivicin [1]. The in vivo restoration of enzymic activities might involve, in part at least, new enzyme synthesis.

Effect of Acivicin Injection on Hepatoma Ribonucleotide Pools. The ATP and UTP concentrations were unchanged. The GTP pool at 2 h after injection decreased to 56% with the lowest value of 32% being reached at 6 h; the pool was still significantly decreased at 24 h, returning to normal range by 48 and 72 h. The CTP concentration was more rapidly and more profoundly affected; at 30 min after injection, it decreased to 46%, and at 2 h reached its lowest level of

Table 6. Effect of acivicin on nucleoside triphosphate pools in hepatoma 3924A[a]

Time after acivicin injection	ATP	GTP	CTP	UTP
Control	793 ± 87[b]	204 ± 23	90 ± 8	190 ± 51
	100	100	100	100
10 min	103	93	120	127
0.5 h	85	78[c]	46[c]	123
1 h	70	69[c]	42[c]	104
2 h	70	32[c]	2[c]	103
6 h	77	32[c]	11[c]	122
12 h	64	42[c]	18[c]	106
24 h	65	57	19[c]	77
48 h	117	118	89	98
72 h	108	130	89	120

[a] The concentration of nucleotides is calculated in nmol/g wet wt, and also expressed as percentage of 0.9% NaCl solution-injected controls. The acivicin dose was 25 mg/kg i.p.
[b] Mean ± SE of four or more rats in each group
[c] Significantly different from controls ($p < 0.05$)

2%, the pools remaining at less than 20% for 24 h and returning to normal range by 48 and 72 h (Table 6). In the host liver, the pattern was similar, but the decrease in GTP and CTP pools occurred to a minor degree, corresponding to the lesser impact of acivicin on the enzymes of the liver [15]. The decline in the pools of GTP and CTP that occurred subsequent to the reduction in CTP and GTP synthetase activities is in line with the high sensitivity of these enzymes to acivicin.

Effect of Acivicin Injection on Hepatoma Deoxynucleotide Pools. In this experiment, an acivicin dose of 5 mg per kg i.p. was given and the animals were treaιed as outlined above. The results in Table 7 show that the most rapid and marked depletion occurred in the dTTP pools which rapidly decreased, reaching a low level of 18 and 9% of controls at 12 and 24 h after injection. The dTTP concentration remained at 40% of control, even after 72 h of treatment. The next most strongly affected was the dCTP pool which also reached the lowest concentrations of 22 and 18% at 12 and 24 h after injection, with 72 h values still only at 40% of controls. The dGTP pool decreased to 55 and 42% at 12 and 24 h and returned to normal at 72 h after injection. The dATP pool was significantly decreased only at 12 and 24 h, when concentrations were 6 and 3% of controls. At other time periods, the dATP pools were not affected (Table 7) [13].

The decrease in the deoxynucleotide concentrations is attributed to a decrease in the nucleoside diphosphate pools although these were not measured in these studies. The low specific acitivity of ribonucleotide reductase [10] compounded by the decrease in the concentrations of GDP and CDP due to acivicin action, might explain the decline in the pools of dGTP and dCTP. The reason for the rapid and profound depletion of dTTP pools is not clear, but it might be attributed, in part at least, to the decrease in the CDP pools. However,

Table 7. In vivo effect of acivicin on dNTP pools in hepatoma 3924A

Time after acivicin injection[a]	dNTP pools: % control			
	dATP	dGTP	dCTP	dTTP
Control (saline)	46 ± 8 100	12 ± 0.6 100	129 ± 11.6 100	11.7 ± 1 100
10 min	113	111	101	32[b]
30 min	95	98	43[b]	22[b]
6 h	83	64	60[b]	30[b]
12 h	6[b]	55[b]	22[b]	18[b]
24 h	3[b]	42[b]	18[b]	9[b]
48 h	80	78[b]	28[b]	28[b]
72 h	54	95	41[b]	39[b]

[a] Acivicin dose 5 mg/kg i.p., four rats/group
[b] Significantly different from control ($p < 0.05$)

since the decline in dTTP appears to precede that of the dCTP by 20 min, there may be other factors involved. It is also apparent that the thymidine kinase salvage activity could not restore dTTP concentrations.

Role of Salvage Enzymic Activities in Determining the Impact of Acivicin Action

It is noteworthy that although acivicin injection decreased the activities of amidophosphoribosyltransferase and GMP synthetase, the levels of GTP decreased, but those of ATP did not. In pyrimidine biosynthesis, the activities of both carbamoyl-phosphate synthetase II and CTP synthetase were inhibited, but this was manifested only in the decline of CTP pools with no change in UTP concentrations. The selective decrease in CTP and GTP pools and the protection of ATP and UTP pools were attributed to the contributions of the pyrimidine and purine salvage pathways (Fig. 4). This hypothesis was tested in the following experiments [13].

In hepatoma 3924A tissue culture studies, we showed that the salvage nucleosides, uridine, deoxycytidine, and guanine together protected the cancer cells from acivicin toxicity [7, 17]. Thus, apparently there were high activities present for the salvage enzymes, uridine and deoxycytidine kinases and guanine phosphoribosyltransferase.

Further investigations demonstrated that the activities of all purine and pyrimidine salvage enzymes were higher than those of the rate-limiting de novo enzymes in the solid hepatoma 3924A [14]. Consequently, inhibition of carbamoyl-phosphate synthetase II activity was counterbalanced by the high uridine kinase activity which at the prevailing tissue uridine level was sufficient to maintain the UTP pool in the acivicin-treated animals. Similarly, inhibition of amidophosphoribosyltransferase activity was counterbalanced by the high adenine phosphoribosyltransferase activity which was able to maintain the ATP

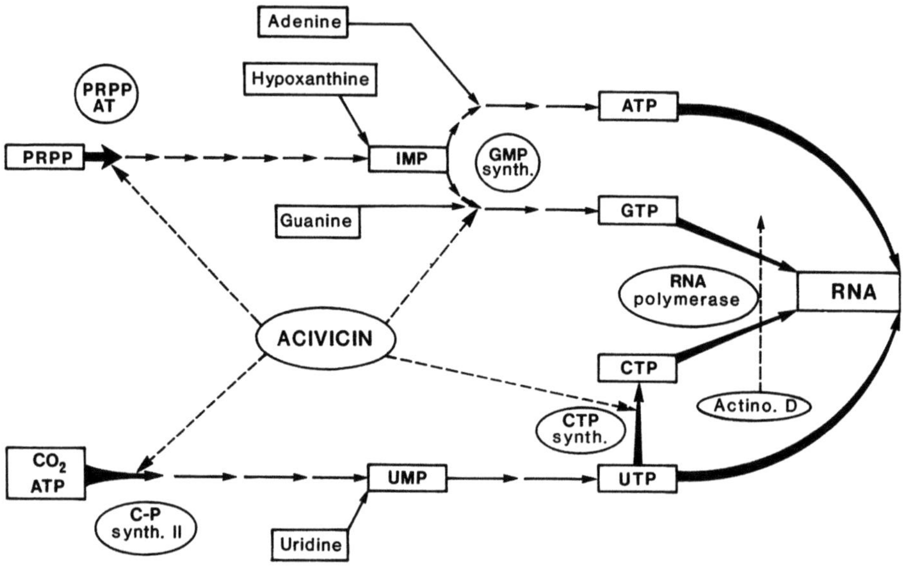

Fig. 4. Enzymic and nucleotide targets in the synergistic action of acivicin and actinomycin and the role of salvage pathways. *AT* amidotransferase; *GMP synth.* GMP synthetase; *CTP synth.* CTP synthetase; *Actino. D* actinomycin D; *C-P synth. II* carbamoyl-phosphate synthetase II

pool. The decline in the GTP pool was attributed to the inhibition of GMP synthetase activity which apparently could not be relieved by the guanine phosphoribosyltransferase activity, perhaps because of low concentration of tissue guanine or low affinity of the enzyme for guanine. The decrease in the CTP pool was an expression of the marked inhibition of CTP synthetase, which in the solid tumor apparently could not be overcome by the activities of cytidine and deoxycytidine kinases, possibly because of the low tissue concentrations of cytidine and deoxycytidine.

Combination Chemotherapy with Acivicin and Dipyridamole

To gain insight into the role of salvage nucleosides in overcoming the action of acivicin, we tested the action of dipyridamole, an inhibitor of nucleoside transport. Our investigations showed that dipyridamole in concentrations of 10^{-6}–10^{-7} *M* effectively blocked the incorporation of purine and pyrimidine nucleosides into macromolecules in hepatoma 3924A cells [17]. Dipyridamole by itself killed the hepatoma cells incubated for 7 days with an $LC_{50} = 20$ μ*M* as determined by clonogenic assays [17]. When acivicin or dipyridamole was incubated with hepatoma cells, survivals of 64 and 65%, respectively, were found. The combination of the two drugs yielded a synergistic kill with only 11% surviving. These results support our hypothesis that combination chemotherapy with an antimetabolite of the de novo pathways, such as acivicin, and an inhibitor of the salvage pathways, such as dipyridamole, should provide

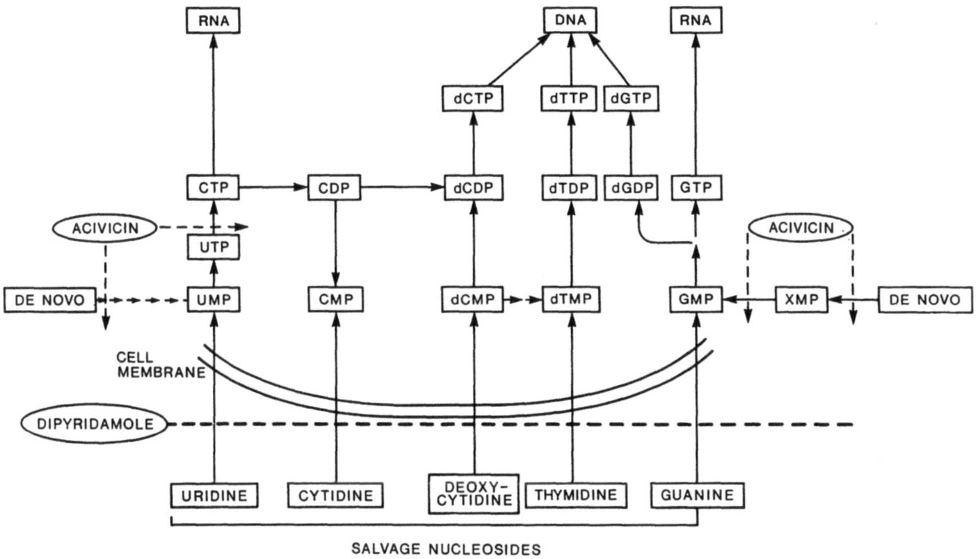

Fig. 5. Attacking points of acivicin and dipyridamole in pyrimidine and purine biosynthesis

synergistic anti-cancer cytotoxic action. A synergistic response on the ribonucleotide pools has also been observed in studies in vivo in transplanted hepatoma 3924A. The attacking points of acivicin and dipyridamole in pyrimidine biosynthesis are shown in Fig. 5. These results should have clinical relevance, since dipyridamole is a licensed drug in the United States and acivicin is in clinical trials [16].

Combination Chemotherapy of Acivicin with Actinomycin in Hepatoma 3924A

The action of acivicin in the depletion of GTP and CTP pools did not achieve a complete cure of the transplantable hepatomas in the rat although in appropriate concentrations it eradicated the hepatoma cells in tissue culture. For combination chemotherapy, we chose actinomycin because it inhibits RNA polymerase activity precisely in the utilization of CTP and GTP. Since RNA polymerase activity was markedly increased in hepatoma 3924A [4], this was a further argument for the use of actinomycin with the objective of inhibiting the increased polymerase activity (Fig. 4).

In in vitro studies the LC_{50} values for acivicin and actinomycin were 1 and 0.01 μM, respectively, and the two drugs together were synergistic in killing the hepatoma cells [15]. In determining the nucleotide concentrations, the evidence showed that in these cells acivicin, as we had already established, decreased GTP and CTP pools, whereas actinomycin caused an accumulation of these nucleotides.

Table 8. Synergistic in vivo action of acivicin and actinomycin on hepatoma 3924A[a]

Treatment	Body wt (% of initial)	Tumor volume (% of control)	Tumor/body ratio (% of control)
Control (0.9% NaCl solution)	100	100	100
Actinomycin	82[b]	58[b]	78[b]
Acivicin	101	58[b]	67[b]
Acivicin + actinomycin	74[b]	16[c]	22[c]

[a] Means of 10 rats in each group are given as percentage of controls. Treatment schedule and doses are as outlined in text
[b] Significantly different from controls ($p < 0.05$)
[c] Synergistic response ($p < 0.05$)

The in vivo synergistic action of acivicin and actinomycin was tested in rats inoculated with hepatoma 3924A. Treatment with acivicin or actinomycin or with a combination of the two drugs was started 24 h after the tumor cells were inoculated and tumor volumes were followed by measuring tumor diameters. The results demonstrated that combination chemotherapy yielded a synergistic inhibition of tumor growth as measured 14 days after tumor inoculation [2] (Table 8). The clinical relevance of these studies is in the fact that actinomycin is well known for its anti-cancer action in choriocarcinoma and our recent discovery of its cytotoxicity against hepatoma cells may provide a new approach in combination chemotherapy with acivicin for the treatment of hepatocellular carcinomas and other tumors.

Acknowledgements. The research work outlined in this paper was supported by grants from the United States Public Health Service, National Cancer Institute, Grant Nos. CA-05034 and CA-13526. The excellent technical assistance of Mrs. Sergia Calderon is gratefully acknowledged.

References

1. Aoki T, Sebolt J, Weber G (1982) In vivo inactivation of carbamoylphosphate synthetase II in rat hepatoma. Biochem Pharmacol 31: 927–932
2. Cooney DA, Jayaram HN, Ryan JA, Bono VH (1974) Inhibition of L-asparagine synthetase by a new amino acid antibiotic with antitumor activity: L-(alphaS,5S)-alpha-amino-3-chloro-4,5-dihydro-5-isoxazoleacetic acid (NSC-163501). Cancer Chemother Rep 58: 793–802
3. Denton J, Lui MS, Aoki T, Sebolt J, Weber G (1982) Rapid in vivo inactivation by acivicin of CTP synthetase, carbamoyl-phosphate synthetase II and amidophosphoribosyltransferase in hepatoma. Life Sci 30: 1073–1080
4. Duceman BW, Jacob ST (1980) Transcriptionally active RNA polymerases from Morris hepatomas and rat liver. Biochem J 190: 781–789
5. Jackson RC, Lui MS, Boritzki TJ, Morris HP, Weber G (1980) Purine and pyrimidine nucleotide patterns of normal, differentiating and regenerating liver and of hepatomas in rats. Cancer Res 40: 1286–1291
6. Jayaram HN, Cooney DA, Ryan JA, Neil G, Dion RL, Bono VH (1975) L-(alphaS,5S)-alpha-amino-3-chloro-4,5-dihydro-5-isoxazoleacetic acid (NSC-163501): a new amino acid antibiotic with the properties of an antagonist of L-glutamine. Cancer Chemother Rep 59: 481–491

7. Lui MS, Kizaki H, Weber G (1982) Biochemical pharmacology of acivicin in rat hepatoma cells. Biochem Pharmacol 31: 3469–3473
8. Neil GL, Berger AE, Bhuyan BK, Blowers CL, Kuentzel SL (1978) Studies of the biochemical pharmacology of the fermentation-derived antitumor agent, (alphaS,5S)-alpha-amino-3-chloro-4,5-dihydro-5-isoxazoleacetic acid (AT-125). Adv Enzyme Regul 17: 375–398
9. Neil GL, Berger AE, McPartland RP, Grindey GB, Bloch A (1979) Biochemical and pharmacological effects of the fermentation-derived antitumor agent, (alphaS,5S)-alpha-amino-3-chloro-4,5-dihydro-5-isoxazoleacetic acid (AT-125). Cancer Res 39: 852–856
10. Takeda E, Weber G (1981) Role of ribonucleotide reductase in expression of the neoplastic program. Life Sci 28: 1007–1014
11. Tso JY, Bower SG, Zalkin H (1980) Mechanism of inactivation of glutamine amidotransferase by the antitumor drug L-(alphaS,5S)-alpha-amino-3-chloro-4,5-dihydro-5-isoxazoleacetic acid (AT-125). J Biol Chem 255: 6734–6738
12. Weber G (1977) Enzymology of cancer cells. Parts I and II. New Engl J Med 296: 486–493, 541–551
13. Weber G (1983) Biochemical strategy of cancer cells and the design of chemotherapy: GHA Clowes Memorial Lecture. Cancer Res 43: 3466–3492
14. Weber G, Lui MS, Natsumeda Y, Faderan MA (1983) Salvage capacity of hepatoma 3924A and action of dipyridamole. Adv Enzyme Regul 21: 53–69
15. Weber G, Prajda N, Lui MS, Denton JE, Aoki T, Sebolt J, Zhen Y-S, Burt ME, Faderan MA, Reardon MA (1982) Multi-enzyme-targeted chemotherapy by acivicin and actinomycin. Adv Enzyme Regul 20: 75–96
16. Weiss GR, McGovren JP, Schade D, Kufe DW (1982) Phase I and pharmacological study of acivicin by 24-hour continuous infusion. Cancer Res 42: 3892–3895
17. Zhen Y-S, Lui MS, Weber G (1983) Effects of acivicin and dipyridamole on hepatoma 3924A cells. Cancer Res 43: 1616–1619

Subject Index

MIX
Papier aus verantwortungsvollen Quellen
Paper from responsible sources
FSC® C105338

If you have any concerns about our products,
you can contact us on
ProductSafety@springernature.com

In case Publisher is established outside the EU,
the EU authorized representative is:
Springer Nature Customer Service Center GmbH
Europaplatz 3, 69115 Heidelberg, Germany

Printed by Libri Plureos GmbH
in Hamburg, Germany